6140107900

CMl3 000 188
12/13
£25.00
338.92707/EVA

Advance praise for

1 Week Loan

This book is due for return on or before the last date shown below

University of Cumbria

D1471025

"Thi ... ... hich is
abstr ... ty, but
to be ... werful
than ... post-
mod ... ices of
real ... ship in
critic ... y and
ecop ... educa-
tiona

... cation,
... nooth;
... *entury*
... *Person*
... *cation*

"Pro ... *ccupy*
Educ ... libera-
tory

... iction,
... *cation*
... *agogy*

"Fac ... ower,
it is e ... neous
pursu ... *tion*
Tina ... s in
which education and social learning can enl ... localization, and promote sustainability, and ... sion
course the global economy has set us on. It d ... ators
and activists alike."

—Peter Newell, Professor, University of Sussex;
Author of *Globalization and the Environment*

LIBRARY INFORMATION
AND
STUDENT SERVICES
UNIVERSITY OF CUMBRIA

OCCUPY
education

GLOBAL
STUDIES IN
EDUCATION

A.C. (Tina) Besley, Michael A. Peters,
Cameron McCarthy, Fazal Rizvi
General Editors

Vol. 22

The Global Studies in Education series is part of the Peter Lang Education list.
Every volume is peer reviewed and meets
the highest quality standards for content and production.

PETER LANG
New York • Washington, D.C./Baltimore • Bern
Frankfurt • Berlin • Brussels • Vienna • Oxford

tina lynn evans

OCCUPY
education

Living and Learning Sustainability

PETER LANG
New York • Washington, D.C./Baltimore • Bern
Frankfurt • Berlin • Brussels • Vienna • Oxford

Library of Congress Cataloging-in-Publication Data

Evans, Tina Lynn.
Occupy education: living and learning sustainability / Tina Lynn Evans.
pages cm. — (Global studies in education; vol. 22)
Includes bibliographical references and index.
1. Environmental education. 2. Sustainable development.
3. Critical pedagogy. I. Title.
GE70.E93   338.9'27071—dc23   2012029818
ISBN 978-1-4331-1967-5 (hardcover)
ISBN 978-1-4331-1966-8 (paperback)
ISBN 978-1-4539-0907-2 (e-book)
ISSN 2153-330X

Bibliographic information published by **Die Deutsche Nationalbibliothek**.
Die Deutsche Nationalbibliothek lists this publication in the "Deutsche
Nationalbibliografie"; detailed bibliographic data is available
on the Internet at http://dnb.d-nb.de/.

The paper in this book meets the guidelines for permanence and durability
of the Committee on Production Guidelines for Book Longevity
of the Council of Library Resources.

© 2012 Peter Lang Publishing, Inc., New York
29 Broadway, 18th floor, New York, NY 10006
www.peterlang.com

All rights reserved.
Reprint or reproduction, even partially, in all forms such as microfilm,
xerography, microfiche, microcard, and offset strictly prohibited.

Printed in the United States of America

*This book is dedicated to the ninety-nine percent
and especially to those who are working diligently and with love
to make societies more sustainable.*

Contents

Part One:

*Looking Back and Looking Around at an
Unsustainable World:
How Did We Get Here and Why?*

Part Two:

The Road Ahead: Setting Guideposts for Living and Learning Sustainability

Part Three:

The Critical Role of Sustainability Education

Foreword

Objection Sustained:

Revolutionary Pedagogical Praxis as an Occupying Force

Peter McLaren
The University of California, Los Angeles and
The University of Auckland

The biosphere is disappearing into itself, and it is no coincidence that those of us living in regions of the geopolitical center, in the very locations where the forces of exploitation are most acutely developed, will be able to resist (with the help of the war economy) this collapse for a longer duration than those laboring in the peripheral countries. This means we are obligated to use this interregnum for political mobilization. And it is here that *Occupy Education* can serve as an important guide. With this bold new work by Tina Evans, we are provided with important conceptual tools to create a new language out of which new epistemologies of liberation can emerge. These can be connected to new engines of class struggle and new pedagogies capable of addressing the ecocrisis of our era.

Motivated by the sustainability crisis and emboldened by the courageous activities of the Occupy Movement, Evans has not only put together a powerful argument on how to deal with the crisis of sustainability but also offers a very timely and important contribution to critical pedagogy and community action struggling amid looming resource shortages, climate change, economic instability, and ecological breakdown.

Evans' critical pedagogy is marshaled in the interest of socio-ecological sustainability and sustainability-oriented social change. In short, her work is a profound demonstration of an effecacious integration of the social and ecological justice movements. In this sense, *Occupy Education* is very much a critical pedagogy of convergence and integration, as the work of European sustainability scholars and activists is brought into dialogue with powerful emergent voices from *las Americas* to work towards a vision of what a world outside of the ravages of neoliberal capitalism might look like.

What initially strikes the reader as a key theme of Evans' project is the way she establishes the wider context of her point of departure, where the author utilizes place-based sustainability theory and action in the varied and multiple contexts of practical lived experience, experience that has been inestimably impacted by the ravages of neoliberal capitalist globalization and sustained opposition to it. Evans' point of departure emerging from this context is precisely the suffering of the planetary oppressed, a departure she makes under penalty of losing herself to the very system which she has been trying so valiantly to overcome.

Evans rejects a reformist discourse and its hegemonic apparatuses and instead chooses to construct a pedagogy of sustainability that can be used as a strategic instrument for liberation, one that is education-oriented but none-the-less maintains a position of extraordinary political effectivity. The upshot of this is the creation of what Richard Kahn in his excellent Afterword calls a "counterhegemonic bloc of ideological alliance" among environmental educators, indigenous scholars, nonacademic knowledge workers, and political activists of various and sundry stripe—or what Kahn in his own pathbreaking work has called "the ecopedagogy movement."

Evans' work is built upon in-depth theories about the nature and purposes of sustainability itself, and Evans is acutely aware that the politics of sustainability can easily be co-opted by the guardians of the state who make empty promises to manage the crisis in the interests of the public good (really in the interests of private greed).

The politics of sustainability is a discourse that can be hijacked by the very interests that Evans is out to unmask. Understanding how such hijacking takes place can be further assisted by engaging the works of the decolonial school that has charted out the conflictual terrain known as the "coloniality of power" (*poder de patron colonial*) whose scholars and activists working in the areas of decolonizing epistemologies and praxis include Ramon Grosfoguel, Anibal Quijano, Linda Smith, Enrique Dussel, Sandy Grande, and others. In addition to addressing the coloniality of power, a revolutionary critical pedagogy of sustainability is as much about creating what Kahn calls a "revitalized ecology of body/mind/spirit" and the struggle for "planetarity" as it is a praxiological undertaking to achieve specific, cumulative goals.

Occupy Education carries with it the implicit but powerful lesson that we need to talk about the future and to ignore those who tell us that normative considerations and utopian thinking are inappropriate for revolutionary critical pedagogues. This would be, in Marx's view, a self-refuting statement as "what will be" is always inscribed within the "what is." To talk about different futures is desirable as long as such reflection is grounded in reality. Normative statements about the future are inescapable for any revolutionary. The elements of the future are contained within the very structure of the present. But we need to have more than a vision of the future, we need to be committed to a vision that arcs towards the justice that eludes us under the ironclad thrall of capitalism.

There is no metaphysical springboard from which to propel ourselves into the future; rather, we propel ourselves from where we are, from being energized by the truth effect of our own commitment to a praxis of liberation—what we may consider a concrete universal—and our full fidelity to such a praxis (Zizek and Milbank, 2009). While we have no original source from which to act (we act from a position of exteriority beyond the totality of social relations), and from which to accept the entreaty of the oppressed, that should not stop us from participating in the struggle to build the world anew. We act not from some divine fiat but from our own compassion, from our love for our brothers and sisters and nonhuman animals, from our thirst for justice, from our desire to end such needless suffering in the world.

Yet the struggle will not be easy. On this path we are threatened by our own human frailty, by those who would betray us, and the principles of revolutionary communalidad, by those who would use us for their own ends, and by the faux revolutionaries who wish to be part of the struggle without sacrificing their own positions of power and privilege. It is these individuals who will take us down the path of working in "collaborative partnerships" with statist institutions all too eager to co-opt limited environmental resources, using what Kahn calls "public relations alchemy."

Evans' work recognizes an important theme that runs through my own work: that the problems associated with global capitalism are not self-standing, they form an organic unity. In capitalist societies such as ours, self-alienating subjectivity is always already social alienation linked to the social relations of production, to racialized and gendered antagonisms, to the normative constraints of the global "power complex" that reduces everything to production and consumption—and it is this alienation that generates the self which remains isolated from its Other.

Grosfoguel (2007, 2008) reminds us that it is a complicated power complex that we are facing with an ignominious history—a history that provides an important context for Evans' work. It was not just economic colonization that visited *las Americas* in 1492 but multiple antagonisms that included: a global class formation where a diversity of forms of labor (slavery, semi-serfdom, wage labor, petty-commodity production, etc.) co-existed and became organized by capital as a source of production of surplus value through the selling of commodities for a profit in the world market; an international division of labor of core and periphery where capital organized labor in the periphery around coerced and authoritarian forms; an inter-state system of politico-military organizations controlled by European males and institutionalized in colonial administrations; a global racial/ethnic hierarchy that privileges European people over non-European people; a global gender hierarchy that privileges males over females and the system of European patriarchy over other forms of gender relations; a sexual hierarchy that privileges heterosexuals over gays and lesbians; a spiritual hierarchy that privileges Christians over non-Christian/non-Western spiritualities institutionalized in the globalization of institutionalized Christianity; an epistemic hierarchy that privileges Western cosmology and systems of intelligibility over non-Western knowledge and cosmologies, and institutionalized in the global university system; a linguistic hierarchy between European languages and non-European languages that privileges Eurocentric knowledge as true communication and rational knowledge/theoretical production yet denigrates indigenous knowledges as 'merely' folkloric or cultural and not worthy of being called theoretical (Grosfoguel, 2007, 2008).

While, as Grosfoguel points out, the racial/ethnic hierarchy of the European/non-European divide transversally reconfigures all of the other global power structures, and while the idea of race and racism becomes the organizing principle that structures all of the multiple hierarchies of the world-system (i.e., coercive [or cheap] labor is done by non-European people in the periphery and "free wage labor" in the core), I would still emphasize that it is the exploitation of human labor in the global capitalist system (i.e., global capitalism's endemic crisis, the social relations of production and the political class conflicts taking place within these relations), that sustains the conditions of possibility of all these other antagonisms, which is not to reduce them all to class.

My own approach to revolutionary critical pedagogy is not so much theoretically multiperspectival as it is dialectical, emerging from the Marxist humanist tradition and beginning with the works of Marx himself. While Marxist educators need to include an ecological dimension in their work (in the discourse of predatory capitalism, is not the exploitation of human labor and endless consumption a logical corollory of the extermination of indigenous peoples?), ecology activists need to engage Marx. Marx is mostly known for his critique of political economy that helped guide his devastating critique of capitalism. But Marx can also be used to help find some markers for charting out what a post-capitalist future might look like.

As Peter Hudis (in press) formulates it in his pathbreaking book, *Marx's Concept of the Alternative to Capitalism*, when labor is determined by necessity and external expediency ends, that is, when we exist outside the social universe of value production and are no longer defined by material production, and our tribulations as human beings seeking to survive the world of vampire capitalism are no longer measured by labor time, then, and only then, are we able to take the first real steps towards freedom. This is because production and consumption will be based, according to Hudis, "on the totality of the individual's needs and capacities." Drawing our attention to Marx's storied phrase, "From each according to their abilities, to each according to their needs," Hudis corrects those who might interpret this phrase as some kind of a quid pro quo. Here we need to understand that Marx is not saying that needs are met only to the extent that they correspond to the expression of a given set of abilities. This is the case because it would mean that human relations are still governed by material production. But the true realm of freedom lies beyond material production. Even when we move from socially necessary labor time to actual labor time, we still are outside of the realm of freedom; entering the realm of freedom only occurs when actual labor time also ceases to serve as a standard measure, and labor serves as an end in itself, as part of an individual's self activity and self development. As Hudis makes clear, free development for Marx could not be possible when human activity and products acquire an autonomous power and limits are externally imposed on the range by which individuals can express their natural and acquired talents and abilities. Marx went so far as to stress the elimination of the basis of both modern capitalism and statist 'socialist' alternatives to value production.

And Hudis gives us something else to consider. He writes that the subjective development of the individual is, for Marx, a crucial precondition of a truly new society; in fact, it was for Marx as significant as such objective factors as the development of the forces of production. Here he took the position that human subjective activity should never be constrained by the forces of its own making. He went so far as to argue that it is not the means of production that create the new type of man, but rather it is the new human being that will create the means of production. Marx understood that there was no way progressive political forces could just "will" a new society into being by a force of the imagination. Any new society would have to come into existence imminently from womb of the old society, with its specific conditions of capitalist production and reproduction and the forces in play that challenge such conditions.

Marx stressed the development of the forces of production (in part, because he did not live to witness the most destructive power in the forces of production) whereas, as Hudis notes, we are witnessing today the need to limit the destructive power of much of these forces before they overtake us completely. Time is running out on the effort to save the planet from capital's vicious self-expansionary nature, and this is where Tina Evans' timely volume, *Occupy Education*, can provide us with a crucial intervention.

We cannot have market freedom, hierarchical harmony or authentic democracy within the social universe of capital—this monstrosity of monopolistic imperialist capitalism—that is unable to distribute overproduction and unable to function even minimally without the extraction of surplus value. We must not be deceived. We must reject liberal pluralism and methodological individualism, as it only serves to bolster neoliberalism and the capitalist state. With this bold new work by Tina Evans, we are provided with important conceptual tools to create a new language out of which new epistemologies of liberation can emerge. These can be connected to new engines of class struggle and new pedagogies capable of addressing the ecocrisis of our era.

What Tina Evans is able to accomplish so skillfully, is to chart out a comprehensive critical pedagogy of sustainability. To this marvelous accomplishment, I would only add that in order to have a critical pedagogy of any kind, we first need to develop a philosophy of praxis, which requires that we recognize that all philosophy is determined by its dialectical relationship to praxis. And I would emphasize that this relationship between philosophy and praxis *is imminently ethical* in that *it is manifested in a preferential option and thematic priority to be given to the oppressed to present their counterstories and testimonios of resistance.* It is also *imminently pedagogical*, in that *it recognizes that the languages and discourses of the oppressed have been domesticated if not destroyed by the pedagogical practices of the state and that new languages of resistance are often coded in the interstices of popular struggles.* It is *imminently transformative* in that it *adopts a class position in solidarity with the oppressed and remains united in popular, ideological, racial, gender and cultural struggles.* As a philosophy of praxis, revolutionary critical pedagogy in the service of ecosustainability will need to be critical, self-reflexive, ethical and practical. Such a praxis is self-relating, it is immanent, it is an inscription into the order of being, a pulsion toward alterity and it is also connected to the larger language of multiplicity and the historical traditions that can help guide it. It is an arc of social dreaming, a curvature of the space of the self as it is inscribed in our quotidian being. I emphasize this feature of revolutionary critical pedagogy *as a way of life* to distinguish it from the critical pedagogy as a methodology, or a set of instructions for effective practice. We know, for instance, that the self is generated by alienation, and that to a large extent we are defined by our own failure to arrive at any fixed identity, as the self returns to itself from the rubble of relations of alienation and in doing so retroactively posits is own presuppositions through the self-sublation of contingency. We are, in other words, invented after the fact. But this should not stop us from presupposing a new world, a better world, a less exploitative world, a world that surges forth from the integuments of the old, through the expressive revelation of the people, through the practical action of service, and through class mobilization.

I do not know at this time of capitalist cholera, if critical pedagogy is the outcome and expression of historical necessity or a contingent force that will be erased by the sands of empty, unproductive time. That is, it is unclear whether critical pedagogy is the result of the constitution of a deeper historical praxis needed at this historical moment or is merely the contingent construction of such a praxis. Surely critical pedagogy is the externalization of an idea—the construction of a contingent singularity that can only be comprehended retroactively. Thus, we can never offer any guarantees as to the way people will be attracted to its principles and its 'truth effect". And we must live with this realization, as difficult as it might be. In truth, we have no choice but to live with it.

Unscrolling before our bloodshot eyes is a vast panorama of decay, as if the sun has been leeched of its radiance and turned into a visual rendition of Bataille's *The Solar Anus*, towards which Benjamin's Angel of History is careering out of control, not so much propelled by a storm from heaven now threatening to capsize the entire planetary system but sucked back into historical time by the black hole of the waste of endless wars and wars to come. Living within the state of planetary ecocrisis so aptly characterized in Richard Kahn's Afterword as constituting "geographies of genocide, ecocide, and zoöcide" (and I would add, epistemicide, the wholesale 'disappearance' of indigenous knowledges and practices by the guardians of Eurocentric knowledge production), we cannot experience our self-presence except through the anamorphically distorting mirror of capital.

As oblivion advances threateningly, and as the profound and irredeemably incoherent discourse of neoliberal privatization secures itself as the generative grammar of our generation, we need to recognize capitalism as a fundamental structuring barrier to democracy. *Occupy Education* can be used to combat neoliberalism and stand united against the structural violence of the state and the globalized economy.

References

Grosfoguel, R. (2007). Descolonizando los universalismos occidentales: El pluri-versalismo transmoderno decolonial desde Aimé Césaire hasta los Zapatistas. In S. Castro-Gómez & R. Grosfoguel (Eds.), *El giro decolonial: Reflexiones para una diversidad epistémica más allá del capitalismo global* (pp. 63–77). Bogota, Colombia: Siglo del Hombre Editores).

Grosfoguel, R. (2008). Para descolonizar os estudos de economia política e os estudos póscoloniais: Transmodernidade, pensamento de fronteira e colonialidade global. *Revista Crítica de Ciências Sociais*, number 80 (March 2008), 115–147.

Zizek, S., & Milbank, J. (2009). *The Monstrosity of Christ: Paradox or dialectic?* Cambridge, MA: MIT Press.

Author's Preface

Why "Occupy" Education?

The title of this work deserves some explanation regarding relationships drawn to the Occupy Movement (a.k.a. the Ninety-nine Percent Movement). Although I cannot predict for certain what this movement might become over time, to date, it has purposefully remained a leaderless movement that has challenged many forms of entrenched power globally. What began in New York City as a protest against Wall Street's pillaging of investors and ordinary citizens in the United States has become an international movement protesting and challenging entrenched and oppressive powers in diverse social contexts. "Occupy" has come to mean many things to many people. There are no "official" proclamations that I or anyone else can point to as *the* quintessential expressions of the character and purposes of the Occupy Movement. And yet, through the actions of those involved and the many voices contributing to the discussion of the meaning of Occupy, the movement has emerged as a clearly counterhegemonic force.

While I cannot claim that this book grew directly out of the Occupy Movement, my work has emerged from the same historical groundswell that has given birth to that movement. My work converges with the Occupy Movement in that I critique the domination and oppression of others that are the embodied expression of concentrated wealth and power. I do not pretend to speak for the movement. I only hope to contribute to it by extending its critique of concentrated social power explicitly into the realm of sustainability-oriented theory, education, and praxis. I see this work as one among many formative statements that both comprise and, at the same time, engage in dialogue with the Occupy Movement.

Acknowledgments

The completion of this book would not have been possible without the support, inspiration, and encouragement of many people. I would like to thank Rick Medrick, Randall Amster, Mark Seis, and Peter McLaren for the time and attention they devoted to reviewing my drafts and helping me to do my best work. Even more importantly, I would like to thank each of you for believing in the value of my work to me as an educator and to other sustainability-oriented educators. In many ways, I have built upon the visions of sustainability that you have shared with me.

Many thanks are also due to the late Richard Douthwaite and to James Fitzgerald, both of whom assisted me in developing key ideas that grew into chapters in this book. Jim, you are a continual inspiration for engaging with local food as a counterhegemonic and life-enriching process. Richard will be sorely missed by the sustainability community. I can't say enough about how his writings on economics and sustainability have inspired my work on enforced dependency and localized sustainability.

I would also like to thank the Prescott College community as a whole for creating the space for sustainability education and my work to thrive. Within that community, I would especially like to thank my cohort of Ph.D. students (Bill Crowell, Jordana DeZeeuw-Spencer, John Gookin, Esmeralda Guevara, Koh Ming Wei, Jessi Kidder La Porte, Kathy Lamborn, Aaron Morehouse, and Mike Wood). You have been and will continue to be an amazing source of inspiration and encouragement.

I would also like to thank my family and friends who have encouraged me along the way. Your support has been an essential to my completing this work. Knowing that you truly care about my work has been such an important source of my strength and drive.

My students in The End of Oil also deserve my deepest gratitude for their willingness to engage in a clear-eyed, critical assessment of the converging sustainability crises of our time, for considering how to respond to these crises, and for engaging in sustainability-oriented action. They have been a constant source of inspiration for me as an educator and writer.

Most importantly, I would like to thank my husband Dennis Lum, my best friend, first editor, truest colleague, and the love of my life. Your patience and guidance have been invaluable in my work, and will continue to be so. I could not have written this book without you.

Notes on the Cover Art

The cover image is adapted with permission from a poster developed by students Whitney Blakstad, Carolyn Cohen, Cortney Kayl, and Adam Wilkes as a visual aid for their group project presentation for the Value of Place. The Value of Place consisted of two upper-division college courses taught in summer 2006 by the author and her dear colleague Kate Niles.

The courses focused on place as a nexus of human/nature relationships. Through classroom learning and educational travel and service in the Four Corners region of the U.S. Southwest, participants explored how people and communities are intimately embedded within the natural environment, whether we recognize these relationships or not. We also explored how the character of our relationships with the places we call home can influence the growth of healthy mutualism or, conversely, how these relationships can drive the diminishment and destruction of both people and the environment.

The image as a whole underscores the meaning of reciprocity as a balance between taking from and giving back to one's community and one's place. To live according to a principle of sustainable reciprocity would mean that human communities and the environments in which they are embedded would exist in healthy interdependency.

The yin yang symbol at the center of the image indicates wholeness and the interdependency of seeming opposites, both within ourselves as beings and within our world and the universe at large. The symbol is depicted in blue and green colors, the dominant colors of the planet as observed from space, the green representing land and the blue representing water. The yin yang symbol indicates a need for balance, and in this particular context, the need for reciprocally nurturing relationships between humans and nature.

The circle outlining the diagram depicts Western industrial society moving in a circular, counterclockwise motion through time. This circle of history circumscribes the globe/yin yang symbol, indicating that humans greatly influence the world and can influence the history of all other life forms on earth. The counterclockwise movement of Western history depicted here indicates that industrial society needs to relearn some of the social philosophies and practices that characterize(d) more sustainable traditional societies.

The ideas and strategies listed within the green portion of the yin yang symbol signify the social sources of the pressing socio-ecological problems we face at this point in history. The ideas and strategies listed in the blue portion of the yin yang symbol represent opportunities for changing the current "taker" mentality of Western industrial culture to create a more sustainable society based on principles of balance, reciprocity, and interdependency.

The graphical axes on which the yin yang (globe) and the course of Western history rest mark a potential point of balance between giving and taking in people's relationships with each other and the natural world. This point of balance represents healthy interdependence and the potential to live according to a principle of reciprocity.

The flames represent where we are now as a global society: as far out of balance as possible on the taking axis of the graph, but perhaps at the very beginning of social movement in a new direction. Western industrial society has largely renounced reciprocity and has lost the ability to see the self in the face of the other, be that other another person, a group of people, or nature. Industrialized society is therefore experiencing fires of destruction resulting from the lack of reciprocity that permeates all facets of modern life, and as depicted in the diagram, we have the potential to jettison ourselves off of the circle of history and life entirely and into a cycle of continually deepening environmental and social crisis. If we are able to continue long enough on the circular path of history toward healthy mutualism and wholeness, we might re-establish reciprocal relationships between modern humans and the natural world.

Part One:

Looking Back and Looking Around at an Unsustainable World:

How Did We Get Here and Why?

Introduction and Overview of Contents

Confronting the Sustainability Crisis

Every day we hear the news: the planet cannot sustain life as we know it. Subjected to attack by the global growth economy on all fronts—the oceans, the land, water, the atmosphere—the earth's systems are breaking down. Species extinction rates now match those of major planetary die-off events, ocean fisheries are in decline from overfishing and other forms of human disturbance, and a warming climate threatens planetary scale, permanent dislocations of human and nonhuman life and the radical alteration of the earth's productive cycles. Meanwhile, the global economy continues to satisfy the appetites of rapidly growing numbers of people globally who are now entirely dependent upon it. Ironically, this dependence creates widespread allegiance as well as ample opportunity for powerful interests to profit from scarcity. And so, the story of Western industrial capitalism's conquest and rule over a diminished world continues.

Humans have unleashed a juggernaut of self-perpetuating and self-reinforcing systems of power and exploitation, and that juggernaut is destroying the diversity of human societies along with the diversity and resilience of other life forms and entire ecosystems. If continued unchecked, human-created systems of power and exploitation will ultimately bring an increasing death toll while also extensively and irreversibly altering the course of all life on earth. Addressing this crisis in human institutions and systems of power is therefore an ethical and survival imperative. We must understand why this destruction is occurring and develop ways of being in the world that respect the lives of other species with which we share this place and time. Socio-ecological sustainability must be the highest priority for human civilizations.

This book encapsulates my attempt to theorize and practice forms of education that rise to the occasion in which we find ourselves at the opening of the twenty-first century. Many of us are truly up a creek without a paddle, and even worse, some of us are paddling society along with the mainstream current in blissful ignorance that we are in fact headed toward a devastating watershed of socio-ecological collapse. What is an educator to do in these times in an effort to help students and communities avert disaster—or to help prepare ourselves and others to engage in sustainability-oriented action in the wake of disturbing, if not devastating, changes in our world? I offer my theories and the example of my practice in answer to this question.

In doing so, I recognize the many deeply insightful, articulate theorists and activists whose ideas and actions have confronted, in their current and earlier stages of development, the very destructive forces that I confront today. I recognize that, despite their well-articulated arguments, their theories, up to this point in time, have proven inadequate to permanently, seriously challenge the power structure of modern, industrial society. Still, their ideas live inside me and in countless others and serve as a lens for novel ways of seeing, understanding, and being in the world. While I cannot claim that my pedagogical theories and practice, however revolutionary, are up to the task of remaking society, my work does invite others to join with me in the process of moving our communities and our world toward sustainability—and maybe, just maybe the time is finally ripe for widespread, sustainability-oriented social change. I hope that, together, we can teach, learn, and live sustainable societies into being. Our lives and so much more depend on it.

Content Overview

This book addresses what is perhaps the central question for sustainability educators today: in this time of converging socio-ecological crises and a rapidly closing window of time in which to act effectively to mitigate or prevent devastating forms of socio-ecological collapse of modern societies, exactly what should sustainability educators teach and how? This is a question I have been working with for a number of years as an inter- and transdisciplinary educator of undergraduate students at a small, public, liberal arts college. In the broader terms of asking what one needs to know and do to create a better society, I have been working with this question for most of my life. I have long been at work as a student and citizen developing a big picture understanding of the world and working toward socio-ecological justice. This process of questioning, learning, and acting is a life's work that is never complete. It represents my dialectical and dynamic engagement with the world as I know it. It entails an opening of both my mind and heart to the harshness and the beauty of the human condition and to the beauty and decline of the ecologies that give humans and other beings life. It is a painful process that is, at the same time, laden with the joy of meaning-making, work that makes my own life worth living, even in the face of mounting crises that may, in the end, prove unsolvable.

My focus in this book is higher education. I develop a theory of critical pedagogical processes that can engage students, faculty members, and the community in sustainability-oriented praxis. This transformative praxis departs from dominant modes of higher education today in that it involves students in the process of naming the world and defining desired action. It seeks to (re)integrate our fractured modern identities and worldviews. It is counterhegemonic in orientation so that it directly confronts the political economy of late capitalism and its means of production as primary drivers of the sustainability crisis. It takes a transdisciplinary approach to integrating the academic disciplines and seeks to heal dichotomous and destructive fractures within the modern worldview such as the separation of humans from nature. It seeks to authentically reconnect people with each other and with the land. It embodies sustainable forms of leadership and entails educational processes and content that encourage personal and community engagement and that foster sustainable living.

Since the pedagogy I advocate is also the pedagogy I practice, I draw upon student reflections on their experiences in my End of Oil course as one means to indicate what might be possible in terms of student learning and praxis within the context of sustainable pedagogy. It is my hope that my theories and practical experience will prove useful to other sustainability educators who choose to adapt my work to their specific historical and socio-ecological contexts. Though I strongly advance my theories, I see myself as engaged in a conversation with others about what we need to know and do to be responsible educators and people. In no way do I propose to have the last word in this conversation. I only hope that my thoughts and the examples of my actions will prove useful in some way to others who share my concerns and hope to make a difference in our world.

My work emerges in many ways from the work of a diverse set of critical social theorists that include both Western and indigenous theorists who have focused their attention on various embodiments of perhaps the central social contradiction of modern societies: domination and oppression—a contradiction that must be resolved in order to actualize social justice, which is in turn a fundamental constituent of sustainability. My theory of sustainability in important ways hinges upon Gramsci's (1971/1999) theory of agency and praxis so that my theory of sustainability is a praxiological theory. It is born in history, and it is also capable of influencing history through its lived application. My pedagogical theories build upon the works of critical educators, including Paulo Freire (1970/2000) and Peter McLaren (2005; 2007; McLaren & Farahmandpur, 2005; McLaren & Houston, 2005; McLaren & Jaramillo, 2007; McLaren & Kincheloe, 2007; McLaren & Kumar, 2009) who advocate counterhegemonic forms of education and agency as means to confront and alter the destructive and unsustainable trajectory of capitalist society and its central dynamic of domination and oppression. My counterhegemonic stance is rooted in an analysis and critique of globalized, neoliberal political economy and its overarching tendency to create and enforce unsustainable forms of dependency among people everywhere. I see combating these socio-ecologically destructive forms of dependency as central to counterhegemonic struggle for sustainability, and I draw upon the social and economic theories of Richard Douthwaite (1999a; 1999b; 2004) and others to argue that reducing and ultimately eliminating unsustainable social dependencies associated with provision of basic needs is of paramount importance. In my focus on the sustainable, localized provision of basic needs, my pedagogical theories draw upon the work of place-based educators and diverse theories of place as well as on the work of those advocating sustainable, localized forms of agriculture as embodiments of counterhegemony in service to sustainability. My work weaves these threads of related theory together to form a theory of a critical pedagogy of sustainability that is rooted in social critique, analysis and critique of globalized political economy, recognition of the importance of place as a context and construct for sustainable living, sustainability-oriented local food action, and counterhegemonic sustainability education praxis.

The pedagogy I advocate is a form of praxis rooted in Gramsci's theory of agency. It involves inter- and transdisciplinary engagement with students and community members in an effort to catalyze the conceptualization and living of sustainable practices. The study of food production, food security, and food sovereignty are central to this critical pedagogy. In addition to developing a critical pedagogy of sustainability for higher education, I draw upon examples from my own teaching to illustrate how engaging students in sustainability-oriented action projects, and in reflection on that engagement, can create an important nexus for sustainability education praxis.

Organization of the Book

In this book, I develop theories about the nature and meaning of sustainability-oriented social change, articulate the form and content of a critical pedagogy of sustainability, and promote this pedagogy as a means to reorient higher education toward addressing the converging crises of sustainability. I begin in chapters one and two by developing a critical social theory of sustainability. This theory informs my critique of globalized political economy and the construction of a second theory that is central to the development of the critical pedagogy of sustainability: the theory of enforced dependency as a pillar of the late capitalist system. This theory is developed in chapters three and four and is used as a launching point for chapter five, in which I argue that (re)inhabitation can serve a vehicle for confronting and combating enforced dependency and for actualizing sustainability. Seeing localized food production and consumption as especially important aspects of (re)inhabitation, in chapters six and seven, I argue for local food sustainability as a strategy for resisting enforced dependency and for promoting personal and community autonomy. I follow this discussion by developing in chapter eight a critical pedagogy of sustainability in higher education and arguing for its widespread application as a form of sustainability education praxis. This pedagogical theory builds upon and integrates all of the theories developed in previous chapters. In chapters eight and nine, I provide examples of how I have implemented my pedagogy in my engagement with students and community members in- and outside the classroom. In chapter ten, I offer my conclusions on the pedagogy and praxis of sustainability in these challenging times.

Chapter 1:

The Critical Social Theory of Sustainability:

Critique of Domination as a Foundation for Sustainability-Oriented Education and Praxis

People thought they could explain and conquer nature—yet the outcome is that they destroyed it and disinherited themselves from it.

—Václav Havel

I begin this chapter with the assumption that the globalized growth economy is on a collision course with the natural limits of earth's systems and with the knowledge that, despite widespread understanding of the declining health of the biosphere, little is being done to address the challenges we face. Within this context, I develop a conceptual framework for education and action to address the sustainability crisis. I hope that my critical social theory (CST) of sustainability will provide insights into contemporary social power structures and cultural processes that inhibit sustainability-oriented social change. I also propose that these insights serve as a springboard for practical, sustainability-oriented action.

My CST of sustainability is not only a critical theory of market systems, and although it builds upon earlier work in CST, it is distinct from the critical theory of the Frankfurt School. My CST of sustainability focuses on agency as a vehicle for addressing the sustainability crisis and draws upon bodies of theory not typically associated with CST. Though I do draw upon Marx's insights, my work cannot be pigeonholed as a Marxist or neo-Marxist analysis. My CST of sustainability recognizes that growth-oriented industrial economies, whether socialist or capitalist, are incompatible with sustainability.

My CST of sustainability is rooted in a critique of social domination as unsustainable. I develop that critique by drawing upon and synthesizing in new ways diverse social critiques of domination and oppression in modern societies including:

- Spretnak's (1997) and Merchant's (1996) critiques of the modern Western worldview as an epistemology and ontology that frames systems domination and oppression;
- theories of the Frankfurt School of critical theorists, particularly Marcuse (1964), regarding complex cultural forms of domination and oppression in modern societies;
- Marx's (1844/1964) theory of the alienation of labor;
- Polanyi's (1944/1957) critique of industrial economies as having made people subservient to the economy rather than the reverse arrangement which had typified pre-industrial societies;
- Proudhon's (1890/1966) theory of property as a form of theft;
- Gramsci's (1926–1943/1996; 1971/1999) theories of cultural hegemony and passive revolution;
- ideas offered by indigenous authors including Armstrong (1995), Martinez (1997), and Salmon (2000), and others who critique industrial society and the modern Western worldview and offer alternative ways of seeing and being in the world;
- deep ecology developed by Næs (1973/2008) and Devall (1980/2008);
- the ecofeminism of Gomes and Kanner (1995);
- the environmental justice theories of Bullard (1993/2008);
- ecopsychology (see Rozak, Gomez, & Kanner, 1995) as a both critique of psychological disconnection from the natural world and a vehicle for healing the human/nature divide conceptualized by modern societies; and
- the systems-theory-based social critique of Homer-Dixon (2006).

As a response to the Western notion of humans living outside of nature and beyond natural laws and limits, in chapter two, I develop an ontology that recognizes humans as embedded within and part of nature. This ontology forms a foundation for sustainability-oriented praxis that pursues human and ecological health as an inseparably intertwined whole. This ontology is the basis and wellspring for distinctly action-oriented aspects of the CST of sustainability. These aspects are rooted in Gramsci's (1926–1943/1996; 1971/1999) notions of agency and praxis, and they embody fitting and practical responses to the deep social critiques embedded within my CST of sustainability. My theory of sustainability praxis provides a foundation for hopeful engagement with the world in the realms of education and society more generally.

Why Choose Critical Social Theory as a Point of Departure for Sustainability-Oriented Theory and Praxis?

As my understanding of the social roots the global sustainability crisis has grown, I have found it frustrating to hear calls to change our way of being in the world through individualized and isolated attempts at personal development and spiritual enlightenment while the societal aspects of human life on earth—those aspects that delineate our shared experience and outline the possibilities for our collective treatment of others and the earth—have remained largely unquestioned "realities." Within the hegemonic social systems of modern-day global capitalism, the material, cultural, political, and economic aspects of our lives become increasingly circumscribed so that we have little choice but to participate in systems that undermine the very kind of personal and spiritual growth necessary for creating systems of healthy, sustainable living. I refuse to believe that a broad segment of society freely chooses to support these systems. I believe, instead, that our participation in collectively violent, even self-destructive systems derives in large measure from the design and functioning of interlocking systems of social power that characterize the modern, industrial, global-capitalist world. My belief mirrors the central thesis of critical social theory.

For many of us, our modern experience is like riding a runaway horse headed for a cliff. We feel powerless to change direction and fearful of leaping off, even though many of us are well aware of the ultimate peril we are in if we stay the course. We're aware of the perils of the warming climate that is a result of industrial production and industrial-style living; we're aware of collapsing fisheries and dying coral reefs, of poisonous pollutants that harm ecosystems and our own bodies, and of the myriad other socio-ecological problems that have their roots in the industrial capitalist growth paradigm—but at least to this point in time, we have not collectively engaged in widespread, significant reconfiguration of our ways of life. The juggernaut of continued globalization and economic growth rolls on. Still, many of us do realize that, without changing social power relationships, we have little hope of realizing true personal and spiritual growth that is most possible in community. We need systemic change so that community life, as lived in particular places, can recapture center stage as the appropriate vehicle for personal fulfillment, learning, and sustainability-oriented change.

Given CST's central focus on critiquing and changing social power relationships characterized by entrenched and self-perpetuating forms of domination and exploitation, I see the potential for CST to serve as a foundation

for (re)creating sustainable societies.[1] Critical theorists attempt to comprehend, critique, and alter social structures and phenomena that embody
features of oppression, domination, exploitation, and injustice. Through
critiquing and altering these systems, critical theorists hope to change or
eliminate these structures and phenomena and extend the scope of freedom,
justice, and fulfillment. Oppression, domination, exploitation, and injustice
figure prominently in today's sustainability crisis, and these prominent
aspects of the unsustainable paradigm provide openings for using critical
theory as a vehicle for sustainability-oriented change.

Since critical theorists work to reveal the dynamics of entrenched power
as it shapes social systems, CST as a framework for inquiry and action also
presents distinct opportunities for developing empathy for those who find
themselves trapped within interlocking social systems of domination and
oppression. Cultivating such empathy can itself open possibilities for
personal and spiritual growth through recognizing one's material, economic,
social, political, and spiritual relationships to others—including countless
others hitherto virtually unseen and unrecognized as having any bearing on
the meaning and lived experience of one's life. So, most certainly, I do
advocate spiritual growth, but growth that is a shared experience, not a
jealously guarded, solitary journey. The spiritual growth I advocate as most
important to realizing sustainability derives from deep understanding of
systems of domination and oppression as these have manifested in a wide
variety of social and ecological settings. In the process of spiritual growth,
this understanding is further cultivated through a growing empathy toward
people and nature caught up in these destructive systems, and this deepened
understanding is fused with practice. When engaging in this process, one is
moved toward living a life in which his/her values and actions form a
coherent whole that strives toward sustainability and justice.

I argue that critical social theory offers an appropriate lens for discovering and analyzing the sources of the sustainability crisis. CST can provide
educators and change agents working toward sustainability with the conceptual and analytical tools for understanding and synthesizing the many diverse
aspects of the sustainability crisis in such a way as to reveal its central

---

[1] I place "re" in parentheses here and in other places throughout this text in recognition that
place-based cultures remain alive in diverse places globally. For members of these cultures,
the "re" in such words as reinhabitation and relocalization may not apply, and there may be no
need to re-create sustainable cultures. The parenthetical "re" also recognizes that, for some
peoples and places, examples have never existed or no longer exist to draw upon for place-
based living, meaning that localized systems of sustainable living must be created for the first
time from scratch.

organizing themes. In this chapter, I build upon the foundations of CST that focus attention on social power as constitutive to creating and perpetuating socio-cultural and socio-economic systems characterized by domination and oppression. I demonstrate how domination and oppression of people mirrors domination and oppression of nature; and, from this premise, I formulate the critical social theory of sustainability. This theory, an extension of CST, is explicitly ecological in its orientation and explicitly oriented toward praxis. The CST of sustainability is also situated within an analysis of fossil fuel depletion and the impacts of that depletion on the global growth economy.[2] I argue that the critical social theory of sustainability can illuminate vital and appropriate avenues for change through education and action. It is my hope that the critical social theory of sustainability will facilitate sustainability praxis rooted in a worldview that sees domination and oppression as the central socio-ecological problem of our time.

We now turn our attention to examining how and why CST serves as an appropriate vehicle for both comprehending and acting to avert the converging socio-ecological crises of late capitalism.[3]

---

[2] The energy depletion and overall economic context for the CST of sustainability is developed in chapter three.

[3] I use the term late capitalism explicitly to imply several things. I propose that the capitalist system is nearing its logical conclusion because it is both consuming the resource base necessary for its functioning and concentrating wealth in fewer and fewer hands. Intense concentration of wealth in the global economy increases the suffering and discontent among the dispossessed, and it spurs their resistance to the system, thereby threatening capitalism's indefinite continuation. Concentration of wealth also impedes the circular flow of money required for continued consumption within the ever enlarging and overproducing capitalist system. This flow is being impeded to such an extent that the global economic system itself is nearing a breaking point. We cannot have continued growth in production in the face of depleting resources such as oil, fresh water, and arable land, and we cannot have continued economic growth at the same time that the ability of consumers to purchase goods is rapidly eroding. The system is starving and bankrupting itself. I do *not*, however, imply that we need only sit back and wait for the system to come apart of its own weight. In fact, such passivity in the face of impending collapse would virtually guarantee chaotic responses in which the uninformed and unprepared would lash out in desperation against many who had contributed little to creating the disaster. We must prepare for the end of capitalism by building alternatives to it, especially concerning the provision of basic needs. We must work to take the system apart and also be prepared for its possible sudden collapse. This rationale for using the term "late capitalism" draws on the works of Mandel (1972/1975) and Jameson (1991).

Why Are We Witnessing a Sustainability Crisis, and
How Can Critical Social Theory Help Us Understand and Address It?

These questions shape the conceptual framework of this chapter. For some time in my teaching and learning, I have been developing a set of theories about why and how globalized, capitalist societies are characterized by what I call *enforced dependency* of individuals, communities, and entire societies on unsustainable social systems and institutions. My emerging theory is not entirely new. It shares a great deal with the critical social theory developed by the Frankfurt School beginning in the 1930s. Perhaps what *is* new about my theories is the direct and deliberate application of critical social theory to the rapidly converging socio-ecological and political/economic crises of our day, including global warming and the current or near-future peaking of global oil production. Others are working to explicitly extend critical theory into the realm of sustainability, but these efforts are recent and not widespread.

Critical theory is useful in macro-analyses of unsustainable systems and institutions and how these reproduce themselves in societies. According to Morrow and Brown,

> [Critical theoretical] analysis at the level of *system integration* may involve concepts involving functional-type part-whole relations. This involvement entails the macrosociological assumption that society, as a contradictory totality, must be analyzed structurally as a process of reproduction and transformation of agency/structure relations over time. But system integration here is understood in terms of an interpretive structuralism that rejects the analogy of organic systems in favor of open, historical social formations. (1994, p. 269)

Here, Morrow and Brown, in describing one thrust of critical social theory, elucidate a framework for analysis that I find useful to understanding and altering social sources of unsustainable systems. They note how CST focuses attention on the structures of society, built and perpetuated through systems of social power. They emphasize that human agency can alter these structures over time and that social structures are not analogous to laws of nature and are, therefore, open to challenge and change. Thereby, Morrow and Brown imply that CST-based critique, as a response to historically and culturally specific social formulations, does not attempt to uncover absolute truths. CST theories and actions based upon those theories must always remain contingent and open to emerging histories and interpretations of those histories.

As a vehicle for analysis at the system integration level, CST can be used to address the following questions: *In spite of the development and dissemi-*

nation of a great deal of credible evidence pointing to a need to mitigate converging socio-ecological crises, why do large scale social systems exhibit an inertia that seems to preclude making needed changes? How do systems of social power reproduce and reinforce themselves over time, and how are these systems related to the inertia present in unsustainable social systems? What roles do individuals and groups, operating at various levels and in various capacities, play in reinforcing or resisting systems of social power?

These are questions that can lead to *understanding* important social and historical forces at work in the current unsustainable social paradigm. This understanding can be articulated as social theory, as I do in this chapter. Note that the form of theory building in which I engage recognizes that systems, in their formation, functioning, and possible alteration or destruction, exist within history as the cause/result of specific conditions and the actions of individuals and groups. These same systems, therefore, can be altered through individual and group agency: people both create systems and are, in many ways, created by them. This chapter explores this dynamic.

In order to articulate why and how CST can serve as a vehicle for sustainability-oriented understanding and action, and in order to create a foundation for explicitly extending that body of theory into the realm of sustainability, I must first grapple with the meaning of *sustainability*. I must then elucidate the conceptual framework of CST to this point in time. Doing so will help me to identify the threads of CST that most effectively support the extension of critical theory into the realm of sustainability education and praxis.

A Working Definition of Sustainability

A central argument of this chapter—that sustainability must be the top priority for human societies—first requires an examination of the meaning of sustainability. The first widespread use of the term "sustainable" derived from its definition in the 1987 report of the World Commission on Environment and Development headed by Gro Harlem Bruntland: "Sustainable development is development that meets the needs of the present without compromising the ability of future generations to meet their own needs" (quoted in Allen, Tainter, & Hoekstra, 2003, p. 25). Many have critiqued the anthropocentrism of this definition. For our purposes here, the expressed concern for the long-term future is useful, but this definition does not take us far in discussing with any precision the sustaining of human ecology, the sustaining of broader ecological systems, or the relationship between the two.

Our definition here must integrate humans and environment: to conceptualize humans as separate from environment and nature is unacceptable because this dichotomy is part and parcel of the current unsustainable paradigm's power structures and systems of exploitation. This division forms a central theme within unsustainable, late capitalist, global reality. Here, I use *theme* in the way that Paulo Freire (1970/2000) defines it in *Pedagogy of the Oppressed*: in order to understand and change the current unsustainable paradigm, people oppressed by it must be able to "*name* the world": to conceptualize themes of oppression as a key activity toward transforming the world and ending oppression (p. 88).

In order to conceptualize themes of oppression, it is important to identify who and what is oppressed. I argue that the systematically impoverished, all those whose cultures and places are being destroyed by globalization, along with the middle classes—who live in a state of enforced dependency upon unsustainable economic and social systems—are oppressed under the current paradigmatic system. Nature is also oppressed in that, in a pattern that parallels the oppression of people in society, its purposes and very being are subjected, controlled, and systematically destroyed to create profits and economic growth. Those who reap the profits and hold the increasingly concentrated wealth and power of the world are the oppressors, though they too will become oppressed under the converging ecological and social crises of our time, if these crises converge with the full force of their potential for destruction. Even now, both oppressors and oppressed suffer dehumanization under the current paradigm. According to Freire, "As the oppressors dehumanize others and violate their rights, they themselves also become dehumanized" (1970/2000, p. 56).

If humans and the natural world operate as an integrated system, if the denial of an integrated human/nature complex is at the heart of many converging crises in the world today, and if social justice is necessary for ecological sustainability—all points argued here—we need a definition of sustainability that calls us to analyze and potentially transform human/nature relationships. James Pittman's *living definition of sustainability* (2007) offers us such an integrated vision. Pittman defines living sustainability as "the long-term equilibrium of health and integrity maintained dynamically within any individual system (organism, organization, ecosystem, community, etc.) through a diversity of relationships with other systems."

Pittman's living sustainability describes what I will call socio-ecological living sustainability. I use *socio-ecological* in addition to *living* because *ecology*—the study of the earth household—is inherently concerned with relationships—with systems views—and these views are central to comprehending power and exploitation. Using the term *ecology* highlights the fact that the systems referenced include both living and nonliving components of earth systems. This definition is *socio*-ecological because society is embedded within ecology, and the definition is *living* because, like all life, it is open to change driven by historical and natural forces. This openness to change highlights an important aspect of the definition: it is place specific; what might be sustainable in a given context is not necessarily so in every context. The appropriateness of changes to and adaptations of this definition is therefore place-specific in the same way that the appropriateness of life adaptations is in many ways determined by the specific context. This definition, to remain viable, must take form and evolve in living situations. *Sustainability*, then, is a set of lifeways lived within specific historical circumstances. Within these lifeways, considerations of the "long-term equilibrium of health and integrity" remain the central focus for communities.

Systems characterized by sustainability thusly defined would contrast sharply with the global capitalist world of today within which both humans and nature are used as tools to translate the health and wealth of the world into vast riches for a few and servitude and suffering for many. According to this definition of socio-ecological living sustainability, a sustainable society and world must prioritize the health and integrity of all community members and must foster relationships that create and support their well-being and recognize their intrinsic value. This definition of sustainability invites both justice-oriented critique of the world as we know it and construction of alternatives to the current reality.

This concept of sustainability recognizes that all possibilities for sustainability *depend* upon our abilities as individuals and communities to effectively and radically transform our world within a framework of intra- and intergenerational social justice. In such a transformed world, we would extend the concept of community membership to both human and nonhuman nature in order to (re)establish community and individual lifeways based on respectful reciprocity. As in many indigenous societies (Armstrong, 1995; Cajete, 2001; LaDuke, 1999; Martinez, 1997; Nelson, 1983; Salmon, 2000; Sveiby & Skuthorpe, 2006), healthy, reciprocal relationships among people and between humans and nature would replace relationships characterized by unhealthy co-dependence between the dominant and the oppressed—a form of human to human and human to nature relationship that pervades global capitalism.

Throughout this text, the term *sustainability* refers to the concept of socio-ecological living sustainability, and my work here is directed toward sustainability-oriented transformation within the life systems of the world. In order to lay groundwork for developing a critical social theory of sustainability, we will now explore the history and intellectual roots of CST in an effort to identify those threads of existing theory most appropriate to extending CST explicitly into the realm of sustainability theory and action.

Intellectual and Historical Roots of the Critical Social Theory of Sustainability

Critical theory is closely associated with the Frankfurt School of social theorists whose ideas represent an articulation and extension of Marxist-derived social theory for twentieth-century historical contexts. I draw from and integrate works both within and outside Frankfurt tradition in developing the critical social theory of sustainability that serves as the foundation for my critical pedagogy of sustainability.

Overview of Critical Theory

Critical theory is closely associated with the Frankfurt School of social theorists (Max Horkheimer, Herbert Marcuse, Eric Fromm, Walter Benjamin, Theodor Adorno, and others) whose ideas represent an articulation and extension of Marxist-derived social theory applied to twentieth-century historical contexts. According to Bentz and Shapiro,

> [CST] attempt[s] to understand, analyze, criticize, and alter social, economic, cultural, technological, and psychological structures and phenomena that have features of oppression, domination, exploitation, injustice, and misery. They do so with a view to changing or eliminating these structures and phenomena and expanding the scope of freedom, justice, and happiness. The assumption is that this knowledge will be used in the process of social change by people to whom understanding their situation is crucial in changing it. (1998, p. 146)

Bentz and Shapiro further summarize the nature and purposes of CST: "The purpose of [critical theory] inquiry is to change oppressive social conditions and to educate some or all of the public about these conditions and the possibility of changing them" (p. 157).

In critical theory, context is of the utmost importance in understanding and transforming social systems (Bentz & Shapiro, 1998, p. 146). Critical theory examines how large-scale social systems manifest themselves in and reproduce themselves through specific phenomena and, reciprocally, how these phenomena contribute to the construction and perpetuation of the larger system.[4] In critical theory analyses, systems are historically situated and can only be grasped as products of and active agents within particular histories. Critical theory draws upon the concept of the dialectic developed by Marx and Hegel in that it calls for analyses of historical phenomena, both small and large scale, in terms of their internal contradictions (Bentz & Shapiro, 1998, p. 147). Additionally, critical theory is concerned with the agency/structure dialectic of society: possibilities for proactive change or *agency* are paired in a unity of analysis with existing power structures in society that mitigate against changes and, thereby, reinforce and reproduce oppression (Morrow & Brown, 1994, p. 228).

Generation of critical theory involves using reflexive methods based in negotiation and argumentation. These methods can include historical analysis and interpretation, textual analysis, self-reflexive analysis, and metatheoretical argumentation, such as epistemological critique of research methods and heuristics (Morrow & Brown, 1994, p. 232). CST engages in immanent critique and ideology critique as central processes for generation of theory. In immanent critique, institutions and societies are analyzed according to their ability to keep their word. Such critique uses the society's or institution's own standards as the measure of success rather than critiquing the institution or society from the outside. For example, if a society claims to be *free*, immanent critique could serve as a means to determine the extent to which that society lives up to its own conceptions of freedom. Ideology critique is similar to immanent critique in that it deals with rhetorical contradictions. Ideology critique of a society or institution focuses on contradictions between official stories (ideologies) and experience (Bentz & Shapiro, 1998, p. 148). The widely believed notion that any person in America has real potential for living the "American dream" serves as a good example of an official story that could be a focus for ideology critique. In probing this notion, one could ask both how and why the American dream became a widely held notion as well as how and why the story of the American dream differs from the lived experience of many Americans.

A central premise of critical theory is that a more just world is an intrinsically valuable goal for human societies, and that a more just world would be one in which social relationships of domination and oppression were continually reduced, and ultimately eliminated (Bentz & Shapiro, 1998, p. 146). Critical pedagogy, as we shall see in chapter five, represents the work of realizing these goals through a praxis anchored in a critically informed view of the world (Freire, 1970/2000, chap. 1). According to critical theorists

---

[4] These self-replicating and self-reinforcing aspects of globalization underpin the concept of enforced dependency developed in chapters three and four.

we must uncover, critique, and engage in praxis to eliminate oppressive power relationships. This work is necessary precisely because oppressive power is at the heart of all social injustice (Freire, 1970/2000). The ultimate goal of critical theory is widespread praxis, ultimately resulting in liberation of the oppressed and oppressors alike (Freire, 1970/2000, p. 88; Morrow & Brown, 1994, p. 158). The freedom experienced in such a world would be both negative (freedom *from*) and positive (presence of opportunity). CST strives toward freedom *from* the heavy burdens and constraints of oppression and freedom *to* realize one's humanity in healthy, mutual relationship with others. Both of these freedoms are heavily constrained within the globalized capitalist paradigm. I will argue in chapter two that the collectively violent[5] principles upon which this paradigm operates systematically concentrate wealth and power in fewer and fewer hands while also extending and deepening dependence on the system.

The scope of CST is vast, especially when the critical project is applied to the current sustainability crisis. Understanding the systemic and ideological factors that create and perpetuate oppressive human to human and human to nature relationships is a daunting task that includes understanding both the functions and composition of natural systems (of which humans are part), understanding political economy and geopolitics, understanding the use and abuse of power in human systems, and understanding the historical and ecological legacies that inform unequal power relationships in global systems. Though grand in scope, CST does not offer a grand-narrative version of history and does not subscribe to drawing direct analogies between human social systems and natural history/systems. Because of its insistence on the role of historical specifics in creating current realities, CST avoids insinuating the inevitability of the world as we know it today.

A Brief History of Critical Social Theory

In this section, I discuss themes and threads of critical theory that are most important to developing the CST of sustainability.[6] I build upon some of the central tenets of CST, integrate these with other important and related socio-cultural and socio-ecological theories, and thereby create a solid platform for extending CST into the realm of the sustainability crisis. First, I will briefly revisit the history of CST in order to chart its trajectory and envision how

[5] See Summers & Markusen, 1992/2003, p. 215. These authors define collective violence as actions by large numbers of people that inflict harm on people and/or environment.

[6] More thorough and detailed histories of the Frankfurt School and critical theory are available elsewhere (Dant, 2003; Held, 1980; McLaughlin, 1999).

that trajectory might usefully be extended into the future as an effective vehicle for sustainability-oriented change.

The Institute for Social Research of the University of Frankfurt, later dubbed the Frankfurt School, was founded in the early 1920s (McLaughlin, 1999, p. 111). Under its first director, Carl Grünberg—an Austro-Marxist— Marxism became the theoretical basis of the school's program (Held, 1980, p. 29). In 1930, Max Horkheimer became the director. Under his direction, theorists associated with the Institute worked to adapt Marx's theories of political economy to the contemporary context. These theorists recognized that the economic determinism developed by Marx had not accurately predicted the capitalist world of the twentieth century. Though they found much of value in Marx's original analyses, these theorists working in 1930s Germany contended with questions of why conditions of political economy had not triggered the socialist revolution predicted by Marx and, most importantly, why fascism was on the rise within the capitalist context. In exploring these questions, Frankfurt School theorists turned their attention to socio-cultural aspects of modern life. They contended that aspects of culture serve to entrench the powerful in their positions of advantage and to create momentum and inertia within socio-economic systems in ways that simultaneously reproduce and extend the capitalist sphere while also undercutting impetus toward socialist revolution. This particular thread of CST, powerfully articulated by Marcuse in his *One-Dimensional Man* (1964), is of particular importance to developing the CST of sustainability as a theory that effectively grapples with how hegemony manifests in society as socially and environmentally destructive collective behavior. This thread of CST helps us comprehend how and why such behavior continues, even in the face of widespread knowledge about poverty, social and environmental injustice, and widespread ecological decline.

Another important theoretical development of the Frankfurt School that dates from the 1920s is the melding of the Marxist theoretical tradition and its insights concerning political economy with Freudian psychoanalytic theory regarding identity formation and social repression of instinctual urges (Held, 1980, chap. 4). This melding created a springboard for the Frankfurt theorists' critiques of culture as a central vehicle for development and reproduction of the administered society and its capitalist political economy. This synthesis also created openings for exploring individual malleability and susceptibility to ideological control.

In 1933, the Institute was moved to Geneva as the Nazis came to power in Germany, and in 1935, it was moved to Columbia University in New York (Held, 1980, p. 34). Particularly since the school's exile to the United States, theorists of the Frankfurt School and their disciples such as Jürgen Habermas have created a diverse body of theory that has touched many disciplines. It can hardly be said that the critical theorists of the Frankfurt School tradition agreed in all aspects of their work or that they were even working within the same field of study at any given time. Therefore, several models of critical theory with divergent foci emerged (Held, 1980, p. 34). According to Held,

> The themes covered by the Frankfurt School [during its exile from Germany] ... are extensive. They include discussions of theories of capitalism, of the structure of the state, and of the rise of instrumental reason; analyses of developments in science, technology and technique, of the culture industry and mass culture, of family structure and individual development, and of the susceptibility of people to ideology; as well as considerations of the dialectic of enlightenment and of positivism as the dominant mode of cognition. As always, it was the hope of Horkheimer and the others that their work would help establish a critical social consciousness able to penetrate existing ideology, sustain independent judgment and be capable, as Adorno put it, "of maintaining its freedom to think things might be different." (1980, p. 38)

In the post–World War II era, Horkheimer and Adorno maintained their focus on extending the valuable contributions of Marx and other leftist thinkers into the contemporary context while, at the same time, attacking Soviet Marxists. Accordingly, they pleased neither "conservative authorities nor radical thinkers," and their politically independent positions led to challenges from all sides (Held, 1980, p. 39).

Marcuse's work became highly popular with the New Left in the 1960s and 1970s due to his commitment to politics and to contemporary radical social struggles. During this time, he became perhaps the most prominent theoretician of the left. Through Marcuse, the critical work of the Frankfurt School in the areas of culture and authoritarian/bureaucratic society became well known, and it was Marcuse that began to expand the work of critical theory into the realm of socio-ecological problems (Held, 1980, p. 39).

After the 1970s, as the social struggles of the past two decades died out or manifested in less directly confrontational settings and efforts and as neoliberal political economy and the radical individualism of the so-called me generation moved into full swing, interest in critical theory in academe and politics slowed to a trickle. Currently, at a time in which the divide between rich and poor has never been wider and in which the very mechanisms that once seemed capable of driving an ever more rapid expansion and deepening entrenchment of the capitalist project seem themselves to be breaking down—an opportunity is emerging to build upon the rich body of critical work developed and instigated by Frankfurt School theorists from the

1930s through the 1970s. Some of their insights may apply equally well, if not better, to interpreting today's societies.

Early CST, with its roots in Marxist analysis, had a great deal to say about society and economy and very little to say about human relationship to nature, though its critiques of society and political economy are highly relevant to sustainability. As sustainability-oriented theorists, educators, and activists come to see with increasing clarity that oppression of people and domination and destruction of nature are two sides of one coin representing the same underlying exploitive values and practices, CST-oriented analysis is being extended to include human relationship with environment (Gruenewald, 2003; Kahn, 2010; Kovel, 2002; Leonardo, 2004; Merchant, 1999, 2008; O'Connor, 1991/2008). This more inclusive analysis of domination and oppression is not entirely new (Marcuse, 1972/2008), but CST-based socio-*ecological* critique is only recently gaining strength.

Toward a Critical Social Theory of Sustainability

In the remainder of this chapter, I draw upon themes and ideas within CST and other critical theories, selecting those that I deem to offer the strongest and most appropriate foundations for confronting the converging sustainability crises of the modern world and for encouraging sustainability-oriented agency. I include the work of theorists not typically associated with the CST of the Frankfurt School, but whose theories articulate well with the central premises and purposes of the CST of sustainability. Some contemporaries of Frankfurt School theorists together with some theorists active since the 1970s (when interest in CST began to wane) share central values and tenets with CST. The work of a number of these theorists is immensely helpful to developing a critical social theory of sustainability. I also situate my own work within the CST tradition as an extension of that body of theory into the current global context of peak oil, climate change, and limits now in sight for global growth capitalism. I do not offer a comprehensive or fixed articulation of the critical social theory of sustainability, but one that works historically and provisionally, and one that serves as an appropriate foundation for praxis in my college and community.

Questioning the Modern Worldview

Critical social theory is historically specific, rooted in the context of modernity as a system of beliefs and values that evolved out of the Enlightenment. Many theories offered by critical theorists hinge upon a critique of the intellectual systems that define modernity. *Dialectic of Enlightenment* by

Max Horkheimer and Theodor Adorno (1944/1972) is an obvious example. In order to form a basis for understanding and interpreting CST, we begin with a discussion of modernity as an intellectual and social construct.

In her book *Resurgence of the Real: Body, Nature, and Place in a Hypermodern World*, Charlene Spretnak (1997) offers an insightful analysis of the conceptual and normative foundations of modernity. The values system Spretnak elucidates derives from the eighteenth-century European period of the Enlightenment and the birth of the scientific method as a mode of inquiry. Specifically, she critiques the dualisms and hierarchies that characterize modernity.

According to Spretnak, within the modern worldview, humans are viewed as primarily *economic beings*. It follows that the satisfaction of human material needs and desires is of primary importance, although, as Spretnak states, the emphasis on accumulation typical of capitalist societies actually worsens the material foundation of the natural world that is the foundation of all human systems (1997, p. 219). Spretnak claims that the values, beliefs, and conceptual orientations inherent in modernity reinforce the likelihood that modern humans will conceive of themselves and behave as *homo economicus* (Spretnak, 1997, p. 219). Among these values, beliefs, and conceptual orientations are:

- progressivism, "the belief that the human condition progresses [in a linear fashion] toward increasingly optimal states as the past is continually improved upon";
- objectivism, "the belief that there is a rational structure to reality, independent of the perspectives of any particular cultures or persons, and that correct reason mirrors this rational structure";
- rationalism, the conception that "knowledge, belief, and the basis for action are properly derived solely from reason...'untainted' by emotions";
- a mechanistic worldview, the belief that "the physical world is composed of matter and energy, which operate in various constellations of cause and effect according to 'laws' of nature [so that] occurrences of creative unfolding and complex interactive responses in nature have no place" in our conception of the world;
- reductionism, the idea that "understanding physical entities...is achieved by breaking them down into smaller and smaller parts";
- scientism, "the belief that all fields of inquiry can attain objective knowledge by modeling their practices after the investigative methods of science";

- emphasis on efficiency through standardization, bureaucratization, centralization of power and decision making, and hierarchical institutional systems;
- anthropocentrism, the belief that humans are what matters most in the universe;
- emphasis on instrumental reason, "modes of thinking used to achieve desired ends rather than to determine values";
- opposition to nature that includes a conception of "nonmodern societies … as having been 'held back' by unproductive perceptions of holism and by conceptualizations of human culture as an extension of nature with reciprocal duties";
- compartmentalization, where life is "considered to exist in discrete spheres…such as family life, work, [and] social life";
- rationalism, valued as a vehicle for throwing off the chains of religion and superstition; and
- "the shrinkage of the cosmological context, the sacred whole, to the scale of humans" (Spretnak, 1997, pp. 219–221).

Spretnak notes that modern values, beliefs, and orientations are also highly gendered—that modernity is "hypermasculine" and patriarchal because valued ideas and practices are identified as characteristically male while devalued ideas, practices, and characteristics—such as emotionality, empathy, and identification with the earth (as in earth-centered, Pagan or indigenous belief systems)—are identified as characteristically female (Spretnak, 1997, p. 221). This gendered bifurcation of values parallels the false dualism of the human/nature construct. Once such dualisms are internalized, a hierarchical arrangement of differentiated ideas and practices follows, and that which is seen as other within the power structure is devalued.

The emergence of the modern worldview is further supported by Christianity's emergence as the dominant religion of Western Europe and its colonies. In her essay "Reinventing Eden: Western Culture as Recovery Narrative" (1996), Carolyn Merchant states that "the story of Western civilization since the seventeenth century … can be conceptualized as a grand narrative of fall and recovery." She continues by stating that "three subplots organize its argument: Christian religion, modern science, and capitalism" (p. 133). Rather than focusing on the biblical theme of human stewardship of the earth, Merchant focuses on human dominion as a biblical theme with regard to human/nature relationship. Although stewardship themes are also clearly present in the Bible (Barbour, 1993, pp. 75–77),

Merchant correctly identifies dominion as the biblical theme that has facilitated both scientism and capitalism as systems of power in the modern age. The modern recovery narrative Merchant explicates is that of female, fallen nature being subdued and civilized through conceptually male systems of rational belief and action.

Ironically, within the construct of modernity, scientific rationality itself becomes myth. According to Dant:

> Critical theorists argue that the form of myth actually resurfaces within the adoption of the [rational] methods of science and technology as the only adequate mode of knowledge. What appears as rationality ... begins to operate as myth in the sphere of culture.... As a cultural system it becomes mythic through rigidifying the processes of nature, treating them as predetermined and beyond the power of human will. It is the unitary and unbending form of reason in modernity that lends itself to domination, both of nature and of human will. (2003, p. 25)

As *the* epistemology of the modern era, rationality is reified and, therefore, becomes irrational, making possible the use of reason to serve ends that destroy nature and that are, thereby, ultimately self-destructive. Therefore, the CST of sustainability critiques aspects of the modern worldview that inform unsustainable socio-cultural systems.

The Critique of Domination

A central focus of the CST of sustainability is critique of domination within socio-ecological systems. This section draws upon critiques of domination and oppression as formative dynamics of globalized industrial capitalism and capitalist culture. Marcuse's theories (1964) anchor this discussion because they effectively address why socio-ecological problems have reached the crisis stage while penetrating critiques of the sources of the sustainability crisis have been available for decades. My critique also draws upon indigenous and ecofeminist thought and includes voices from the environmental justice movement. These critical traditions extend the critique of domination into areas highly relevant to developing the critical social theory of sustainability. I also draw upon Marx's (1844/1964) theory of alienation of labor as a process that informs domination and oppression of common people. Because Karl Polanyi (1944/1957) situates his critique of industrialism within the context of the human history of community in a way that speaks to the importance of (re)establishing reciprocal, sustainable relationships between people and between people and place, his work also provides an important foundation for my critique of domination. I also draw upon the work of Proudhon (1890/1966) for insight into how government has contributed to enforcing the dependency of the masses upon those who dominate global society.

The critique of domination lies at the very center of CST. It lays bare the exploitive and destructive mechanisms of globalized capitalism. It is through this centrally organizing critique that CST can be usefully extended into the realm of sustainability. The critique of domination helps us to comprehend and contribute to reversing the destruction of people and nature—both of which serve as fodder for the growth economy and are emblematic of the extension of capitalism into every corner of the globe. The destruction of nature is at the center of the sustainability crisis. This destruction coincides with the obliteration of diverse human cultures developed within diverse ecosystems—indigenous cultures that embody both the means and meaning of living in unmediated relationship with nature as the source of all life. The CST of sustainability focuses on domination and oppression in the realms of both society and nature, and it is particularly concerned with how the domination and oppression manifested in both realms articulate with and reinforce one another.

According to Marx (1844/1964, pp. 106–119), alienation derives from the capitalist mode of production through which human work is fragmented and the relationship of that work to nature as an integrated system is abstracted and obfuscated (p. 114). According to Marx, work and production are natural to humans and are done in the absence of physical or material need (p. 113). Work is a meaningful activity through which humans express their species being and contemplate their relationship to and meaning within nature. Through work, humans change nature and thereby change their own nature. In order for this process to be a meaningful expression of humanity, workers must control the processes and the products of their work. According to Marx:

> It is just in his work upon the objective world ... that man first really proves himself to be a *species being*. This production is his active species life.... The object of labor is, therefore, the *objectification of man's species life*: for he also duplicates himself not only, as in consciousness, intellectually, but actively, in reality and therefore he contemplates himself in a world that he has created. In tearing away from man the object of his production, therefore, estranged labor tears from him his *species life*.... (p. 114)

Wage labor and the introduction of the detailed division of labor in the nineteenth century reduced workers to a commodity—labor—and robbed them of control through the deskilling of artisans.[7] Marx also recognized that these changes in production transferred the locus of work from the home to the factory (Dant, 2003, p. 61). This dislocation and the resultant abstraction of the market economy further centralized power and control in the hands of management and created the conditions for widespread commodity fetishism. Factory production and the centralization of control inherent in the capitalist mode of production systematically undermined self-sufficiency and created a system of enforced dependency of workers on the free market distribution of commodities.

Similarly to Marx, Polanyi cites factory production as the basis for a mode of production and a self-regulating market that would undercut and eventually virtually eliminate the ability of people to work and support themselves independent of the capitalist system:

> Although the new productive organization was introduced by the merchant...the use of elaborate machinery and plant involved the development of the factory system and therewith a decisive shift in the relative importance of commerce and industry in favor of the latter.... The more complicated industrial production became, the more numerous were the elements of industry the supply of which had to be safe-guarded. Three of these...were of outstanding importance: labor, land, and money. In a commercial society their supply could be organized in one way only: by being made available for purchase.... The extension of the market mechanism to the ele-ments of industry—labor, land, and money—was the inevitable consequence of the introduction of the factory system in a commercial society.... As the organization of labor is only another word for the forms of life of the common people, this means that the development of the market system would be accompanied by a change in the organization of society itself. All along the line, human society became an ac-cessory of the economic system.... (1944/1957, p. 75)

Polanyi concludes that "improvements" of industrialism were "bought at the price of social dislocation" (p. 76).

Writing a century before Polanyi on the question "what is property," Proudhon (1890/1966) makes points similar to those of Polanyi (1944/1957) regarding the commodification of the factors of production. He concludes that property is theft 1) because it is a means of eliminating traditional access by common people to the basic factors of production and 2) because these factors were not created by property owners but were stolen from the natural commons, the birthright of every person. While Polanyi (1944/1957) and Marx (1844/1964) point to the mode of production as a central causative

---

[7] Similarly, Kropotkin (1902/1989) argues that systems of mutual aid, free association, and decentralization of social power and control—that characterized tribal societies, village life, and the medieval cities of Europe during the era of the craft guilds—represented forms of social organization characterized by nonalienating production.

factor in the alienation of labor, Proudhon (1890/1966) emphasizes government's role in creating and defending private property rights as a central driver in the alienation of producers from the factors of production. Both forms of alienation and the concentrated powers that produce them combine to enforce dependency of common people on goods and services produced by systems over which they have little control.

The forms of material and psychic domination that characterize global growth capitalism have their roots in the alienated dependency fostered by the capitalist economic system and its means of production as well as by governments that historically created and maintained the rules of the capitalist system. The CST of sustainability recognizes that the organization of labor and production combined with widespread dependence on the free market as a provisioning system are central means through which the capitalist order dominates and controls individuals and societies globally. Therefore, alternate means of production and consumption that reduce dependency and that foster community control of economy and the restored subservience of economy to society should inform social critique and action toward sustainable living.

Critical theorists extend Marx's critique of the mode of production by analyzing how domination, characterized by both social control of the individual and dependency of the individual on the system, is perpetuated and entrenched through culture. Critical theorists critique work as a form of capitalist production and reproduction that extends beyond the realm of industrial production and into all aspects of modern life. Dant (2003) summarizes the thinking of Adorno and Horkheimer on this point:

> As enlightenment thinking has taken hold of modernity, technology has led to a systematic organisation of work that employs tools and machinery in the domination of nature. That systematicity has extended to the domination of workers, whose work life…is driven by the rhythms and demands of machines. What the critical theorists argue is that these processes extend into the cultural life of modern society. Just as the worker is dominated at work, so (s)he is at home and in leisure, which leads to a restriction of private life outside the process of production. (p. 46)

According to Adorno and Horkheimer, the culture industry produces "consumers as so many automatons, all thinking and acting in the same way" (Dant, 2003, p. 47) so that amusement, as part of the reproductive mechanisms of late capitalism, is an essential element of work that sustains the capitalist order (Dant, 2003, p. 47). Similarly, Baudrillard claims that, in both work and non-work, individuals participate in the circulation of signs, symbols, and values that reinforce and perpetuate the system of late capitalism. Dant summarizes Baudrillard's critique in stating that "Marx's analysis of the mode of production has collapsed into the sphere of consumption" (2003, p. 57).

Continually expanding the sphere of domination and exploitation of people and nature is unsustainable. While domination of landscapes and human others seemed for a time to benefit the dominant, we are now witnessing diminishing returns on this strategy of subjugation for production and profit as we confront natural limits (Meadows, Randers, & Meadows, 2004), and perhaps the limits of human tolerance for abuse (Homer-Dixon, 2006, pp. 204).

As Joseph Tainter noted in *The Collapse of Complex Societies* (1988), societies become vulnerable to collapse as an outgrowth of an internal dynamic of increasing social complexity. Late capitalist industrial society is the most complex system in human history. According to Tainter, social complexity—characterized by "more parts, different kinds of parts, more social differentiation, more inequality, and more kinds of centralization and control"—is a problem-solving strategy (p. 37). Early on, societies typically receive high returns on investments in complextiy in terms of social adaptation to stresses. These investments have costs in terms of money, effort, energy, etc., and growing complexity eventually yields diminishing returns (Homer-Dixon, 2006, p. 223). It is important to note that investment in further complexity as a means to solve social problems benefits some more than others—typically those with disproportionate social power. Eventually, according to Tainter (1988), for growing numbers of people, the returns on investments in further complexity become small or even negative when compared to the alternative: collapse of the complex society into a reduced state of complexity (chap. 4), perhaps into a form in which "small, internally homogeneous, minimally differentiated groups [are] characterized by equal access to resources, shifting, ephemeral leadership, and unstable political formations" (p. 37). Late capitalist society is likely approaching a point of collapse as continued investments in technologies, bureaucracies, economic integration, and political control themselves create a plethora of problems that require social response.

Living as they did prior to both the obvious convergence of socio-ecological crises of sustainability and the rapid upswing in the pace of globalization, and also having been touched directly by the rise of fascism in Europe, the founders of critical theory focused a great deal of attention on destructive collective behavior, exploring how and why individuals participate in extensive collective violence and self-degrading behavior instigated by authoritarian leaders or by an overarching social system pervaded by authoritarian logic. Individual humans have adapted and bent to the needs and judgment of community since time immemorial (Tar, 1977, pp. 88–89), but critical theorists recognized something qualitatively different in the bending of human will and desires to fit the capitalist mode of production and capitalist cultural reproduction. They also noted technology's central and catalyzing role in fostering capitulation to the logic of the system (Marcuse, 1964, p. 158).

Critical theorists, immersed in the intellectual tradition of the Enlightenment (though they directly confronted and challenged Enlightenment thinking and constructs) (Tar, 1977, p. 88), tended to focus on the individual as the rightful center of sovereignty and the focus for considering questions of human mental and social health. With their partiality to Freudian analysis, critical theorists tended to interpret the domination of the human sphere in terms of individual repression and to focus less attention on the destruction of indigenous and place-centered communities as a characteristic phenomenon of the capitalist machine. Critical theoretical focus on the individual derives from an emphasis on autonomy. According to Maeve Cooke,

> Critical social theory is not individualist in the sense of asserting the priority of individual goals at the expense of communal and social values or in the sense of conceiving of human beings atomistically, as self-contained centres of ethical value. Indeed, critical social theories frequently appeal to an idea of social solidarity and understand ethical value in intersubjective terms. Nonetheless, this kind of theory is individualist in two important senses. First, in the sense that it prioritizes individual human flourishing over that of the collective. Second, in the sense that it stresses the need for the individual herself to be able freely to accept a given conception of human flourishing as the best.... The concept of autonomy articulates its commitment to individual freedom in this sense. (2005, p. 383)

Focus on individual repression is as relevant today as it was during the formation of critical theory. In particular, Marcuse's concept of repressive desublimation of individuals within late capitalist society remains a highly relevant critique of a complex form of domination that continues to pervade industrial societies. Marcuse elucidates this concept in his *One-dimensional Man* (1964) where he argues that individuals in late capitalist industrial societies participate in their own oppression as they are coerced by the dominant culture to surrender their liberty in exchange for material objects, comfort, and the sensuous consumption of products of the culture industry. In Freudian terms, desublimation would mean an end to repression of sensuous individual desires. Marcuse argues in his development of the oxymoronic concept of repressive desublimation that, when the satisfaction of these desires becomes the central organizing principle of society, possibilities for democracy shrivel, and tyrants are allowed to shape the lived experience of the masses. For Marcuse, under late capitalism, desublimation for the masses ironically leads to their captivity and repression in all facets of life, other than the satisfaction of material needs.[8] Marcuse argues that uncovering the contradictions of advanced industrial society has become incredibly challenging because entrenched powers have done a masterful job of creating a society in which the requirements for their hegemony have been made congruent with the perceived needs of the many. In such a society, opportunities for critique are undermined even as domination and oppression persist.

Marcuse (1964) implies that democracy can only exist absent the coercive and self-reinforcing repressive desublimation that characterizes late capitalist society: "The rights and liberties which were such vital factors in the origins and earlier stages of industrial society yield to a higher stage of this society: they are losing their traditional rationale and content" (p. 1). He continues:

> The concept of alienation seems to become questionable when the individuals identify themselves with the existence which is imposed upon them and have in it their own development and satisfaction. This identification is not illusion but reality. However, the reality constitutes a more progressive stage of alienation. The latter has become entirely objective; the subject which is alienated is swallowed up by its alienated existence. There is only one dimension, and it is everywhere and in all forms. The achievements of progress defy ideological indictment as well as justification; before their tribunal, the "false consciousness" of their rationality becomes the true consciousness. (p. 11)

In Marcuse's analysis, technological and material progress, associated sensual comfort and pleasure enjoyed by growing numbers of people, and pervasive messages of the culture industry that coerce participation and identification with the system conjoin to overwhelm critique of domination and integrate the vast majority of people as willing participants in the capitalist project. Control of the messages of the culture industry by powerful elites has intensified since Marcuse's time with the concentration of media ownership triggered by industry deregulation (McChesney, 1999). The system and its participants adopt identical goals, attitudes, and logic that manifest a self-reinforcing and self-perpetuating way of life. Resistance has been rendered futile and is perceived as contradictory to the well-being of the individual.

---

[8] Marcuse's concept of repressive desublimation is similar to Antonio Gramsci's (1971/ 1999) concept of cultural hegemony discussed later in this chapter, but social control through repressive desublimation requires that individuals attain a relatively high level of material comfort, while hegemonic social control extends to all or nearly all social classes.

For critical theory, Marcuse sees a perpetually negative role: "The critical theory of society possesses no concepts which could bridge the gap between the present and its future; holding no promise and showing no success, it remains negative" (Marcuse, 1964, p. 257). Still, Marcuse sees some chance that critical theorists and the most exploited within the capitalist system may serve as agents of social change. With regard to critical theory itself, he states that it "wants to remain loyal to those who, without hope, have given their life to the Great Refusal" 1964, p. 257). [9] For Marcuse, critical theorists may play a role in social change in that they keep alive a deep critique of domination, one that leaves them virtually without hope.[10]

Regarding the most exploited within the capitalist system, those for whom repressive desublimation itself is out of reach materially, Marcuse writes: "Their force is behind every political demonstration for the victims of law and order. The fact that they start refusing to play the game may be the fact which marks the beginning of the end of a period" (1964, p. 257).

Regarding the potential of the exploited masses to employ critical theory in realizing the radical overthrow of the capitalist order Marcuse states, "The chance is that, in this period, the historical extremes may meet again: the most advanced consciousness of humanity, and its most exploited force. It is nothing but a chance" (1964, p. 257). Now, more than 40 years after Marcuse wrote *One-dimensional Man* (1964), the dispossessed within the globalized capitalist system have grown dramatically in both number and proportion, including within highly industrialized societies. Labor control through the division of labor and mechanization of production (and resultant job loss) combine with the concentration of wealth and the cannibalizing dynamic of the system (which pays workers/consumers ever lower wages while requiring them to consume more). As the global economy confronts limits to growth and teeters on the edge of collapse, there exists a unique opportunity to bring together the historical extremes of which Marcuse spoke, but doing so to the effect of successfully challenging the capitalist order would mean that people would both comprehend and act upon a far-reaching critique of the domination and destruction embodied in the capitalist order.

---

[9] When Marcuse wrote *One-dimensional Man* (1964), there appeared to be no end in sight to technological progress and concomitant increases in the material well-being of industrial society. Therefore, he argues for a "great refusal" to participate in a wholly unfree society characterized by repressive desublimation.

[10] Marcuse closes *One-Dimensional Man* with these words: "At the beginning of the fascist era, Walter Benjamin wrote: It is only for the sake of those without hope that hope is given to us" (1964, p. 257).

Such a conjoining of thought and effort based on a critical assessment of society could catalyze a humane outcome to the collapse of global growth capitalism, but it is perhaps far more likely that collapse will come with few prepared to make sense of it and that, even for those who are prepared, actions in crisis may be irrational. The concepts of autonomy and individual rights are central to Marcuse's argument, but the solutions to the social crisis of a one-dimensional society must be collective, and we have little time to cultivate both understanding of the system and collective action appropriate to realizing a freer and more just society.

In forming an appropriate basis for sustainability-oriented action, the critique of social domination must go beyond the basic recognition that domination is, in part, culturally constructed. It is important also to recognize that capitalist forms of domination share a lineage with racism and sexism that arose within complex societies prior to the advent of industrialism. Like all manifestations of social domination, racism and sexism derive from a worldview that fragments socio-ecological wholes and ranks them. This worldview is far too narrow, exclusionary, and conflictive/aggressive to serve as a useful foundation for (re)creating sustainable societies.

In "The Rape of the Well-maidens: Feminist Psychology and the Environmental Crisis," Gomes and Kanner (1995) analyze gendered aspects of oppression that inform unsustainability. According to these authors, understanding the acculturation of male children within Western society is important to understanding the socio-ecological crises of our time. They argue that Western society's high valuation of radical autonomy and the association of that autonomy with males are highly destructive to developing healthy interpersonal and human/nature relationships. According to Gomes and Kanner (1995), to be male—or to secure a position of power within society (whether one is male or female)—is to separate from and appear not to depend upon others and nature (p. 113). They conclude that "radical autonomy is a cultural ideal that does not allow for other forms of growth, especially those based on relationship and connection" (p. 113), and "domination becomes a way to deny dependence" (p. 115) which is seen as a weak, feminine quality.

Within the modern capitalist order, male-identified domination is directed at both human others and nature, and it thereby prevents the dominant from developing holistic, sustainability-oriented relationships with people and nature. According to Gomes and Kanner (1995),

> By acknowledging our dependence, we allow gratitude and reciprocity to come forth freely and spontaneously. This is especially true when power in a relationship is fairly equal.... When we deny our dependence on another person, we threaten not

only to engulf them but to feed on their strength and vitality, often until we have used them up.... A striking parallel is seen in the physical destruction of ecosystems.... (p. 115)

Gomes and Kanner argue that perpetuating systems of gender inequity is inherently self-destructive and unsustainable. Within such systems, women are socialized as nurturing caregivers while men are socialized to seek radical autonomy, and both men and women may seek social power through cultivating the appearance of radical autonomy from others and nature. Since radical autonomy is not actually attainable, those who seek to assert their complete independence may be driven to "dominate the world so thoroughly that the autonomy of all else is wiped out" (Gomes & Kanner, 1995, p. 114).

Gomes and Kanner (1995) contribute an important aspect to the critique of domination as a foundation for the CST of sustainability. They emphasize that gender-based discrimination and domination contribute in important ways to the unsustainability of modern social systems, and they point to the reduction and eventual elimination of these forms of discrimination and domination as important aspects of developing sustainable socio-ecological systems.

Robert Bullard's "Confronting Environmental Racism" (2008) offers a critique of domination based on race. Bullard offers an excellent analysis of racism as an important factor influencing the siting of dirty industries and environmentally damaging activities of all kinds. In his analysis, Bullard demonstrates that racism operates as a clearly distinguishable factor apart from class status.

Racism infused colonialism (Pagden, 1982/1988), and Bullard draws important connections between 1) colonialism as an extractive enterprise within far flung empires and 2) continuing "internal" colonialism within countries.[11] Colonialism, internal colonialism, and globalization, with their racist foundations, form the foundation for the environmental justice movement. This movement seeks an end to discriminatory placement of hazardous production and waste facilities near communities whose ethnic and economic history and status disadvantage them in terms of social power (First National People of Color Summit, 1991/2008).

---

[11] In his book *This Sovereign Land: A New Vision for Governing the West*, Daniel Kemmis (2001) makes a similar argument about the West within the U.S. serving as an internal colony for the entire nation. Bullard's argument also closely relates to Alan Miller's article "Economics and the Environment" (1999). Miller's discussion of world-system theory highlights how colonialism has never really ended; it has only been transformed into economic colonialism that continues to manifest in the dependency of economically peripheral nations on the centers of power.

Critique of racist domination is important to developing the CST of sustainability for similar reasons that critique of gender discrimination is important to developing such a theory: sustainability means an end to destructive domination and engulfment of the other—be the other women, ethnic/racial groups, or nature.

The critique of domination also extends to language. According to Horkheimer, "The spoken word cannot deny its collective coinage, for language is a true reflection of social structure" (as quoted in Tar, 1977, p. 92). In a culture of domination and oppression, we tend to speak the language of subjugation.

In her article "Keepers of the Earth" (1995), Jeannette Armstrong, a member of the Okanagan tribe of British Columbia, discusses linguistic foundations of human domination of nature. Throughout her article, she discusses differences between Okanagan and English in how each represents relationship. Armstrong shows how the subject/object construction of English and other Western European languages (which nearly always represent humans as the subjects acting upon objects) creates a mental construct of humans as separate from environment and thereby fosters an orientation toward domination of nature rather than reciprocity. By contrast, in Okanagan, *human* and *earth* share the same root syllable. Humans are conceived of as the "land dreaming capacity" (p. 321); they *are* the earth. Okanagans embody their homeland—and all that shares that homeland with the tribe. They are made of and directly related to that which the land provides them (p. 323). As Armstrong shows, there are ways of being in the world and with nature that are not characterized by domination, though globally dominant cultures face serious linguistic challenges to conceptualizing them.

The critique of domination outlined here lies at the heart of the CST of sustainability. This critique builds upon the foundations of Marx's critique of the alienation of labor and Polanyi's critique of modern industrial society's subservience to the free market economy. It incorporates the insights of Adorno, Horkheimer, and Marcuse regarding the extension of capitalist domination into the realm of culture, and it highlights the important role of otherization based on gender, race, and ethnicity as contributing factors in the sustainability crisis. We now turn to a discussion of how (re)inventing human-to-human and human-to-nature reciprocity can form a foundation for sustainability praxis.

Reciprocity as an Organizing Principle for Sustainability

The critique of domination provides us with useful insights into how and why industrial society finds itself on the brink of socio-ecological disaster, but it does not provide a roadmap leading us back from the brink. According to sociologist Robert J. Antonio (1981), "Critical theory has emphasized primarily the negative moment of the dialectic. It has attacked domination, rather than describing explicit, determinate possibilities for new social formations" (p. 341). Critical theorists effectively point to the sources of the human/nature destruction we are currently witnessing provide openings for sustainability-oriented social change, but the question remains for us to answer as we face the sustainability crisis: *how exactly can humans live fulfilling lives while also reconciling their activities with natural limits and with the health of the natural world?*

The character and processes of the collective life of humans lie at the center of questions of sustainable living. Chapters three and four of this text will focus on ideas and examples that can usefully inform sustainable community life. For now, suffice it to say that two problems of unsustainability must be addressed: the vast scale of capitalist society and the near complete lack of human-to-human and human/nature reciprocity as a basis for building meaningful relationships. As a number of authors note, individuals adapting their desires and actions to serve community need not result in domination of individuals by the collective (Armstrong, 1995; Berkes, 1999; Martinez, 1997; Polanyi, 1944/1957). In fact, individuals can grow intellectually and spiritually and develop meaningful relationships with people and places *through* engaging in community building work (Loeb, 1999). The vast scale of globalized society interferes with building or even recognizing reciprocal relationships, and the capitalist system integrates individuals into its logic through instrumental relationships in which people serve as means to ends rather than centers of experience and value in their own right.

Even in the face of the momentum of the capitalist system, however, we should not assume that meaningful, reciprocal relationships are somehow unnatural in human society. As Polanyi aptly demonstrates in his classic work *The Great Transformation* (1944/1957), self-regulating communities, founded upon principles of reciprocity, have been the norm for all but the most recent human history. It is through the process of creating free markets that communities and the lives of individuals have been made secondary and subservient to a self-regulating free market. Polanyi states:

> The outstanding discovery of recent historical and anthropological research is that man's economy, as a rule, is submerged in his social relationships. He does not act so as to safeguard his individual interest in the possession of material goods; he acts

so as to safeguard his social standing, his social claims, his social assets. He values material goods only in so far as they serve this end. Neither the process of production nor that of distribution is linked to specific economic interests attached to the possession of goods; but every single step in that process is geared to a number of social interests which eventually ensure that the required step be taken. These interests will be very different in a small hunting or fishing community from those in a vast despotic society, but in either case the economic system will be run on noneconomic motives. (p. 46)

According to Polanyi, the free market's influence on the shape and character of modern societies marks a radical historical departure that places people in service to the economy, rather than the reverse. This development is a substantial force behind forms of domination and oppression analyzed by critical theorists. Polanyi continues:

The explanation, in terms of survival, is simple. Take the case of a tribal society. The individual's economic interest is rarely paramount, for the community keeps all its members from starving unless it is itself borne down by catastrophe, in which case interests are again threatened collectively, not individually. The maintenance of social ties, on the other hand, is crucial. First, because by disregarding the accepted code of honor, or generosity, the individual cuts himself off from the community and becomes an outcast; second, because, in the long run, all social obligations are reciprocal, and their fulfillment serves also the individual's give-and-take interests best. Such a situation must exert a continuous pressure on the individual to eliminate economic self-interest from his consciousness to the point of making him unable, in many cases…, even to comprehend the implications of his own actions in terms of such an interest…. The premium set on generosity is so great when measured in terms of social prestige as to make any other behavior than that of utter self-forgetfulness simply not pay. (1944/1957, p. 46)

Derek Jensen (2000), drawing on the work of anthropologist Ruth Benedict, makes similar points. He explains how many indigenous societies "eliminate the polarity between selfishness and altruism by making the two identical" (p. 212). According to Jensen, in a society which operates on principles of mutualism and reciprocity, to behave selfishly would be considered insane (p. 212). Indigenous authors and authors who have studied indigenous societies also emphasize the importance of social relationship and reciprocity. By contrast to modern industrial societies, many traditional indigenous societies are not inherently oppressive but are, instead, governed by mutuality and reciprocal relationship within the community and between people and nature (Armstrong, 1995; Berkes, 1999; Cajete, 2001; Kropotkin, 1902/1989; Martinez, 1997; Nelson, 1983; Polanyi, 1944/1957, chap. 4; Salmon, 2000; Sveiby, & Skuthorpe, 2006).

Aggressive and conquering, armed with advanced technology in the form of weapons, and aided in conquest by disease (Diamond, 1999; Mann, 2002), Western society has overwhelmed nearly all precapitalist indigenous societies. But, the sustainability crisis will signal an end to global growth capitalism as we have known it (Astyk, 2008; Campbell & Strouts, 2007; Guggenheim, 2006; Heinberg, 2005; Homer-Dixon, 2006; Intergovernmental Panel on Climate Change [IPCC], 2007; Li, 2008; Meadows, Randers, & Meadows, 2004; Simmons, 2005), and to live well beyond the crisis, or perhaps even survive, we will need to (re)learn life in reciprocating society.

Examples of indigenous lifeways can help us conceptualize the principles and processes at work in a reciprocating society. Jeannette Armstrong (1995), comments on her life lived in two worlds: the Western world dominated by oppressive industrial capitalism and the lifeworld of her traditional indigenous culture. She describes her experience of Okanagan socio-cultural and environmental destruction within the dominant society: "I have always felt that my Okanagan view is perhaps closer in experience to that of an eyewitness and refugee surrounded by holocaust" (p. 317). Through reading "Keepers of the Earth" (1995), we learn that the Okanagan people see themselves as one with their land, as incarnations of the land itself (p. 324). The land gives the Okanagan people their language—their entire system of meaning making (p. 323). As noted above, in Okanagan, people are earth, and they have a duty to take care of the earth as source of all life (p. 324). We also learn that, Okanagans conceive of the individual as a melding of four selves: the physical self, the emotional self, the thinking/intellectual self, and the spiritual self. According to Armstrong, all four selves are important and must exist in balance with each other as complementary parts of a whole person (pp. 320–322). This view differs from modern Western thinking in which the rational intellect is more valued and trusted than other parts of the person. In Okanagan culture, a leader is not characterized above all else by his/her intellect or by her/his ability to dominate others or systematically extract surplus value from others and nature within rationalized systems of production. Instead, leadership ability is judged by a person's emotional ability to connect with others (p. 321).

Like Armstrong, Dennis Martinez (1997), of O'odam and Crow tribal heritage, emphasizes how indigenous peoples have lived in reciprocating relationship with the land. According to Martinez, many indigenous societies have not only avoided inflicting long-term damage to ecosystems, they have developed cultures of care for the earth that make their presence on the land an ecological benefit. Martinez (1997) explains how Indian burning and indigenous harvest of plants and animals benefited the long-term health of native homelands. He also explains that, when the U.S. Forest Service prohibited Native American use of some lands in an effort to protect the land and species harvested by indigenous tribes, the ecosystems declined due to the absence of human participation in ecological processes (pp. 116–118). Charles Mann, makes similar points in his article "1491" (2002) in which he discusses evidence that the Americas were highly populated prior to the arrival of Western colonial conquerors. According to Mann, the human imprint on the land was everywhere, and human presence was in many cases nondestructive, even beneficial. He claims that modern societies, in order to live sustainably, will need to "find it within themselves to create the world's largest garden" (p. 53).

Referring to preindustrial, precapitalist human history, Martinez states:

> There had to have been a way for people to have lived sustainably, because these populations, smaller than we have at present, were here for a very long time. People could have exhausted a resource in any given generation very easily, had they not had a fundamental restraint and a fundamental notion of reciprocity—what to give back to that system. (1997, p. 110)

Martinez makes it clear that he is not promoting the myth of the noble savage (p. 110), a myth that conceptualizes indigenous people as half-human creatures, not even intellectually capable of doing harm. For indigenous societies past and present, learning to live sustainably on the land and developing cultures centered around doing so was and is a practical matter for sustaining all aspects of the human being. Living sustainably in the long term has not meant absence of mistakes made by individuals and cultures, and it has not meant that indigenous cultures are and were devoid of conflict or violence. It has meant that many indigenous societies developed cultures that allowed them to carry on for long periods of time in the specific places where their cultures served to articulate sustainable connections between people and place. Such societies are described by Enrique Salmon (2000) as "kincentric."

The critical theory of sustainability articulated here does not promote the notion that the sole aim of environmental activism should be to maintain pristine landscapes, untouched by human presence. Instead, it seeks to develop ideas and practices that restore health to the interaction between humans and environment. What restoration means with regard to particular places and human uses of environments must be decided on a case-by-case basis, and the decisions themselves must remain contingent over time as changes occur on the land and in human societies. The CST of sustainability recognizes that, in the words of Martinez (1997), "There are very few places on this globe that one could adequately describe as pristine" and that "the anthropogenic landscape" has existed for a very long time (p. 109). The CST of sustainability subscribes to the view articulated by Martinez that "biological diversity and cultural diversity are linked. You cannot have one without the other" (p. 109).

Given precapitalist human history as conveyed by Polanyi (1944/1957), Mann (2002), and others, and given that indigenous societies offer examples of living in healthy relationship with nature (Armstrong, 1995; Berkes, 1999; Cajete, 2001; Martinez, 1997, Nelson, 1983; Salmon, 2000; Sveiby & Skuthorpe, 2006), a central focus for the critical social theory of sustainability should be restoring systems of reciprocity and mutuality within contemporary lifeways. In the words of Dennis Martinez, "Do you think you can go on and on, decade after decade, taking and taking and taking and not expect something bad to happen? It's impossible. It's a complete violation of Natural Law" (1997, p. 119).

Gramsci's Concepts of Cultural Hegemony and Passive Revolution as Bases for Analysis and Action

Like the critique of domination and the notion of reciprocity, the concept of cultural hegemony deeply informs the critical social theory of sustainability. The concept of hegemony discussed here is articulated by twentieth century political theorist Antonio Gramsci (1971/1999, pp. 57–58).[12] Gramsci's conceptualization of bourgeois cultural hegemony highlights cultural barriers to acting in opposition to entrenched systems of social power that damage the environment and reduce quality of life for many. Gramscian Marxism is concerned with why revolutionary movements have failed in Western countries (Salamini, 1974, p. 363) and why fascism arose in Europe— concerns Gramsci shared with the critical theorists of the Frankfurt School.[13]

According to Gramsci (1971/1999), "Hegemony designates a system of social control, and specifically the control of the subaltern classes and groups, without the preponderant use of force/coercion." It is a system

---

[12] This section deals with modern hegemony as an expression of bourgeois social control. It is important to recognize that Gramsci advocates a form of hegemony of the masses within which the contradictions of society would be eliminated and the necessity for Marxist praxis would cease to exist (Salamini, 1974, p. 372).

[13] Both the German theorists and the Italian theorist were intellectually active during the rise of European fascism. While imprisoned by Mussolini's fascist regime, Gramsci wrote perhaps his most famous work, *The Prison Notebooks* (1926–1934/1996).

characterized by domination whereby the oppressed assume the values and worldview of their oppressors and, thereby, engage in their own oppression (pp. 57–58; see also Persaud, 2001, p. 37). Hegemony has cultural and political components. Political hegemony, the control of decisions of society, may occur through force, as in political dictatorship, or it may result from deep social penetration of cultural hegemony. Gramscian analysis focuses on the latter form of hegemony, in particular how hegemony serves as a platform for bourgeois capitalist domination in modern societies. According to Salamini (1974), "Gramsci conceives of hegemony as an ideological phenomenon first, and only secondly as political fact" (p. 368). Ideological hegemony is the social "leadership" Gramsci sees as a precondition to achieving political leadership in modern capitalist societies and a requirement for maintaining it:

> A social group can, and indeed must, already exercise "leadership" before winning governmental power (this is indeed one of the principal conditions for the winning of such power); it subsequently becomes dominant when it exercises power, but even if it holds it firmly in its grasp, it must continue to "lead" as well. (Gramsci, 1971/1999, pp. 57–58)

Gramscian cultural hegemony functions as an internalized colonization in which subaltern classes actively participate in their own oppression in cultural, economic, and political life. Freire eloquently describes this system in action:

> At a certain point in their existential experience the oppressed feel an irresistible attraction towards the oppressors and their way of life. Sharing this way of life becomes an overpowering aspiration. In their alienation, the oppressed want at any cost to resemble the oppressors, to imitate them, to follow them. This phenomenon is especially prevalent in the middle-class oppressed, who yearn to be equal to the "eminent" men and women of the upper class. (1970/2000, p. 62)

In a Western world replete with racism, sexism, and classism, conceptual divisions wherein the myth of meritocracy is largely assumed truth, many of those oppressed by dominant groups internalize belief in their own inferiority along with belief in the corresponding superiority of the dominant. Modern capitalist society is a hegemonic society, and cultural hegemony is a powerful elixir that unites people—even those whose interests are not well served—in reinforcing the trajectory of history led by dominant groups and institutions. Accordingly, we can speak of the oppressed who believe they *choose* to participate in capitalist, consumerist societies.

Gramsci's concept of passive revolution (1971/1999, pp. 105–120) is also important to understanding how cultural hegemony is perpetuated and violent revolution averted by hegemonic blocs. Passive revolution becomes necessary when subaltern groups recognize that their interests differ from those of dominant groups and when this recognition and the social turmoil it engenders threaten to rupture the hegemonic order. In order to maintain hegemonic control, political leaders may undertake passive revolution in which they act in opposition to their own short-term interests by engaging in alliances and pursuing top-down reforms that mitigate the sources of dissatisfaction among the subaltern classes. Passive revolution, by avoiding a rupture in the overarching socio-economic and political order, serves the long-term interests of the powerful. It also deepens cultural hegemony by creating the illusion that dominant and subaltern groups share important common interests.

The concept of cultural hegemony is highly compatible with critical theory whose proponents contend that dominant culture serves to entrench the powerful and to reproduce and extend the capitalist sphere while also undercutting impetus toward socialist revolution. As already noted, Marcuse's *One-Dimensional Man* (1964) is a prime example of CST in this vein. In *One-Dimensional Man*, Marcuse develops a far-reaching and sophisticated analysis of repressive desublimation as a form of hegemony he observes in American society. Although Marcuse never cites Gramsci in this work, his arguments build a bridge for uniting CST with Gramscian analysis of hegemony. In *One-Dimensional Man*, Marcuse compellingly and clearly analyzes the interests and mechanisms that create, perpetuate, and deepen social, economic, and political hegemony. Marcuse's work articulates directly with that of Gramsci regarding cultural hegemony and passive revolution as social phenomena central to defusing the potential for revolution in capitalist societies.

However, it is important to recognize that a fairly high level of material comfort must be enjoyed by many people as a precondition for widespread repressive desublimation (Marcuse, 1964). By contrast, cultural hegemony can pervade all social classes. Therefore, Gramsci's concept of cultural hegemony is more effective than Marcuse's notion of repressive desublimation for developing a CST of sustainability applicable to socio-economic contexts of extreme poverty and to societies existing on the periphery of the globalized world.

The hegemony spawned within colonized populations persists today and supports global capitalism as a world-system. Citizens of poor and indebted "developing" nations are led to believe that they will one day enjoy the material comforts of their former conquerors, and many internalize this myth. Idealized images of Westerners and Western lifestyles portrayed in media such as Hollywood films (Henzell & Rhone 2002), magazines, and advertisements contribute to the seduction of the oppressed worldwide into the hegemonic system. Hegemonic belief systems operate to further entrench the powerful within the world-system as well as within individual nations.

Helena Norberg-Hodge describes hegemony in action in Ladakh in the film *Ancient Futures* (International Society for Ecology and Culture [ISEC], 1993) where she documents the social dislocation and environmental damage that occur when Western culture and global capitalism descend upon a subsistence culture. She attributes the Ladakhi people's willingness to participate in these changes that damage the local environment and disrupt belief systems and practices of reciprocity to a "psychological pressure to modernize." According to Norberg-Hodge, much of this pressure stems from Western media and advertising as well as contact with Western tourists.

Cultural hegemony as a critical lens shares much with CST critiques. In questioning hegemonic cultural reinforcement and perpetuation of oppressive capitalist society, CST admits that we have been lied to—and that we lie to ourselves and others about deeply important things. CST argues that many of our hopes and beliefs about our culture, our politics, and the workings of the world are in fact built upon lies. Some of the lies give rise to alluring and comforting fables in the form of official stories about our institutions and societies: that we and our modern culture are on a linear and upward path to ever-growing and improved knowledge and material well-being (and the corollary belief that all pre-modern and non-modern societies were/are lesser, backward, ignorant), that we as free people collectively chose our current reality because it was the best choice possible, that we moderns are freer than any other people has ever been, that we live in a meritocracy rather than a class-based society, that everyone in the world wants to (and should want to) be like us (and the corollary that those who are not like us are somehow inherently defective) (Bennet, 2007; Clark, 2005; Jarecki, 2006; Spretnak, 1997). Social systems that effectively perpetuate hegemony successfully create such illusions of freedom and choice along with ideologi-cal myths that become so deeply interwoven in the social fabric of a nation, a people, a culture that to unravel the rotten threads would threaten the very integrity of the social fabric. Successfully hegemonic systems also impercep-tibly require people to weave the threads of official lies into their own identities so that a threat to hegemony is perceived by individuals as a threat to personal integrity. Therefore, if taken seriously, a CST-based critique calls for a rather painful assessment of what lurks behind the façade of modern culture as well as an assessment of the shadow parts of ourselves that articulate with oppression.

Cultural hegemony is also present in critiques about the utility of CST as a lens for analysis and praxis toward sustainability. For members of dominant cultures, it is quite natural that the kind of deeply probing analysis generated by a CST-based inquiry informed by notions of Gramscian cultural hegemony would lead to deeply disquieting emotions: a sense of betrayal, guilt, sadness, rage (Bennet, 2007; Clark, 2005; Jarecki, 2006; Jensen, 2004). Such emotional awakenings can prompt us to ask: isn't CST too depressing, incapable of inspiring and motivating for change, and just too uncomfortable to be worth it? Even those who are comfortable with many forms of deep cultural critique represented by such schools of thought as deep ecology and ecopsychology may feel repulsed by the notion that we are pawns and tools of powerful interests. It is natural in a hegemonic culture to sense that there must be something wrong with ideas that question the very foundations of society. Such critique, after all, is—and explicitly aims to be—destabilizing in that its ultimate goal is the remaking of society itself.

Because they shed light on processes of domination and exploitation active within unsustainable hegemonic societies, Gramscian notions of hegemony and passive revolution create points of departure for critical pedagogy based on the CST of sustainability. Recognizing that these processes are entrenched in modern societies clarifies that engaging in counter-hegemonic critical pedagogy is a very challenging prospect. Still, because the sustainability crises will increasingly compel us to critically examine our social forms and practices in an effort to develop sustainable lifeways, opportunities may increase for counter-hegemonic thinking and education to play increasingly prominent roles in creating a sustainable future.

In chapter two, I will develop a theory of sustainability praxis that derives from the critique of domination developed in this chapter. Together, these two chapters articulate my overarching CST of sustainability. The theory of praxis developed in chapter two informs my critical pedagogy of sustainability and can also inform hopeful, sustainability-oriented social engagement that addresses the sustainability crisis and its causes.

Chapter 2:

Agency, Epistemology, and Ontology in Sustainability Praxis

You act not to achieve a certain outcome; you act because it is the right thing to do.
                              —Paul Wilson reflecting on the life of Václav Havel

To this point, I have articulated a CST of sustainability informed by an overarching critique of domination as a pervasive theme in modern industrial societies and by indigenous worldviews and systems of reciprocity. Even if we find the CST of sustainability accurate and convincing, though, without a concomitant theory of praxis that locates both critique and agency within the process of history, we are left with an idealist foundation for action. According to the idealist formulation, conceptual knowledge is the foundation for social change. If we assert, as I suggest here, that both the materialist and idealist conceptions of consciousness and being are operative in any one person or context—that consciousness determines being while being also determines consciousness—we have begun to articulate a foundation for sustainability praxis. We can work to open people's minds to sustainability-oriented critiques of society *and* work to build on-the-ground examples of sustainable living capable of influencing people's consciousness and driving social change. This is an essentially Gramscian conception of agency (Gramsci, 1971/1999, pp. 333–334).

In this section, I probe a central tenet of CST: that social formations are historical and, therefore, open to change through human agency—that nothing in human history is absolute or immutable because all history is contingent upon past and future action. It is important to remember that CST not only calls for us to identify the sources and processes of hegemony and domination; it calls upon us to *liberate ourselves* from oppression. The CST of sustainability rests upon the conviction that people have the capacity to transform the world. Otherwise, why theorize about the need for transformation? And as educators, why teach sustainability? The purpose of this section is to elucidate the concept and role of praxis in sustainability-oriented education and social change.

Gramsci articulates a philosophy of praxis that derives from historicist foundations. His notion of praxis serves exceptionally well as a springboard for the philosophy of sustainability praxis, an essential component of the CST of sustainability. Gramsci's Marxism is highly original in that he focuses on the role of critical consciousness building as an important and necessary subjective aspect of revolutionary change. Salamini summarizes Gramsci's view on the role of subjective consciousness in realizing radical social change:

> The domination of a class over another is always the domination of a given [worldview] over another; consequently any revolutionary movement, if it is to be successful, has to be preceded by a profound intellectual and cultural reform of human consciousness. (Salamini, 1974, p. 378)

With regard to the role of consciousness, Gramsci's work strongly parallels the pedagogy of Paulo Freire (1970/2000).

According to Salamini (1974), Gramsci's philosophy of praxis arises from the premise of hegemony as a cultural and ideological fact (p. 370). Since the hegemony of the bourgeois effectively assimilates the masses into supporting the interests of the ruling classes, Gramsci sees "ideological revolution as a precondition for political revolution" (Salamini, 1974, p. 368). According to Gramsci, in order to effect revolutionary change, people must develop a critical self and social consciousness that allows them to step outside hegemonic culture and recognize how they have been manipulated into supporting a society that does not serve their best interests. Salamini (1974) summarizes Gramscian theory regarding the role of critical consciousness in informing radical social change:

> In historical situations where the power resides formally in the hands of ruling classes, the working classes, politically organized and conscious of their role, can and must exercise an ideological hegemony by subtracting themselves from the bourgeois ideology and progressively attracting into their orbit all other subaltern classes. Ideological hegemony (defined as "intellectual and moral direction") is a preliminary condition for the actual seizure of state power and the creation of a new state. The proletariat, Gramsci contends, can and must become a *dominant* class before becoming a *ruling* class. (p. 368)

The Gramscian conception of the role of consciousness in building a hegemony of the masses that has the power to move society toward liberation from social contradictions closely parallels Marcuse's call for people to engage in "the great refusal" to collude with the capitalist project (1964, pp. 255–257).

This conception also strongly parallels the Freirean (1970/2000) notion of concientization as a process through which individuals become capable of "naming the world" (p. 88)—of clearly comprehending their own situation and interests—rather than allowing hegemonic interests to name the world for them.

Gramsci (1971/1999) sees the development and recognition of the concept of hegemony itself as a major step in reasserting the roles of human consciousness and agency in praxis:

> Critical understanding of self takes place therefore through a struggle of political "hegemonies" and of opposing directions, first in the ethical field and then in that of politics proper, in order to arrive at the working out at a higher level of one's own conception of reality. Consciousness of being part of a particular hegemonic force (that is to say, political consciousness) is the first stage towards a further progressive self-consciousness in which theory and practice will finally be one. Thus the unity of theory and practice is not just a matter of mechanical fact, but a part of the historical process, whose elementary and primitive phase is to be found in the sense of being "different" and "apart," in an instinctive feeling of independence, and which progresses to the level of real possession of a single and coherent conception of the world. This is why it must be stressed that the political development of the concept of hegemony represents a great philosophical advance as well as a politico-practical one. For it necessarily supposes an intellectual unity and an ethic in conformity with a conception of reality that has gone beyond common sense and has become, if only within narrow limits, a critical conception. (pp. 333–334)

The concept of hegemony places Marxist thought and action within the realm of human strategies for social change. This conceptual framework informs Gramsci's reassertion of Marxist humanism through which he reclaims Marxist thought from the hands of scientific Marxism with its theories of "necessary, constant, or immutable economic laws" (Salamini, 1974, p. 372). In Gramsci's thinking, "any law of automatism admitted in the analysis of sociohistorical phenomena tends to stifle human will and human creativity and tends to mystify and alienate human consciousness" (Salamini, 1974, p. 372). Adherence to belief in automatism creates in individuals a dangerously passive response to modern capitalism, one that can serve to reinforce the hegemony of bourgeois society (Salamini, 1974, p. 363).

Though he emphasizes the role of consciousness in realizing social change, Gramsci sees idealism as incapable of uniting theory and practice (Salamini, 1974, p. 364). His philosophy of praxis is absolutely historicist. Salamini (1974) elucidates the humanist and historicist foundations of Gramsci's philosophy of praxis:

A basic historicist assumption is that any idea exists and develops in dialectical relationship with praxis. Praxis is history; it develops and transforms itself with history.
Marxism as a philosophy then is history becoming conscious of itself, and since
history is conscious human activity, man becomes the central focus of reality (humanism). In sum, in its very essence Marxism is "absolute historicism" and "absolute humanism." Failure to lay the active and conscious human will at the base of a
Marxist [worldview] inevitably leads, Gramsci warns, to solipsist theories (subjective idealism) for which the self is the only object of knowledge or mechanistic theories (positivism and scientific Marxism), which posit the existence of necessary laws
and principles in the historical development. In this respect, Marxism is the process
of historicization of human thought, which relocates ideas and ideologies in their
specific and concrete historical framework, the process of relativization of existing
social structures and social arrangements. (p. 371)

For Gramsci, ideas and theories are themselves historical and function
historically in the ongoing creation of reality. He states:

This is the central nexus of the philosophy of praxis, the point at which it becomes
actual and lives historically (that is socially and no longer just in the brains of individuals), when it ceases to be arbitrary and becomes necessary—rational—real.
(Gramsci, 1971/1999, p. 369)

Still, according to Salamini (1974), Gramsci's conception of ideas as
historical does not "preclude the assessment of the truth and the validity of
such ideas. Their validity is determined…by their capacity to mobilize and
guide the masses toward the attainment of ideological and political hegemony" (p. 372). The hegemony toward which society should aspire is one in
which the contradictions of capitalist society that manifest in domination and
exploitation are ultimately eliminated (Gramsci, 1971/1999, p. 405). Similarly, CST seeks to open up space and processes for people to "name the
world" (Freire, 1970/2000. p. 88)—to tell their own truths from their own
positions. These truths vary, but the process of identifying conceptions of
hegemony and hegomonic powers through critique is not a completely
relativistic exercise. It involves the searching for *themes* of oppression that
are repeated and related in their sources, even though they interplay with
individual and community variance to create a wide variety of specific
incarnations.

Understanding the historical character and the goals of Gramsci's notion
of praxis allows us to draw clear connections between praxis and the social
justice aspect of sustainability. When we contextualize praxis within the
definition of sustainability articulated above, we recognize that a society free
of contradictions cannot be one that destroys humanity and nature. Such
behavior would introduce social contradictions in the form of domination

that restricts the freedom of people and other creatures to realize their own ends. Therefore, Gramsci's conception of praxis—as a process through which humans consciously develop a form of hegemony that eliminates contradictions from society—in fact *embodies* sustainability-oriented social change in its complete socio-ecological milieu. Sustainability praxis must necessarily transcend fragmented, disciplinary inquiry and action and seek to develop an integrated worldview capable of articulating with history in all its dimensions. Such a transdisciplinary epistemology has been a central goal of critical theory from its inception (Held, 1980, p. 34).

Praxis as conceived by Gramsci is self-reflexive in that theory is mani-fested within and through history. Marxist theory is seen as infused with specific class interests that themselves contain social contradictions which Marxism seeks to eliminate. Gramscian Marxism, therefore, foresees its own slow erasure through the process of praxis: "If…it is demonstrated that contradictions will disappear, it is also demonstrated implicitly that the philosophy of praxis too will disappear, or be superseded" (Gramsci, 1971/1999, p. 405). Salamini (1974) elaborates:

> The task of resurrecting and rejuvenating Marxism requires a radical redefinition of Marxism itself. Marx has to be read in a new light and has to be considered as only a phase in the evolution and the elaboration of the "philosophy of praxis".… Marxism as historicism signifies that it is, within history, a theory of history, itself a transitory phase in the history of the development of human thought. (pp.370–371)

Because, for Gramsci, there is no absolute theoretical validity, Gramscian Marxism does not prescribe for itself a universal validity beyond history: "Every theoretical system has validity within the limits of a specific histori-cal context, therefore it is bound to be superseded and deprived of signifi-cance in the succeeding historical context" (Salamini, 1974, p. 371). The validity of praxis can be judged by its demonstrated and living ability to reduce and ultimately eliminate social contradictions. According to Salamini (1974), "The development of subaltern classes and their ascendant movement toward cultural, ideological, and political hegemony is the most fundamental criterion for the analysis of all historical, social, and cultural phenomena" (p. 387). Gramscian praxis is highly compatible with the concept of sustainabil-ity articulated above in which sustainability is viewed as historically, culturally, and place specific. Sustainability is not a set of prescribed rules applicable to any and all situations. Instead, it is a process of praxis that seeks to eliminate socio-ecological manifestations of domination and oppression that circumscribe the freedom of nature and people to realize their own ends. The effectiveness of any sustainability praxis can be judged according to its ability to satisfy these purposes.

The Gramscian conception of praxis provides an intellectual home for sustainability educators and activists living and working in a world of contradictions. Within a sustainability context, many of us are conscious of the harm we do to others and the world by simply living a "normal" industrial lifestyle, but most of us pursue only partial solutions, if any at all. But then, we are also living in a world where our choices are limited by the systems and structures within which we live, and we understandably take actions that make "common sense" within that world because we "must" do so if we wish to remain part of the world as we know it. Within this setting, our consciousness is very much shaped by our being within the world as we know it so that we may act in two contradictory ways: by adapting to the world as it is and by seeking to change the world. Even in this inconsistency, I see potential for a transformative agency. Many of us comprehend that the world as we know it is indeed unsustainable and are taking action toward change, even as we embody contradictions of the unsustainable world within our praxis.

Through his philosophy of praxis, Gramsci creates a sociology of knowledge highly relevant to sustainability. "For Gramsci, objectivity represents an *inter-subjective consensus* among men; that is, objectivity is a *historicized* and *humanized* objectivity" (Salamini, 1974, p. 376). This intersubjective consensus is lived into being within specific contexts and can, therefore, actively reflect a multiplicity of worldviews based in history and place (Salamini, 1974, p. 369). This diversity mirrors that of resilient ecosystems and societies. The CST of sustainability has an explicit values framework rooted in the critique of domination and oppression. From this framework, intersubjectively determined, historically and culturally specific guideposts for sustainability can be set and developed through praxis. In Gramscian terms and according to the CST of sustainability, valid knowledge is praxis: "The validity of sociological research resides not in its scientific function but rather in its ideological function, that is, in its capacity to organize the experiences of the masses" (Salamini, 1974, p. 377).

Those who drain power away from the system by refusing to participate in aspects of society that perpetuate oppressive power relationships are engaging in a form of Gramscian praxis. Marcuse laid the groundwork for praxis that embodies the recognition that opting out of administered society can fundamentally challenge the system (Dant, 2003, p. 62). In his advocacy of the "great refusal" at the end of *One-Dimensional Man* (1964, pp. 256–257), Marcuse offers no clear answer to domination and oppression, but he claims that refusal to play the game, refusal to fit one's life to hegemonic realities, may signal the beginning of the end of the current paradigm. Chris Carlsson, author of *Nowtopia* (2008), demonstrates that many of those who opt out engage in unpaid work in the autonomous sphere of life that simultaneously decreases their reliance on capitalist production, distribution, and wages (chap. 1). According to Carlsson, we need critical, reflexive, creative destruction of late capitalism and its culture in order to create a free and sustainable world. Gramsci (1971/1999) offers a conception of praxis that is particularly apt for these purposes.

Engaging in the great refusal is often misinterpreted as nonaction because to refuse is seen as not constructive—constructive in the sense of actually building something. I would say that engaging in the great refusal *is* constructive. It means living a coherent life by refusing to collude with power in order to benefit personally at the expense of others and nature. Such action is a manifestation of the critical self and social consciousness Gramsci (1971/1999) deemed so necessary to praxis (pp. 333–334). Living this refusal implies taking an unusual path in life by building a lifeworld that lives into being an alternative worldview: the great refusal leads to praxis. In living such a life, one is telling a powerful story to others who may be inspired to also engage in the great refusal. Those who refuse to fit themselves to the current paradigm build relationships among themselves and create ways of living that embody the refusal and that begin to advance a hegemony of the masses that may eventually resolve social contradictions based in domination and oppression. Native American resistance to Western cultural hegemony also represents such (re)creation of alternate lifeways rooted in refusal to assimilate into the dominant paradigm (see for example Armstrong, 1995; Martinez, 1997; Nelson, 1983; Salmon, 2000). In Gramscian praxis, we see that alternatives to oppressive hegemony are in part present in critique itself, which is the spark that ignites praxis (Gramsci, 1971/1999, pp. 333–334).

Homer-Dixon highlights the need for critically informed agency to address the sustainability crisis. He claims that, in a time of disintegration,

> people will want reassurance. They will want an explanation of the disorder that has engulfed them—an explanation that makes their world seem, once more, coherent and predictable, if not safe. Ruthless leaders can satisfy these desires and build their political power by prying open existing cleavages between ethnic and religious groups, classes, races, nations, or cultures. First, they define what it means to be a good person and in so doing identify the members of the we group. Then they define and identify the bad people who are members of the they group. These are enemies such as immigrants, Jews, Muslims, Westerners, the rich, the poor, or the nonwhite, who are the perceived cause of all problems and who can serve as an easy focus of fear and anger. (Homer-Dixon, 2006, p. 279)

Homer-Dixon continues by stating that extremists have an advantage over nonextremists in times of crisis. They tend to be better organized, have a more coherent philosophy, have clear goals, have a clear identity, be dedicated to their ideas, and even have a plan of action. Therefore, they can mobilize quickly (pp. 291–292).

Engaging and encouraging others to engage in consciously informed praxis *now* can reduce the possibilities for demagogues and fascists to seize power during the sustainability crisis. To serve these purposes effectively, praxis needs to be undertaken as pervasively as possible in a wide variety of settings. Small groups and localized communities must become the locus for praxis in order for praxis to remain situated within specific contexts and thereby avoid becoming a failed enactment of ahistorical, idealist formulations of sustainability. Homer-Dixon advocates such activity, especially in the realm of developing community resilience in the basic necessities for life. He states,

> In our communities, towns, and cities, we can use small-scale experiments to see what kinds of technologies, organizations, and procedures work best under different breakdown scenarios.... By experimenting with new ideas about politics, economics, and values, we'll be better advocates of a coherent vision of the future and a plausible way of getting there.... (2006, p. 292)

Sustainability praxis in the realm of consciousness building and creation of alternative lifeways can provide living examples and actual lifelines of resiliency in a time of crisis while also serving to catalyze further sustainability praxis. Chapters four and five will expand upon these ideas.

The struggle embodied in the CST of sustainability is rooted in agency and adheres to the Gramscian conception of praxis. It is always historical. The struggle manifests in history in response to the circumstances at hand; and, therefore, its particular form is always contingent. I believe we are only now approaching a historical moment when large numbers of people might begin to engage in a great refusal of late capitalism, and this refusal might well be driven by both consciousness and the historical circumstances embodied in the sustainability crisis. The Occupy Movement may be early evidence of such an awakening. The sustainability crisis will dictate that modern industrial society cannot continue in its current form. Failure could signal change. Given our personal histories and the form and function of the world we have inherited—characterized as it is by entrenched and self-perpetuating systems of domination and oppression that enforce dependency on the capitalist system—we may be ill prepared for the challenge of remaking societies that are just and sustainable. Still, our historical situation is changing rapidly, and we may soon find ourselves within a socio-ecological context that has more in common with traditional place-based cultures than with modern globalized societies in terms of the energy and tools we have to work with. The CST of sustainability serves as a vehicle for praxis within this new context, but it must have an explicitly ecological orientation to serve effectively in this capacity.

An Explicitly Ecological Framework for Critical Social Theory

As sustainability-oriented theorists, educators, and activists come to see with increasing clarity that oppression of people and domination and destruction of nature are two sides of one coin representing the same underlying exploitive values and practices, critical-theory-oriented analysis is being extended to include human relationship with the environment (Gruenewald, 2003; Kahn, 2010; Kovel, 2002; Leonardo, 2004; Merchant, 1999, 2008; O'Connor, 1991/2008). This more inclusive analysis of domination and oppression is not entirely new (Marcuse, 1972/2008), but a critical theory approach to socio-ecological critique is only recently gaining strength.

Modern consciousness abstracts humans from nature, denying that humans can exist only within nature (Spretnak, 1997). This conceptual human/nature divide is not only a division into two, it is a tiered dualism: humans on top, nature acting in all supporting roles (as tool, as resource, as setting) (Shepard, 1995). The subjugated *other*, first conceptualized as nature itself, is born with this divide. And there have been many *others* as systems of hierarchy have proliferated to encompass gender, races, non-Western cultures, and more. Cultural systems of hierarchy in Western societies and the projection of a hierarchical worldview upon nature itself surely are among the keystone concepts upholding the house of cards that is the unsustainable world of globalized industrialism.

The conceptualization of the human/nature complex as two separate constituents is both a false and destructive construct. Humans are inextricably embedded in ecosystems, though we may so extensively dominate those ecosystems that we no longer clearly recognize our relationships to broader nature. We may stay indoors all day, but the oxygen we breathe still comes from the life activities of plants. We may eat in restaurants far removed from the farm, the oceans, and the rivers, but the food we eat still comes from biological processes dependent upon ecological relationships that we may manage but never fully control. Our intimate, bodily relationships with other living organisms and with ecosystems make us entirely dependent upon the natural world. Whether we recognize our impacts or not, if we live in unhealthy ways, the natural world is damaged, and if the health and resilience of nature are destroyed, people ultimately suffer. Fritjof Capra describes a continuum of life and relationship in his book *The Hidden Connections* (2002). This work highlights the depth of human connection to all life. As Capra shows, we share the opening chapters of the story of our origin, many of our biological needs, and much of our very form and structure with all other life. We emerged through a dependence on earlier life forms, and our dependence on non-human life continues (Capra, 2002, chap. 1).

How does the conception of humans as separate from the environment inform systems of power and exploitation? Instead of focusing attention on our dependence upon and inclusion within ecological systems, we emphasize our ability to control and dominate nature and our ability to seemingly separate ourselves from ecological forces that can threaten our survival as individual creatures. We set up a self-interested priority system of belief and action where human "needs"—which are ever expanding within the capitalist system (Berry, 1987, p. 15; Douthwaite, 1999a; Kovel, 2002, chap. 4)—take priority over nature. In so doing, we have devalued nature relative to ourselves, and we risk destroying that on which we depend (Kovel, 2002). Conceptual separation of humans from environment informs all other systems of thinking and action where domination and oppression of the other are seemingly justified.

Repairing this conceptual rift between humans and nature in Western society is at the heart of achieving ecological sustainability. Working to heal this division differs from preservationist efforts which, recognizing the extent of negative human impacts on environments, typically aim to exclude humans from sensitive or exceptionally aesthetically appealing environments. The environmental justice movement, by contrast, seeks reconciliation between humans and the environment and creation of systems that promote the health of the human/nature complex (Di Chiro, 1996; First National People of Color Summit, 1991/2008). As noted above, this movement can play an important role in reintegrating humans with nature so as to (re)create healthy, sustainable communities. Ecopsychology and deep ecology represent additional areas of thought and work that bridge the conceptual gap between humans and nature in Western culture.

I argue for an explicitly ecological framework for the CST of sustainability, one that conceptualizes humans and nature as an inseparable human/nature complex and one that addresses how the concept of human separability from nature informs Western cultural notions of domination of both nature and other people. I have argued that domination and oppression of people and nature stem from the same worldview and that, ultimately, domination and oppression of the natural world is domination and oppression of people who are part of the natural world and wholly dependent upon nature. The capitalist exploiter sees everyone and everything as a means to an end—a view that negates the intimacy of reciprocity discussed above as foundational to sustainable lifeways.

As noted above in our discussion of reciprocity as a central focus for the CST of sustainability, the modern Western notion of the human/nature dichotomy has not been and is not a feature of every culture. Ideas of human dominance are culturally specific. Many indigenous societies embodying cultural systems of human/nature respect and reciprocity have lived for long time spans in healthy interrelationship with place—(Kropotkin, 1902/1989; Polanyi, 1944/1957, chap. 4). While examples exist of indigenous societies that have misjudged natural limits, indigenous societies that have success- fully lived in healthy interrelationship with place offer us inspiration for new themes that could inform sustainability praxis in modern societies (Arm- strong, 1995; Berkes, 1999; Grim, 2001; LaDuke, 1999; Martinez, 1997; Nelson, 1983; Salmon, 2000; Sveiby & Skuthorpe, 2006). The potential exists to eliminate through praxis the human/nature dichotomy of Western culture. The CST of sustainability focuses attention on the need to do so. The CST of sustainability provides a conceptual foundation for sustainability activists and educators to develop ideas, consciousness, and actions appropri- ate to dispelling the myth of the human/nature divide.

We will now explore the contributions deep ecology and ecopsychology to advancing the CST of sustainability and sustainability praxis.

Deep Ecology and Ecopsychology:
Toward Sustainability Agency

When searching for an effective, overarching socio-ecological critique that would lend itself to environmentally related action, deep ecology and ecopsychology appear at first glance to be more directly relevant to the task than critical theory. CST offers us a lens through which to examine the abuse of power in the "rational" modern world, particularly within a capitalist economy which seems bent upon destroying, not only the foundations for its own survival, but the foundations for life itself. With some notable excep- tions (Marcuse, 1972/2008), however, CST does not situate its analysis explicitly within an ecological context. And although critical theory can address ecological themes (Marcuse, 1972/2008), it typically does not directly critique the human/nature divide that undergirds Western culture. I argue bringing the analyses of critical theory, deep ecology, and ecopsychology into dialog creates a CST that can more effectively address socio-ecological challenges in their full complexity.[1]

It is the conceptual divide between humans and nature in the Western tradition that both deep ecology and ecopsychology aim to heal. These two systems of belief and associated practice open windows upon the divide through which educators and others might glimpse its constitution and character. Still, as is the case with what one can see from any window, the view is informative, even inspiring, but still only partial. Enacting the principles of deep ecology and practicing ecopsychology represent important and useful means to better understand ourselves and our place the world, but they must be complemented by systematic theorizing on political economy—and by action founded upon all of these theories.

In this section, I discuss the contributions deep ecology and ecopsychology can make to conceptualizing a CST of sustainability and to acting on this theory through sustainability education at the college level. I propose that CST, deep ecology, and ecopsychology offer complementary theoretical and practical approaches within the CST of sustainability. I also propose that place can effectively serve as a unifying concept and construct within which these complementary theories can translate most effectively into sustainability-oriented practice.

Deep ecology, founded by Norwegian philosopher Arne Næs, represents a clear attempt to conceptualize humans as part of nature. Næs' first principle of deep ecology exemplifies this goal: "rejection of the man-in-environment image in favor of the *relational, total-field image*" (Næs, 1973/2008, p. 143). With his second principle, "biospherical egalitarianism," (p. 144) he strikes directly at the dualistic hierarchy of humans over nature. The third and fourth principles, "diversity and symbiosis" and "anticlass posture" (p. 144–145) call for elimination of destructive hierarchies and highlight the value of difference as a constituent of healthy mutualism. With his fifth principle, Næs calls upon us to "fight against pollution and resource depletion" (p. 145). Næs enjoins us to counteract the ill effects of the global economy and its requisite economic growth, but he does not offer the kind of deeply systemic analysis of political economy that would help us understand how to combat the problems of pollution and resource depletion at their causal roots.

Deep ecologist Bill Devall (1980/2008) points to deep-seated social and economic sources to ecological destruction and depletion. He writes:

---

[1] Similarly, ecological Marxist James O'Connor (O'Connor, 1991/2008) and eco-socialist Joel Kovel (2002) argue that the present planetary crises call for a conjoining of socialist and ecological vision and action.

Treating the symptoms of man/nature conflict, such as air or water pollution, may divert attention from more important issues and thus be counterproductive to "solving" the problems. Economics must be subordinate to ecological-ethical criteria. Economics is to be treated as a small subbranch of ecology and will assume a rightfully minor role in the new paradigm. (Devall, 1980/2008, p. 158)

Devall envisions an appropriate reordering of social priorities—economy serving socio-ecological ends rather than the reverse—but he does not mention the complexity and difficulty involved in this task. I notice the passive voice—"is to be treated"—and ask: Treated by whom? And how? These questions invite a combination between the principles of deep ecology and CST-based critique of industrial political economy. Deep ecology is rooted in idealism, and CST is rooted in materialism. A brief explanation of the idealist and materialist traditions highlights the suitability of CST as a complement to deep ecology.

Devall's arguments embody an essential idealism that is a source of deep ecology's incompleteness as a theory for guiding needed change. He closes his essay "The Deep Ecology Movement" with the statement: "From the perspective of Deep Ecology, ecological resistance will naturally flow from and with a developing ecological consciousness" (Devall, 1980/2008, p. 159). According to Devall, clearly *consciousness determines being*, an idealist formulation. The historical materialist argument reverses the idealist notion of causality by claiming that *being determines consciousness*. As I have argued above with regard to Gramsci's theory of praxis, both idealism and materialism, taken in their pure form, are incomplete as a foundation for sustainability-oriented praxis.

A recent, striking example of the incapability of the idealist formulation to explain behavior is played out in the film *An Inconvenient Truth* (Guggenheim, 2006). Al Gore relates how he once thought that, once federal legislators in the United States were supplied with clear and reliable information about global warming, they would naturally act to protect the world and future generations by creating policies to solve the problem. In reality, they did no such thing. Each of us who drives a car today also knows s/he is contributing to global warming by doing so, but we still drive. We are conscious of the harm we are doing, but most of us pursue only partial solutions, if any at all. But then, we are living in a world where our choices are limited by the systems and structures within which we live.

Marxist materialist theories are incomplete in important ways as well. Marx believed that the proletariat, who were directly exploited within capitalism, would develop a class consciousness that would involve them directly in political revolution, thereby allowing them to seize control of capital and end exploitation of labor. But such a global revolution has not occurred even though the exploitation of workers by the elite has been brutal and direct. Clearly, something more complex than can be described by naïve idealism or early Marxist materialism is at work.

Marxism is grounded in an analysis of power and exploitation. Understanding the complex processes, both economic and cultural, that today perpetuate and advance globalized economy and concentrate wealth in fewer and fewer hands requires developing an understanding of the use and abuse of social power. Developing this understanding calls for building upon Marx's original analysis in important ways. CST can help us here because critical theorists seek to uncover the processes through which social power is perpetuated and further entrenched in modern society. As noted above, Marcuse's (1964) concept of repressive desublimation and Gramsci's (1971/1999) concept of cultural hegemony are of particular importance for understanding how and why the global economy continues to expand with the seeming support of the masses.

Ecopsychology offers additional perspectives that are useful for developing the CST of sustainability. With its roots in psychology and that discipline's traditional focus on the individual—including focusing attention on realizing self-integration and forming a healthy personal identity—ecopsychology retains a focus on the self, but the self *in context*. Ecopsychology recognizes explicitly that human psychological health is inextricably intertwined with the health of the environment: that when we damage the earth and other creatures, we damage ourselves. Therefore, ecopsychology deals explicitly with psychological trauma resulting from human abstraction from nature. Like deep ecology, it is concerned with healing the human/nature divide (see Rozak, Gomes, & Kanner, 1995).

It is perhaps ecopsychology's concern for individuals that makes it such a welcome complement to deep ecology. Freya Mathews (1992/1999) argues that deep ecology's focus on a relational total field image conception of all beings in nature (Næs' first principle), combined with the deep ecological belief that nature knows best, leaves one to assume that all human action—as part of a nature that knows best—is natural and, therefore, morally unassailable. Mathews argues that concern for individuals, human and nonhuman, is at the heart of human emotional concern with and for nature, even though this concern for fellow beings can at times conflict with the health of larger wholes. Mathews claims that we are neither solely immersed in a larger field image of being nor solely separate beings. She argues that we are *both*—our relationships *and* our individual selves. In this way, according to Mathews, we are like the "wavicles" of quantum physics: we are two irreducible things at once (Mathews, 1992/1994. p. 240).[2]

Ecopsychology's concern with such issues as proper individual education and initiation into reciprocity with others and the natural world (Shepard, 1995, p. 30; Armstrong, 1995) and confronting and processing ecological grief (Windle, 1995), represent a concern for the individual in community while also recognizing larger ecosystemic relational wholes. Ecopsychological thought, since it concerns itself with *the individual in context*, may lead more directly than does deep ecology to emotional connection across the human/nature divide. In this way, it serves as a complement to deep ecology in addressing concerns of ecological health and integrity. Ecopsychology can help us recognize and consciously experience the self within the community of life. It is the denial of such fellowship with creation that has allowed us to pillage the earth through "rational," mechanized means of extraction and production. We need to revive the human sense of the community of all beings, and ecopsychology has a role to play in this revival.

Without a complementary grounding in systemic and structural analysis informed by CST, however, followers of ecopsychology, like those of deep ecology, can fall into the idealist trap. At the extreme, they may believe that ecopsychological health will, on its own, create radically improved conditions of being, thereby bypassing a necessary confrontation with entrenched powers. I disagree. The longer we remain unaware of the powers that, often unconsciously, entrap us in the status quo, the longer we will be manipulated by these hegemonic powers. In a materialist/idealist analysis, we can come to see how our being, as experienced in the modern capitalist world, has fostered a consciousness that oppresses us, others, and nature.

Ecopsychology can help us build healthy emotional connections with nature and identities that recognize our fundamental interrelationship with the natural world. Deep ecology can help us focus on the need to emphasize relational wholes in a modern world that has too long focused on reductive thinking and analysis. Critical theory, rooted in the critique of domination, can help us analyze entrenched systems of power so that we might act effectively to dislodge them, and indigenous worldviews can help us understand the concept and practices of reciprocity among people and between people and nature. All of these strands of thought inform the critical social theory of sustainability. We now turn our inquiry to systems theory as another source for sustainability-oriented thought and action.

---

[2] Physicist Danah Zohar makes a similar ontological argument in much greater detail in her book *The Quantum Self* (1990).

Agency and Systems Theory

Systems theory can contribute effectively to developing a CST of sustainability, but it must be used with care. It is a school of thought that, like ecopsychology and deep ecology, has many adherents among sustainability advocates. The systems theory I refer to in this section derives from the sciences, including chaos theory, quantum physics, and nonlinear dynamics. The new sciences offer us important insights about holism, relationship, and creative emergence within complex systems that help us see that our world and the universe manifest from much more than simple cause-and-effect relationships (Capra, 2002). Since developing a sustainable worldview means conceptual reintegration of humans with ecologies and recognition of the interconnectedness of all life and all natural phenomena, systems theory offers important opportunities to develop more sustainable ways of seeing and being in the world.

But some works of systems theorists offer little direct critique of society. These works instead emphasize comprehending the workings of systems of all kinds—from ecological systems to social systems (Folke, Hahn, Olsson, & Norberg, 2005; Laszlo, 2006). Systems theory extends to systems of all scales and scopes (Laszlo, 2006, pp. 89–109). The theory offers important insights into how systems of all kinds may embody emergent properties and possibilities that are characteristic of systems as wholes—offering a means of understanding why a system really is more than the sum of its parts. When system functioning is described as more or less autonomous, however, direct critique of the uses and abuses of social power tends to slip into the background, along with discussion of human agency. Detailed explanations of principles of system functioning can impart an almost autonomous quality to systems, as though emergence and other forms of system change operate outside human history, only minimally, if at all, influenced by human choice and action. Systems theory that does not directly engage in critique of domination and oppression can only superficially inform action, especially when the level of social analysis remains general and not concrete.

In his book *The Chaos Point: The World at the Crossroads,* sustainability and systems theorist Ervin Laszlo (2006) uses systems theory as a basis for discussing the sustainability crisis and possibilities for averting catastrophic social and ecological collapse. While Laszlo does advocate human agency in history as essential to global society in avoiding catastrophe, his critique is too general and too abstractly idealist in nature to serve as a strong foundation for sustainability-oriented action. The sources of unsustainability remain mystified since Laszlo refrains from analyzing in depth the specific economic, social, and cultural mechanisms that coerce often unconscious collusion by the masses. The form of agency Laszlo advocates is mostly within the realm of consciousness. He emphasizes human thought as an active agent in complex systems, an agent capable of triggering almost spontaneous sustainability-oriented creative emergence (p. 10). He states: "When a society reaches the limits of its stability and turns chaotic, it becomes supersensitive, responsive even to small fluctuations such as changes in the values, beliefs, worldviews, and aspirations of its members" (Laszlo, 2006, p. 10). Here, Laszlo (2006) claims that the *nature* of the social system as a complex system is currently manifesting an opening for consciousness to create a shift in the social paradigm.

Chapter seven of *The Chaos Point* (Laszlo, 2006, pp. 61–82) is titled "What You Can Do Today." In this chapter, Laszlo advocates the following: "shed obsolete beliefs" (p. 61), "adopt a new morality" (p. 70), "dream your world and act on it" (p. 74), and "evolve your consciousness" (p. 76). Laszlo also briefly confronts many of the destructive and unsustainable assumptions of capitalist industrialism such as "nature is inexhaustible" (p. 65), "the world is a giant mechanism" (p. 66), "the market [effectively] distributes benefits" (p. 67), "the more you consume the better you are" (p. 68), "the more money you have, the happier you are" (p. 68), and "economic ends justify military means" (p. 69). Laszlo is calling upon his readers to think differently. He is also calling upon them to make a difference, but unfortunately, he offers little direct social critique that could usefully inform action. In calling us to bring new dreams of a sustainable reality to life, Laszlo states:

> When you see things as they are and ask *why are things the way they are*—which is likely to be very different from the way you dream it—you come across a maze of complex explanations and a tangle of unsolved problems. But if you dare to dream, and share your dream with friends and neighbors and ask, *why are things not the way I dream it*, you will find answers—and ways you can come together to start making a world that resembles your dream. (2006, p. 76)

This call to action assumes that people are capable of dreaming sustainable dreams within an unsustainable system and maintaining those dreams for the long haul—while, in contradistinction, the economic and social incentives and rewards of the current social system flow from securing social power and from exploitation rather than from reciprocity and care for nature and others. His call to action also assumes that somehow people who cultivate a sustainable consciousness will coalesce in harmonious sustainability-oriented action. A CST-based analysis of the challenges to sustainability in a hegemonic society would question such assumptions. Most importantly, though, Laszlo's claims here relate directly to the idea expressed early on in *The*

Chaos Point (Laszlo, 2006, p. 10) that shifts in consciousness can tip the system into a sustainable state. Laszlo implies that this change can be accomplished without deeply interrogating or actively confronting the current system, a position highly contradictory to the CST of sustainability. We do need to dream new dreams, but we need to do much more than this as a strategy for sustainability.

In order to maintain a sustainability-oriented vision and motivate action, I argue that we need a critique that can tell us more about why our socio-ecological systems are nearing catastrophic collapse—a critique that can offer us specific insights about how the institutions and systems of economic globalization enforce the dependency of nations, communities, and individuals on unsustainable socio-ecological systems. We also need a critique that can tell us about how the systems of global capitalism perpetuate themselves through the complicity of ordinary people. We need to critique the cultural aspects of domination and the sources of our own collusion with the system, aspects of unsustainability of which many of us remain totally unaware. Developing an understanding of Marcuse's (1964) concept of repressive desublimation and Gramsci's (1971/1999) concept of cultural hegemony can help us become aware of how we have been colonized by capitalist society and how we also act to perpetuate our own domination and the domination of others and nature. A strong foundation for counter-hegemonic critique of our current society has been effectively laid by critical social theorists, and we can build upon that foundation in addressing issues of sustainability.

When it does not foreground deep critique of social history, a systems theory approach to explaining unsustainable society can offer everyone a way out of being indicted for the problems of our world today. Noncritical versions of systems theory are likely to remain unchallenged by the powerful. Nonspecific critiques that call upon readers to change the world through changing their thinking do not directly confront the domination and oppression that pervade the capitalist system and, thereby, deflect direct confrontation between oppressors and oppressed. Uncritical systems theory is mostly incompatible with a CST-based analysis of social systems and with the call to agency embodied in such analysis.

Versions of systems theory that articulate with CST's firm commitment to deep analysis of society as a historical formulation do exist, however, and these versions offer the possibility for integrating the insights of systems theory into the CST of sustainability. In his book, *The Upside of Down: Catastrophe, Creativity, and the Renewal of Civilization*, Thomas Homer-Dixon (2006) offers a version of systems theory that focuses attention simultaneously in three important areas: 1) the historically rooted systems of domination and oppression, 2) the workings of complex natural systems not created by humans, and 3) perhaps most importantly, the interactions between these two large-scale systems and among the subsystems that comprise them.

Homer-Dixon's (2006) theory derives from the work of well-known ecologist and systems theorist Crawford Holling (Homer-Dixon, 2006, pp. 225–234). Holling, and his colleagues at the Resilience Alliance, have developed a version of systems theory called "panarchy theory" (Homer-Dixon, 2006, p. 226). According to this theory, complex systems manifest themselves in adaptive cycles. These cycles have a growth phase character-ized by a system's rising potential for novelty and rising connectedness and self-regulation. In its growth phase, a system continually gives rise to many differentiated parts that function in tightly connected relationship with one another, leaving little room for redundancy or autonomy that would increase resiliency. While complexity is increasing, a system is declining in overall resilience due to the formation of tightly coupled, complex interrelationships that make the system brittle and increasingly vulnerable to cascading and catastrophic collapse (Homer-Dixon, 2006, pp. 226–228). Following collapse of its growth phase, a system demonstrates reduced levels of differentiation. It is "far less interconnected and rigid" and "far more resilient to sudden shock" (Homer-Dixon, 2006, p. 228). Homer-Dixon concludes that system collapse may, in the end, increase system health and resilience to shocks, but at the great price of destruction of the growth phase incarnation of the system (Homer-Dixon, 2006, pp. 227–228).

Additionally, according to Holling and his colleagues,

> No given adaptive cycle exists in isolation. Rather, it's usually sandwiched between higher and lower adaptive cycles. For instance, above the forest's cycle is the larger and slower-moving cycle of the regional ecosystem, and above that, in turn, is the even slower cycle of global biogeochemical processes, where planetary flows of materials and elements—like carbon—can be measured in time spans of years, dec-ades, or even millennia. Below the forest's adaptive cycle, on the other hand, are the smaller and faster cycles of subecosystems…. (Homer-Dixon, 2006, pp. 229-230)

The entire network of adaptive cycles spans a huge range of scale in both space and time (Homer-Dixon, 2006, p. 230). In the growth phase, while high levels of differentiation and interconnectedness are emerging, systems and their subsystems can become synchronized so that the overall system becomes increasingly vulnerable to collapse triggered from any of many system levels.

According to Homer-Dixon (2006), complex global society may be in-creasingly vulnerable to collapse due to system synchronicity and tight system integration driven by efficiency. To support this claim, he draws on Holling's analysis saying that Holling believes we are on the verge of large-scale, systemic crisis for several reasons. For one, adaptive cycles have become nested together in global systems with many systems and subsystems simultaneously poised on the brink of collapse. This synchronicity has been encouraged by tight interlinking of systems. Increasing global connectivity increases chances for deep collapse triggered by high-level system collapse triggering cascading collapses in lower-level systems (Homer-Dixon, 2006, p. 231). In more resilient systems with looser coupling of parts and reduced synchronization across systems and subsystems, higher and lower system cycles form a buffer for individual adaptive cycles and help keep them from collapsing. Larger, slower moving cycles tend to provide stability, and lower, faster-moving cycles tend to inject novelty that keeps the system flexible and responsive rather than brittle (Homer-Dixon, 2006, p. 230).

Homer-Dixon sees the moment of a system's collapse to a level of re-duced complexity as a time for potential "catagenesis." He defines catagene-sis thusly:

> [The term] combines the prefix *cata*, which means "down" in ancient Greek, with the root *genesis*, which means "birth...." Ecologists use catagenesis to refer to the evolution of a species toward a simpler, less-specialized form.... I retain the idea of a collapse or breakdown to a simpler form, but I especially emphasize the "gene-sis"—the birth of something new, unexpected, and potentially good...in essence, the everyday reinvention of our future. (Homer-Dixon, 2006, p. 22)

Homer-Dixon (2006) foresees a near-term, widespread collapse of tightly interconnected social systems. He also foresees widespread potential collapse of ecological systems on which societies depend for survival and for generat-ing economic growth. The extent and character of collapse for Homer-Dixon, however, remain historically contingent on past, present, and future actions of individuals, organizations, and societies.

It is important to recognize that, according to Homer-Dixon (2006), sys-tem collapse at a particular time, of a particular character, or deriving from particular causes is *not determined as an essential characteristic of the system itself*. A system's history, as determined by agents active within the system, plays a critical role in determining the character and timing of such outcomes. Homer-Dixon states, "A complex system's history turns out to be crucially important because it profoundly shapes what the system becomes, and it can't be rewritten or repealed" (2006, p. 26). This interaction between a system and natural/human history/agency can be described as a form of path dependence (Homer-Dixon, 2006, p. 27). When an individual or a system chooses one path or one solution from among many, later paths or solutions that come into view derive from the original choice. It becomes increasingly difficult to revisit and revise past choices which become assumptions upon which aspects and parts of a system are continually constructed (Homer-Dixon, 2006, p. 27). Systems exist and function historically and not as machines with predetermined products and outcomes. They move through growth and collapse phases *according to the actions of individuals and groups that make up the system and according to natural phenomena that impact system functions.* Still, possibilities for agency within a highly complex system, with its functions based on countless interlocking assumptive premises and resultant choices, does become rigidified and open to an increasingly narrow field of vision and possibility for deep change—barring, of course, the advent of collapse.

Historical contingency must be foregrounded as a central organizing principle of systems in order for systems theorists to avoid giving the impression that systems function according to natural laws that exist outside the realm of human agency. Giving such an impression is dangerous in this time of converging socio-ecological crises, for if human ability to shape the present and the future appears out of reach, people may allow themselves to be swept along by the momentum of business as usual rather than seek system change. And it is crucial to alter current systems in ways that increase socio-ecological resilience—if we wish to avoid a collapse that could flatten the modern era with such force that even those aspects of modern society we might wish to retain would disintegrate. Depending on the force of such a collapse, human societies might cease to exist altogether.

Some additional aspects of the behavior of complex systems help us understand how social and ecological collapse could be so violent and total. The existence of thresholds in complex systems is one such aspect. According to Homer-Dixon (2006), thresholds in complex systems represent nonlinear system behavior. As disturbances or changes in the system accumulate, the behavior of the entire system can cross a threshold—suddenly shift into a "radically new mode" (pp. 24–29). The converging socio-ecological crises of climate change, accumulating pollution, species loss, loss of ecological resilience, peak oil, economic turmoil, and social inequity and injustice represent accumulating and mutually reinforcing changes in our world, some or all of which could contribute to creating a threshold. Once crossed, such a threshold would create socio-ecological conditions radically different from those of the present, conditions under which survival for humans and others creatures could be in question. Like many climate scientists, Homer-Dixon sees accumulating changes in the climate system as a potential threshold:

[One] reason we should be very concerned about our climate future is likely the most important, at least from the point of view of possible social breakdown in coming decades: the prospect of abrupt climate shifts or "nonlinearities." In a world where billions of people are tightly coupled to a steady stream of services from a stable climate—depending closely on regular rainfall to grow their food, for example—a sudden flip to a new climate regime would be a prescription for chaos. (2006, pp. 168–169)

An additional aspect of the behavior of complex systems that points toward possible near-term collapse is synergy. According to Homer-Dixon, "Synergy happens when people, things, or events combine to produce a larger impact than they would if each acted separately" (2006, p. 106). Synergy is facilitated by the tight interlinkages and efficiency that characterize complex systems at the height of the growth phase. Negative synergies and system breakdown become increasingly probable within systems as they become more complex and tightly integrated. According to Homer-Dixon, connectivity increases the likelihood of harmful, self-reinforcing feedback loops, a form of negative synergy. He provides an example that is especially apt for current times: "A stock market crash or financial panic is…a vicious circle, because selling drives down prices, begetting fear in the market and more selling, which lowers prices even more" (Homer-Dixon, 2006, p. 115). Homer-Dixon emphasizes the role that synergy can play in complex system breakdown:

A society is more likely to experience breakdown when it's hit by many severe stresses simultaneously, when these stresses combine in ways that magnify their synergistic impact…, and when this impact propagates rapidly through a large number of links among people, groups, organizations, and technologies. (2006, p. 110)

Homer-Dixon concludes that the growth phase of global capitalist society is not sustainable (2006, p. 253). He states:

As our global social-ecological system moves through the growth phase of its adaptive cycle…it's losing resilience…. Capitalism's constant pressure on companies to maximize efficiency tightens links between producers and suppliers; reduces slack, buffering, and redundancy; and so makes cascading failure more likely and damaging. As well, capitalism's pressure on people to be more productive and efficient drives them to acquire hyperspecialized skills and knowledge, which means they become less autonomous, more dependent on other specialized people and technologies, and ultimately more vulnerable to shocks…. Meanwhile, worsening damage to the local and regional natural environment in many poor countries is fraying ecological networks and undermining economies and political stability. And finally pressure is increasing within both rich and poor societies too—from tectonic stresses like demographic imbalance, growth of megacities, and widening income gaps. (p. 252)

According to Homer-Dixon, "A prudent way to cope with invisible but inevitable dangers is to...build *resilience* into all systems critical to our well-being" (2006, p. 283). He defines resilience as "an emergent property of a system...not a result of any one of the system's parts but of the synergy between all its parts" (2006, p. 284), and he advocates "boosting the ability of each [system] part to take care of itself in a crisis" as a means to boost overall system resilience (2006, p. 284). Self-sufficiency and redundancy both serve to increase system resilience.

In the systems theory of Homer-Dixon (2006), social justice and resiliency reinforce one another and call for agency. Working for social justice would increase system resiliency:

> Our world's capacity to avoid "deep collapse"—or synchronous failure—depends on resilience throughout the system. In practical terms today, this means we must focus our attention on boosting the resilience of the world's weakest societies—those with horribly damaged environments, endemic poverty, inadequate skills and education, and weak and corrupt governments. If we don't, our entire global social-ecological systems will become steadily more vulnerable to the diseases, terrorism, and financial crises that emerge from its least resilient components. (pp. 286–287)

Increasing social justice and resilience would also mean confronting global capitalist ideology:

> In its most dogmatic formulation, this ideology says that larger scale, faster growth, less government, and more efficiency, connectivity, and speed are always better. Slack is always waste. So resilience—even as an idea, let alone as a goal of public policy—isn't found anywhere on the agendas of our societies' leaders. (Homer-Dixon, 2006, p. 287)

In his critique of capitalist society and call for social justice, Homer-Dixon (2006) articulates systems theory that is highly compatible with CST. His work also aids in the application of CST to the context of sustainability in that it helps us understand linkages between human societies and ecological systems that make the globalized world-system brittle and vulnerable to widespread and deep collapse. Critical theorists critique global capitalism's drive for efficiency and its unjust processes of concentrating wealth in fewer and fewer hands. These theorists reject authoritarianism and domination of individuals by a capitalist society that increasingly treats people as things, as cogs in a vast machine. Homer-Dixon (2006) helps us see that the industrialized growth economy should be rejected, not only on the grounds that it impoverishes the quality of human existence and damages nature, but also on the grounds that it contributes in important ways to priming both society and nature for devastating collapse.

Systems theory, as presented by Homer-Dixon (2006), offers an important analytical lens for the CST of sustainability. His theory calls upon us to consider the dynamic nature of systems as constantly changing and open to the influence of multiple agents and circumstances. Perhaps most importantly, it focuses our attention on issues of system resiliency, rather than "stability" which would indicate a more or less static state. Homer-Dixon's (2006) theory helps us see that our socio-ecological systems are ever changing, but it also highlights the need to consider how human agency might affect the ability of local and global systems to respond with resilience rather than catastrophic collapse in the face of stresses and disturbances. Levels of resiliency demonstrated by communities, nations, and socio-ecological systems globally in the face of currently converging socio-ecological crises of sustainability will in many ways determine the capacity of societies, and the ecosystems of which they are part, to respond with creative catagenesis. Those systems that embody low levels of resilience will be—and are at this moment—in jeopardy of deep collapse from which little of current complex socio–ecological interrelationships may survive.

Systems theory can contribute most effectively to a CST of sustainability that serves as a basis for sustainability-oriented education and action. In order to do so, it must highlight the essential roles of history and agency in creating current realities. The systems theory of Homer-Dixon (2006) contributes effectively to a CST of sustainability that grapples with domination and oppression as features of unsustainable, brittle systems lacking in resiliency. His theory can play an important role in adapting CST to the context of the current sustainability crisis—and in pointing the way toward effective sustainability-oriented action.

Conclusions: Toward Sustainability Praxis

Because it can help people comprehend the many obstacles to sustainability embodied in late capitalist society, the CST of sustainability forms an important part of the foundation for sustainability-oriented education and praxis. This theory can also help people understand historical openings for change emerging within the sustainability crisis itself. This understanding can assist sustainability educators and practitioners with focusing their praxis in the most effective ways possible. To these ends, the CST of sustainability asserts the importance of in-depth, transdisciplinary social critique rooted in a critique of domination that explicitly recognizes that the exploitation of both people and nature derive from the same sources within capitalist society. The CST of sustainability recognizes that humans are part of nature and dependent upon nature as the source of all life. This recognition promotes forms of praxis through which human health and ecosystemic health can be pursued as a single end. The CST of sustainability derives from historicist foundations. It recognizes self-reflexively that its critique manifests in and through history and that its validity must be judged according to its ability to inform sustainability-oriented action. The validity of the CST of sustainability must be judged over time by its ability to advance sustainability through continually adapting to a rapidly changing socio-ecological context. It is my hope that this new theory will usefully inform sustainability praxis in a wide variety of social, economic, educational, and ecological contexts. If the sustainability crisis is not addressed through radical changes in thinking and living patterns, the world will continue to lose much of its existing beauty, diversity, and resiliency.

Chapter 3:

The Political Economy of Enforced Dependency

You have to remember, what we know now as the Third World didn't exist. There were a few major powers, each of which had a large empire. We had no voice. We had no presence and just were part of somebody else's power structure. You ask whose interest. I ask the question: who set it up?

—Michael Manley, former Prime Minister of Jamaica, commenting on the Bretton Woods conference.

The collective violence of globalization is keenly felt daily by the exploited peoples of the world. They feed the engines of profit with their labor while surviving in a state of enforced dependency upon the global economy. Tragically, the subsistence lifestyles that are their heritage have been rendered backward and uneconomic when compared to cheap mass production and industrial agriculture (Douthwaite, 2004, pp. 114–116; Polanyi, 1944/1957, chap. 3–5). Ironically, the relatively privileged citizens of industrialized nations such as the United States also experience the psychic and spiritual violence wrought by globalization. They may benefit materially, but they lack many of the meaningful relationships with people and nature that typified more resilient, self-reliant, and localized communities prior to the rapid expansion of globalization (Achbar, Simpson, Bakan, & Crooks, 2005; Armstrong, 1995; Berry, 1987; Martinez, 1997; Nelson, 1983; Polanyi, 1944/1957, chap. 4; Salmon, 2000).

The self-perpetuating and self-reinforcing systems of power and exploitation that are the focus of this chapter embody a global system of collective violence. Collective violence consists of "actions by large numbers of people that contribute to large-scale destruction" (Summers & Markusen, 1992/2003, p. 215). The destruction can be social, environmental, or both. I argue that the political economy of late capitalism *requires* collective violence and systematically entraps people as both willing and unwitting participants. The large scale of globalization and the complex relationships it embodies diffuse responsibility for destructive outcomes and promote a culture of obedience in the face of power.

The enormity of the global economy and its inherent requirement for growth place heavy pressure upon people and nations worldwide to deliver growing numbers of consumers and ever more raw materials and energy to feed the system. This pressure has manifested in the overt manipulation and exploitation of peoples and places of the Global South and in the increasing pace and demands of life and work in the industrial world. Within the world-system, former colonies attempt to carve out their share of wealth at the same time that a specific set of global institutions, relationships, structures, and trends ensure that they remain at the service of the industrialized world (Robbins, 1999, pp. 101–107).

To this point in this book, we have explored the CST of sustainability from a mostly socio-cultural perspective. This theory must also address the systems and institutions of economic globalization in order to elucidate how globalized capitalism intensifies cultural hegemony and repressive desublimation while also fostering increasing dependence of individuals, communities, and nations on the globalized system. This chapter closely articulates with the CST of sustainability, particularly with the Gramscian (1971/1999) notion of cultural hegemony and with Marcuse's (1964) concept of repressive desublimation.

This chapter is rooted in *political* economy. I demonstrate how the neoliberal global economy is a *political* entity that is far from neutral in awarding privileges and inflicting harm. The global economy is not a disinterested and self-regulating machine, and the very idea of the self-regulating market is itself ideological (Polanyi, 1944/1957, chap. 6). Nations, communities, and individuals caught in the global economy's web—virtually everyone everywhere at this point in time—are constrained in their day-to-day and long-term decision making and actions into supporting the global system through their participation in it. Some of the more fortunate participants may be materially rewarded by the system, but even their participation is ultimately self-defeating in that it undermines long-term sustainability and well-being for individuals and communities. But there are few choices for most of us: we participate under threat of duress, even death.

In this chapter, I focus on the post–World War II (WWII) period—late capitalism—which is characterized by the rapid growth of an increasingly integrated global economy, and I examine mutually and self-reinforcing systems of concentrated political and economic power within the globalized world-system. Specifically, I define the concept of enforced dependency, and I analyze the structural and systemic features of enforced dependency in the globalized world. I analyze power structures and practices that embody the momentum and inertia of globalization, and I focus attention on the mechanisms that enforce the dependency of ever-growing numbers of people on an economic system that paradoxically and simultaneously both sustains and depletes them.

I explore how and why the economic and political possibilities of societies narrow as a result of late capitalist, neoliberal globalization so that, even should a nation or community wish to reverse or change course, the late capitalist system enforces continued political and economic dependency on the globalized economy. While the poor quite obviously pay the biggest price for the economic and political gains of dominant societies and classes within the world-system, I argue that everyone will lose in the end as the processes and logic of neoliberal globalization increasingly undermine the resilience of all system participants. Continual loss of community resiliency is an ominous trend in the globalized world because, in order to build more sustainable societies beyond the late capitalist crisis in history, it is precisely socio-ecological resiliency that we will need. Through elucidating in this chapter important drivers of destructive and ultimately self-destructive enforced dependency, I offer a springboard for generating interest in resiliency building as an important aspect of sustainability-oriented education and praxis.

I begin this chapter by highlighting three important sources of unsustainability within the world-system: fossil fuel depletion, ecological breakdown, and the structural crisis of capitalism as a viable economic system. These converging crises point to the possibility of near-term socio-ecological collapse, a possibility that drives my search for alternatives to the current political/economic paradigm. I then define the concept of enforced dependency that serves as a heuristic for the analysis developed in this chapter. This definition is followed by an in-depth discussion of late capitalism as a tightly integrated, multifaceted world-system of enforced dependency. I then summarize how the institutions, ideologies, strategies, and structures discussed in this chapter combine to create and perpetuate the hegemony of global capitalist elites and culture. I conclude this chapter by suggesting some directions for cultures and economies to move toward socio-ecological sustainability.

Converging Crises

We begin our analysis with a brief overview of three global-society-changing crises that frame and inform every aspect of the arguments articulated in this chapter. These crises are ecological breakdown, fossil fuel depletion, and the structural crisis of capitalism as a viable economic system.

Ecological Breakdown

Climate change represents perhaps the most well-known all-pervasive threat to ecosystems planet wide (Feasta, 2008; Guggenheim, 2006; IPCC, 2007; Orr, 2009). Additional drivers of ecological breakdown include widespread pollution that has made its way into all parts of the biosphere; overharvesting of plants and animals that threatens to eliminate possibilities for population regeneration and to decrease species resilience through narrowing genetic diversity; extensive human takeover of habitats worldwide that places further stress on species resilience and biodiversity as well as on genetic diversity within species; and freshwater damming, diversion, and mining that are dewatering some ecosystems and flooding others (Barlow & Clarke, 2002; Booth, 2002, chap. 4–5; Douthwaite, 1999b, chap. 10; Homer-Dixon, 2006; Meadows, Randers, & Meadows, 2004; Reisner, 1993). To make matters worse, multiple sources of ecological damage compound one another (Booth, 2002, chap. 4), bringing species to the point of extinction and ecosystems to the verge of collapse.

Further compounding these problems is the fact that societies everywhere have been shoehorned into the capitalist mode of production and distribution—a form of economic life and culture defined and driven by capital accumulation and *not* by reciprocal duties to others and the natural world. The capitalist system recognizes money as *the* measure of value and security and, accordingly, capitalism focuses its participants on making money as the central, perhaps only, means to achieving well-being. Human dependence upon the capitalist system increases the odds for catastrophic ecological breakdowns by severing cultural recognition of the links between environmental and social health that typified many place-based indigenous societies (Armstrong, 1995; Martinez, 1997; Nelson, 1983; Salmon, 2000). Although globalized capitalism has obscured the relationships between environmental destruction and capital accumulation and between environmental and social health, we are witnessing the limits of this obfuscation as global ecological crises become increasingly evident everywhere (Homer-Dixon, 2006; Meadows, Randers, & Meadows, 2004; Orr, 2009).

As we analyze the political economy of enforced dependency, it is important to remember that, not only is capitalism reaching the limits of its ability to contain its own contradictions, but the biosphere is also nearing its limits to support extractive, wasteful, and toxic capitalist development (Orr, 2009). Not only is capitalism consuming its own base in economic terms by concentrating wealth and power (Wallerstein, 2008a, 2008b), it is poisoning and depleting the ecological base required for the economic growth that is integral to its continuance (Daly, 1999).

Fossil Fuel Depletion

The founders of CST offer theories regarding the cultural dynamics of hegemonic society, and these theories serve as a foundation for the CST of sustainability developed in chapters one and two. But these same theorists did not—in fact, probably could not in their time—consider the future of hegemonic society within an energy depletion scenario. As part of the process of extending CST into the realm of sustainability, the CST of sustainability situates social analysis within a peak oil and gas context and considers how and why this new context creates openings for rapid and pervasive social change. Rooted as it is in praxis, the CST of sustainability seeks to use these openings to advance sustainability. In later chapters, we will explore possibilities for (re)localization of societies and economies as an important sustainability-oriented strategy within the emerging context of energy decline.

Oil, natural gas, and coal fueled the industrial revolution, and they re-main the primary sources of energy driving the global economy. We examine depletion of each of these energy sources, with a focus on oil, in order to understand how fossil fuel depletion threatens to destabilize the global growth economy and, ultimately, render capitalism as we know it unwork-able.

Inevitably and soon, global demand for oil and natural gas will outstrip global extraction and supply of these same resources (Campbell & Strouts, 2007, part I; Clark, 2005, chap. 3; Deffeyes, 2001; Douthwaite, 2004, p. 118; Greene, 2004; Heinberg, 2005, chap. 3; Kuntsler, 2005, chap. 1; Roberts, 2004, chap. 2; Simmons, 2005). The long-term implications of declining fossil energy supplies are immense. To fully comprehend the implications, it is important to understand the complexities of energy supplies and their interrelationships with the global economy, geopolitics, food production, transportation, and more. This work has been done well by others (Campbell & Strouts, 2007, part I; Clark, 2005; Deffeyes, 2001; Greene, 2004; Hein-berg, 2005; Kuntsler, 2005; Roberts, 2004; Simmons, 2005). I offer here only a brief summary in order to shed light on the unsustainability of fossil fuel dependency and to demonstrate how this dependency fosters instability in the global economy.

Oil depletion is a well-known, studied, and documented fact. The first person to study this phenomenon, Dr. M. King Hubbert, a highly respected petroleum geologist, predicted in 1956 that oil production in the U.S. would peak in the early 1970s (Heinberg, 2005, p. 97; Deffeyes, 2001, chap 1). U.S. domestic production data show that domestic oil production in fact did peak in 1970. In hindsight, the idea of peak production makes complete sense. Oil is not renewable. Oil fields are under pressure from the layers of rock, earth, and sometimes water above them. As a result of any penetration of the cap rock, the pressurized, liquid oil will to escape upward from the field, meaning that the effort needed to extract the resource is low at first. As more and more wells are drilled into a new field, oil production increases for a time (and the pressure within the field slowly drops) until production peaks. Thereafter, no matter how many new wells are drilled, production will inevitably decline as the oil becomes harder to extract (Greene, 2004; Heinberg, 2005, chap. 3; Kunstler, 2005, pp. 24–25; Roberts, 2004, pp. 50–51). Declining production is a familiar domestic reality in the United States. Once foremost among petroleum exporting nations, in 2007 the U.S. imported just over 65% of the oil it consumed (United States Energy Information Agency, 2008). Many well-respected petroleum geologists predict a global peak in oil production before 2015 (Deffeyes, 2001; Greene, 2004; Heinberg, 2005, chap. 3). In a lecture given in Cork, Ireland, in the summer of 2008, well-known and internationally respected oil geologist Colin Campbell predicted peak production of all petroleum liquids in that same year.

The arguments of those who downplay or refute the importance of peak oil production coalesce around two central points: 1) that there is a great deal more oil to be discovered/extracted and 2) that human innovation and technology will make limits to oil production irrelevant (see Huber, 2002; Lomborg, 2001; Lynch, 2001, 2003). This cornucopian camp consists mainly, though not entirely, of economists and those whose current political and economic interests are best served by assuming a rosy outlook for future oil production (Heinberg, 2005, pp. 118–136; Kunstler, 2005, pp. 28–29; Roberts, 2004, chap. 4).

Economic growth is predicated upon ever growing supplies of fossil fuel-based energy, and predicting limits in sight to extraction of these resources is neither likely to buoy the consumer and producer confidence necessary for maintaining economic growth nor to garner political support from populations dependent upon the global growth economy for their livelihoods (Campbell & Strouts, 2007, pp. 11–12; Greene, 2004).

Regarding the possibility of future discoveries, worldwide oil discoveries peaked in the mid-1960s and have declined to such an extent that it would likely be impossible for even large discoveries to reverse the trend (Heinberg, 2005, pp. 109, 114). Witness the significant non-OPEC oil sources that are in decline (the United States, Mexico, and the North Sea among them). Recent finds also do not compare in size to earlier finds made in significant oil-producing regions such as the Middle East and Texas (Roberts, 2004, p. 51). In 2008, the world was using more than four barrels of oil for every one barrel discovered (Campbell, 2008). If this trend could be easily reversed, there is little doubt it would have been during the past 40 years. According to Homer-Dixon,

> Despite exploration companies' immense investments of capital and technology, oil discovery in the U.S. has declined steadily since 1930. As the petroleum geologist Colin Campbell notes, "The United States had the money to [discover more oil], it had the incentive, [and] it had the technology, so the fact that discovery reached a peak—and then declined inexorably for...seventy years—is not for want of trying. It was due to the physical limits of what nature gave them." (2006, p. 87)

Peaks in oil production lag decades behind corresponding peaks in discovery. According to peak oil educator Richard Heinberg, at this point, "countries in [oil production] decline account for about 30 percent of the world's total oil production" (2005, p. 115). The late Matthew Simmons, founder and Chairman Emeritus of Simmons & Company, the world's largest energy investment bank, asserted in 2005 that Saudi Arabian oil production was likely at or near its peak (Simmons, 2005, parts three and four). Even if new discoveries could push global peak oil production many years into the future, these discoveries would only delay the inevitable while also serving to further socio-economic dependence on oil-based production, transportation, and development patterns.

The concept of net energy is also important to understand when considering oil production and energy sources that might substitute for fossil fuels. The oil produced from a new field requires little effort to extract, but given time, the effort required to lift oil from a declining field intensifies, so that the energy profit from the endeavor declines. Eventually, if extraction continues long enough, lifting oil from a declining field becomes an energy-losing proposition. Optimists cite the many useful technologies employed in the modern energy industry for discovering and producing oil reserves as evidence of long-term ability to increase production levels or at least hold them flat (see for example Lomborg, 2001, and Lynch 2001). What is not often acknowledged is that these technology investments are also *energy investments* and that the harder we must work with new technologies, the lower our net energy return. Furthermore, even if we were to suddenly be able to dramatically increase flows of petroleum with these technologies, doing so would only deepen the eventual energy crash with regard to this finite resource (Heinberg, 2005, chap. 3–4).

And what about the second argument advanced by the cornucopians: that human ingenuity and technology will provide us with sources of energy that substitute for oil? (see for example Lomborg, 2001). For some energy "sources" such as hydrogen that are cited as potential contributors to a new energy economy, the net energy picture is particularly dim. Hydrogen must be refined from natural gas or electrolyzed from water using electricity. According to the second law of thermodynamics, the hydrogen captured through this process actually has *less* energy available to be applied to work than was available for use from the electricity or the natural gas used to capture the hydrogen in the first place. While hydrogen may serve useful purposes as a storage medium for excess energy generated from renewable sources, it is hardly an energy *source*; it is only an energy carrier (Greene, 2004; Heinberg, 2005, pp. 161–168; Scheer, 2007, pp. 89–94).

Furthermore, many forms of renewable energy cited as potential replacements for oil (including wind and solar power generation, wave and tidal generation, hydroelectricity, and nuclear and geothermal generation) produce electricity. Our global transportation infrastructure runs on liquid fuels, not electricity (Heinberg, 2005, chap. 4; Hirsch, Bezdek, & Wendling, 2005), and transportation remains the biggest challenge in terms of finding replacements for oil. With the global fleet numbering hundreds of millions of vehicles (Heinberg, 2005, chap. 2)—each of which required the equivalent of about 90 barrels of oil to fabricate (Greene, 2004)—and considering the miniscule to nonexistent infrastructure for alternative fuels (Heinberg, 2005, chap. 4; Hirsch, et al., 2005), transportation represents a problem of monumental proportions—and one that will require us to use vast amounts of oil to address technologically. Although electric vehicles can be produced, an incredible energy investment would be required to produce the vehicles and the additional infrastructure needed to transition all automobiles and trains to electricity, and it is doubtful that electricity could drive global shipping and air travel. Though a concerted effort to reduce the use of liquid fuels in cars by investing in electric-powered trains and light rail could theoretically free up liquid fuels for use in long-distance shipping, it seems highly unlikely that nations faced with ailing economies and high debt levels (nations that are now cutting back on basic services) will be able to make these kinds of investments any time soon, at least not on the scale necessary to avoid disruption of international shipping.

Biofuels such as ethanol and biodiesel can be used as liquid fuels to run some of the current transport system, but growing the crops necessary to fuel the vast amounts of transportation required for globalization means competing with food production. Skyrocketing tortilla prices in Mexico have been linked to increased demand for corn-based ethanol in response to rapid oil price increases in 2007 and 2008. Corn producers sought the highest profits for their product, and a shortage of corn for food resulted (Patel, 2007). The European Union's goal of deriving 10 percent of its liquid fuels from plant-based sources by 2015 has been a factor in recent land grabs in Africa where large tracts are being acquired to produce and export biofuel feedstocks from countries where vast numbers of people go hungry (Vidal, 2010).

We can run some existing vehicles on natural gas, but most have not been converted to run on this fuel. Furthermore, production of conventional natural gas in the U.S., where a large percentage of the global fleet is used for day-to-day transportation, peaked in 1975, though a second smaller peak based on unconventional shale gas production appears to have occurred in 2010 (Hall & Klitgaard, 2012, p. 81). The depletion picture for conventional natural gas is also similar to that for oil except that, once peak production passes for a natural gas field, production declines much more rapidly than for an oil field (Heinberg, 2005, pp. 139 and 143). According to geologist Arthur Berman, the picture is even worse with regard to unconventional sources of natural gas, which he believes are vastly overstated and, when produced, tend to deplete very quickly. In the Barnett shale, 25 to 35 percent of gas wells drilled between 2004 and 2006 were sub-commercial (producing gas at a loss) within five years (Berman, 2010). The process of hydraulically fracturing the shale, necessary for production in shale gas fields, also has high energy costs associated with shipping millions of gallons of fracking fluids to and from each well site for use and later disposal. Add to this picture that fracking poses serious risks to drinking water safety and air quality and that earthquakes appear to be associated with deep-well injection of used fracking fluids, and the wisdom of exploiting unconventional gas comes into question (see Pro Publica's series on fracking, 2008–2012).

Given this supply picture combined with the fact that recently built electricity generating capacity in the U.S. is fueled by natural gas, the potential for an energy crisis triggered by natural gas depletion becomes apparent (Greene, 2004; Heinberg, 2005, p. 142). Add to this scenario that natural gas serves as a major feedstock for creating plastics and synthesizing agricultural fertilizers, and the potential for widespread economic impacts and even food shortages also becomes apparent (Greene, 2004; Heinberg, 2005, chap. 5). Some suggest that regions facing declining natural gas production may be able to rely on gas shipped by tanker. This proposition would require heavy infrastructure (and, therefore, energy) investments, not to mention that the process of super cooling and shipping gas by tanker negatively impacts the net energy available from these resources (Darley, 2004).

Oil shale and oil sands as sources of liquid fuels have their own net energy and environmental costs. These sources are inefficient in terms of net energy profit as compared to liquid petroleum. It is likely that exploitation of these deposits will continue as conventional oil supplies decline, but these resources will not make up for conventional petroleum reserves in net energy terms (Heinberg, 2005, pp. 127–128; Hirsch, et al., 2005, pp. 40–42). What is more, the processing of oil sands releases vast qualtities of carbon dioxide, something we do not need more of in our atmosphere according to climate scientists (Guggenheim, 2006; IPCC, 2007; Orr, 2009).

Optimists often cite energy efficiency gains over time, noting that, as technology improves, we can do more work with lower quantities of energy (see for example Huber, 2002, and Lomborg, 2001), but it is important to note that efficiency means little in the depletion picture without reduced total usage. With worldwide population growth and newly industrializing nations, total global energy consumption was rising quickly prior to the 2008 global economic downturn. Rising efficiency has also been correlated with increased energy density of primary energy sources. It is likely that increased efficiency levels will be harder to achieve if we must increasingly rely on renewable energy sources and coal—both of which offer opportunities, but at reduced energy density levels (Heinberg, 2005, chap. 4).

Some experts cite large reserves of coal as a substitute for depleting oil and gas reserves (see Hirsch, et al., 2005, pp. 43–44). Coal can be liquefied to produce synthetic petroleum (Hirsch, et al., 2005, pp. 43–44, 78); but, as with oil shale and oil sands, we have little to no infrastructure globally for refining this liquid fuel (Hirsch, et al., 2005, pp. 78–79). Furthermore, dependence on coal for transportation—even if we could build enough processing plants and filling stations—would require massive mining and contribute heavily to climate change—*and* we would be relying on yet another depletable source of energy for transportation, and one much less energy dense than petroleum. Furthermore, increasing our reliance on coal as a source of transportation fuel and/or for electricity generation would mean almost certain climate catastrophe.

Dense, convenient sources of fossil fuel energy have fueled industrialism, and extensive use of fossil fuels by relatively rich nations and people has contributed to the ever increasing concentration of wealth and power that is destabilizing the global economy. But fossil fuel depletion is likely to play an important role in raising possibilities—even mandates—for social change as economic growth proves unsustainable. Peak oil and gas are among the leading contributing factors to the instability and unsustainability of the late capitalist order.

The Structural Crisis of Capitalism

According to Nobel Prize–winning economist Joseph Stiglitz and sustainability-oriented economist Richard Douthwaite, globalization has created an unstable economy that is highly susceptible to crisis (Douthwaite, 2004, pp. 118–119; Stiglitz, 2002, p. 6). In 2009, political economist Damien Cahill noted:

> In the space of a year the unthinkable occurred. What began as a collapse in one segment of the U.S. housing market has spilled over into a crisis of the global financial system. In response, the governments of two of the world's most powerful capitalist nations, the U.S. and Britain, have nationalized major financial institutions, reversing the privatization trend of the last two decades. French President Nicholas Sarkozy proclaimed, "Laisez-faire is finished." Dominique Strauss-Kahn, head of the International Monetary Fund (IMF)—which requires developing nations to impose neoliberal structural adjustment programs on their citizens in order to qualify for loans—laid blame for the crisis on "not enough regulations or controls." (Cahill, 2009, p. 35)

It seems that the neoliberal project of fostering free trade, privatization, and globalization in what was deemed to be a self-regulating market, instead of promoting stability, has delivered a major crisis that threatens the existence of capitalism as the dominant form of modern political economy.

Deregulated, free-market capitalism is consuming its foundations. Production has largely been outsourced to the Global South while buying power has been concentrated in the U.S. and other industrialized countries (Kaplinsky, 2005, pp. 164, 178–180). Workers who produce products see their incomes eroded by economic pressure for low cost production. Meanwhile, prior to the burst of the housing bubble in 2008, buyers in the U.S. maintained their spending in an atmosphere of declining real wages by accruing ever more debt. According to William Greider,

> The present regime is pathological fundamentally because it broadly destroys consumer incomes while it creates the growing surfeit of goods…. Greater social equity is consistent with and, indeed, required for a sound and expanding economy: when rising incomes are broadly distributed, it creates mass purchasing power—the rising demand that fuels a virtuous cycle of growth, savings and new investment. When incomes are narrowly distributed, as they are now, the economic system feeds upon itself, eroding its own energies for expansion, burying consumers and business, even governments, in impossible accumulations of debt. (Greider, 1997, p. 321)

The fact that global oil and gas producers will be able to capture ever higher proportions of total world spending as these resources become scarce further increases the concentration of wealth and power in the global economy, thereby threatening to destabilize the system by impeding the circular flow of money between consumers and producers.

Several important factors have combined to produce a continuing crisis in the global financial sector and a global economic downturn (see Greider, 1997; Wallerstein, 2008a). These include:

- concentration of wealth and buying power that results in patterns of overproduction and underconsumption that both creates and reflects widespread economic hardship that jeopardizes the long-term growth of the economy;
- deregulation of banks that has allowed excessive risk taking in institutions so large that their failure could destabilize the entire global financial sector;
- over-indebtedness of individuals at the same time that real wages have declined (particularly in the United States) and over-indebtedness of entire nations;
- depletion of resources required for economic activity and growth; and
- shrinking margins for potential profits as globalized competition has reduced prices for commodities and products.

Heavy debt loads borne by consumers and nations, the decline of ecosystems, and the depletion of resources have combined to make the current downturn especially challenging to reverse (Orr, 2009). Keynesian methods of priming the economic pump through government spending are also unlikely to generate the desired effects in a globalized economy because money spent within one nation quickly leaks away into other areas of the global economy (Douthwaite, 2004, p. 119). As suggested by world-system theorist Immanuel Wallerstein (2008a, 2008b), the current crisis may indeed signal the beginning of the end of capitalism itself. According to Stiglitz (2009), it is the "the belief that markets are self-regulating and that the role of government should be minimal" that set the stage upon which the factors of economic collapse have converged.

Using the financial sector as a prime example, Homer-Dixon cites tight coupling among institutions, nations, and subsystems of the global economy as another important source of economic instability:

> Any bank faces a fundamental mismatch between the time frames of its liabilities and assets: although its customers' deposits can be withdrawn quickly, its loans are usually invested for long periods…. [There are also] risks arising from the soaring connectivity, speed, and complexity of the international financial system. Today's communication technologies have so increased the number, tightened the coupling,

and boosted the pace of transactions within globalized markets that once a destabilizing feedback loop—like a stock market crash or a run on a weak—takes hold, it can spiral into a crisis before policy makers can respond, and then it can cascade outward to affect other economies far and wide.... Today, the international financial system resembles a huge, crowded theater that's vulnerable to fire.... Speculative capital can be moved from one economy or currency to another with the click of a computer mouse, which means that everyone can converge on the financial system's exits simultaneously. But only those who escape first win, and investors and speculators are terrified of being left behind with a worthless stake in an imploded currency or economy. (Homer-Dixon, 2006, pp. 182–183)

The large real estate bubble created in recent years in the United States, largely as a result of low interest rates and easy credit, further compounded existing economic instabilities. Home buyers made risky purchases during the bubble years and assumed adjustable rate mortgages with their ability to repay resting on the assumption of rapid and never ending increases in home values. Predatory lending practices contributed to home buyers assuming dangerous levels of mortgage debt that would lead to widespread foreclosures in an economic downturn, as has occurred since fall 2008. Deregulation in the banking sector laid the groundwork for risky mortgages to be repackaged as mortgage-backed securities and widely marketed and sold as stable investments, when in fact these securities were very risky investments indeed. The collapse of the real estate bubble in the U.S. triggered the severe global recession that began in fall 2008.

Global banks have also participated in what amounts to casino capitalism using complicated financial instruments such as derivatives and credit-default swaps to gamble with assets. According to Joseph Stiglitz (2009), in the global casino economy, "the problem is that, with this complicated intertwining of bets of great magnitude, no one [can] be sure of the financial position of anyone else—or even of one's own position." Banks and investors were able to take huge risks for huge returns while the lack of transparency in the system hid the dangers from average citizens and while investment banks and stock brokers encouraged the investment of pensions and other funds critical to social welfare into this temporarily highly profitable but ultimately extremely fragile system.

We will explore below how lending policies of the Bretton Woods institutions also create economic instability within many nations and how this instability contributes to the fragility of the world-system. At this point, it is important to understand that capitalism itself is proving inequitable, unreliable, and unsustainable as a global system of production and distribution (Douthwaite, 2004). Therefore, as more and more people are drawn into dependence on this system for access to the necessities of life, the world's populations find themselves in an increasingly perilous position.

We now turn our discussion to exploring how and why people and entire nations become dependent upon the globalized economy—and why they find it nearly impossible to reverse this dependency once it is established.

The Concept of Enforced Dependency

The concept that I call *enforced dependency* is not entirely new, but as a distinct concept, I believe it can offer opportunities for fresh insights on globalization. The concept of enforced dependency provides a lens for understanding the weak points within the structure of global capitalism, particularly those that relate to its long-term socio-ecological and economic sustainability. In this section I define enforced dependency as it functions in multiple arenas in the globalized world system.

Enforced dependency is a form of reliance upon external resources or externally created conditions. For such dependency to function as enforced dependency, it must, once established, progressively undermine the self-sufficiency and resilience of the dependent person, community, institution, or government, making the dependent party increasingly vulnerable to exploitation. The initial conditions of enforced dependency are often established through colonialism or imperialism. The "enforcement" of enforced dependency derives from the increasingly dangerous and/or destabilizing results that would entail from severing the dependent relationship. Under conditions of enforced dependency, the resiliency of the dependent party decreases progressively over time. Typically, dependent parties are also progressively co-opted into supporting the system of enforced dependency upon which they have come to rely, even as the system progressively robs them of freedom, independence, and resiliency. As the dependency deepens, the social power of dominant people, institutions, or governments increases with regard to dependent parties. Dominant parties may increasingly constrain the decisions and actions of dependent parties in order to enhance their opportunities to gain material and financial wealth and increase their social power. Though dominant parties may gain substantial wealth and power through enforced dependency, over the long term, their own resiliency may be negatively impacted as the socio-ecological capacity of dependent parties to serve as sources of wealth and power for dominant parties declines.

Homer-Dixon describes the process of what I call enforced dependency and articulates how it interfaces with Gramscian cultural hegemony (1971/1999, pp. 57–58) and with Marcusean repressive desublimation in the late capitalist paradigm (1964):

We find it far easier to play by the rules if we actually believe in the legitimacy and reasonableness of the larger system that lays down those rules. We become invested in the capitalist worldview. Without it, our modern world wouldn't make much sense at all: we wouldn't know our social and economic roles, and we'd have difficulty connecting and communicating with people. We realize, too, that it's senseless to challenge openly our economic system's overarching logic because we'd be challenging the source of our own paycheck—the goose that laid the golden egg, so to speak. The basic truth of this economic arrangement is crystal clear to everyone: the interests of business prevail over all others. So our economic system generates pervasive insecurity; this insecurity impels us to play by the rules; our need to play by the rules encourages us to find these rules morally legitimate; and our belief that the rules are legitimate creates a huge obstacle to changing them. For many of us, the denial is entirely rational. (Homer-Dixon, 2006, pp. 217–218)

We can view enforced dependency as a central theme of late capitalism. Freire advocates consciousness building in critical pedagogy through identifying and critiquing oppressive themes that pervade social reality. The process of critiquing these themes serves as an avenue for praxis toward resolving the contradictions of capitalism and creating new ways of producing, distributing, and being in the world (Freire, 1970/2000, chap. 3). The concept of enforced dependency is a tool for both analyzing unsustainable systems and for suggesting alternatives to them.

As globalization extends into new areas of life and new communities, systems of local self-sufficiency and decision making—that are also often comparatively sustainable—give way to systems of economic and social dependency based upon money and the globalized market (ISEC, 1993; Norberg-Hodge, 1991/1992). This widespread dependency of societies upon the very systems that oppress them serves to increasingly entrench hegemonic forces and to increase the momentum of the trajectory of globalization itself (Miller, 1999). Path dependency with regard to past and current choices also contributes to the entrenchment of enforced dependency over time because, once a community or society has made or been forced to make some critical choices—monetizing their economy is one of these—future choices become predicated upon structures created by past decisions. Reversing past decisions becomes increasingly difficult because doing so would create widespread social and economic dislocation. Capitalist global integration that creates and enforces dependency is also characterized by path dependency in that, once a community or society is folded into the global economy, the people find it nearly impossible to extract themselves from the system. The post-war Green Revolution in agriculture provides an excellent example of path dependency.

Elucidating the sources of enforced dependency in the world-system means constructing a critique of globalization that focuses attention specifically on those institutions, practices, cultural phenomena, and ideologies that foster dependency at multiple levels. Stiglitz (2002) defines globalization as "the removal of barriers to free trade and the closer integration of national economies" (p. ix). We begin with this definition while noting that the constituents of globalization as a phenomenon also specifically include all of the dependency enforcing institutions, practices, cultural phenomena, and ideologies discussed below. We now turn our attention to applying the world-system analysis of Immanuel Wallerstein as a vehicle for understanding the overarching phenomena that characterize globalization.

Capitalist Industrial Production as Dependency in the World-System

In this section, I argue that enforced dependency infuses globalized late capitalist systems of production so that the global economy operates as world-system that disproportionally serves the interests of economic and political elites. In his classic work *The Great Transformation*, Karl Polanyi (1944/1957) documents how, during the eighteenth and nineteenth centuries in England, the capitalist mode of production and capitalist political economy replaced traditional means of production and distribution. He notes that, within traditional societies worldwide, economic activity served to build and maintain community relationships and to satisfy community goals. Social cohesion and mutual benefit were secured through practices of redistribution of produced goods. Under the new capitalist mode of production and trade in England, society was made to serve the self-regulating market (chap. 3–6). According to Polanyi, factory production was an essential precursor to this monumental shift (1944/1957, pp. 40–41, 57). It is important to note that the rise of the market as *the* primary focus for social organization and production had influenced colonial relationships as well.

Capitalist ideologues have argued that the impetus for detailed division of labor in capitalist production, as distinct from the social division of labor,[1] as discussed by Braverman (1974, p. 70) and others (see also Ollman, 1971; Wallimann, 1981, chap. 6), is primarily about efficiency, but it is not. It is about control of the production process and the centralization of that control

---

[1] The social division of labor (Braverman, 1974, p. 70) is differentiation of work based upon age, sex, and physical strength. Walliman (1981, p. 89) and Ollman (1971, p. 160) emphasize that the division of labor in capitalist society, unlike the social division of labor, is characterized by a division between mental and material labor. Ollman (1971) further states that "the division of labor whereby people do only one kind of work and rely upon others to do whatever else is necessary to keep them alive is a more inclusive social expression of man's alienated productive activity" (p. 159). Alienated and detailed division of labor enforces dependency in the sense discussed in this chapter.

with a new class of factory owners and managers. Kropotkin (1902/1989) recognized the social power inherent in the craft guilds of medieval European cities wherein artisans directly controlled the production process and also the entry of new artisans into particular crafts (chap. 5–7). In the capitalist paradigm, the work of artisans was broken down into myriad different parts so that anyone could be hired to do it, or better yet, machines could do it independently. This change wrested control away from artisans (Kropotkin, 1902/1989, chap. 7–8), who had been dispersed throughout society, and concentrated it in the hands of capitalists. This great transformation created new forms of social dependency in England that were early incarnations the enforced dependency that appears to have reached its apex in the late capitalist world-system.

According to Wallerstein's world-system analysis (1974, 1976, 2003, 2005, 2006, 2007, 2008a, 2008b), capitalist globalization has produced uneven development consisting of core, periphery, and semi-periphery nations and regions. Core areas are those that have benefited the most from globalization. These nations and regions exert their influence on other parts of the world-system in order to capture disproportionate shares of international wealth and power. Periphery zones typically had been colonies of core nations that extracted natural resources needed for industrial production. As a result of their colonial history, peripheral nations lack strong central governments and diverse economies. Semi-peripheral nations are exploited by core powers while they also exploit the peripheral zones; these nations and regions often embody a tension between a strong central government and a domestic landed elite. Within the world-system of late capitalism, relationships among nations and regions resemble those of the colonial era, but without direct political control of the periphery by the core (Wallerstein, 1974, 1976). In recent decades, *core* is becoming less of a geographic designation as the power of transnational corporations and global finance increasingly eclipses that of nation states (see Harvey, 1989).

Within the world-system, nations and peoples who were conquered during the wave of European colonization remain at a competitive disadvantage (Wallerstein 1974, 1976). The purpose of a colony to an empire is not to instill self-sufficiency and resilience; it is to provide resources for the colonizing power. The power imbalance between colonizer and colonized is intentionally maintained wherever possible by the forces of empire (Miller, 1999). The extractive economies of the colonies represent a more or less direct throughput of energy and matter with little diverse use and reuse of resources and human capital at the local level. Jane Jacobs (2000) emphasizes how diverse uses and reuses of matter and energy are essential to economic diversification and resultant resiliency, just as the diverse use and reuse of matter and energy is critical to development of diverse and resilient ecosystems (pp. 43–63). Economic relationships between colonizers and colonies have offered few possibilities for colonies to develop economic resiliency that would enable them to escape peripheral status. Even after independence, former colonies remained disadvantaged with regard to economic diversification and resilience, industrial development, knowledge of modern technology, infrastructure development, ownership of intellectual property, and more. According to Stiglitz, African aspirations following independence have remained almost entirely unfulfilled (2002, p. 5).The Global South has been expected to compete in the global economy, but the playing field has remained tilted in the direction of core powers (Miller, 1999).

Fossil fuels have played an important role in creating and deepening the divisions among core, periphery, and semi-periphery zones. Beginning with the industrial revolution, energy-dense fossil fuels, created incredible opportunities for increasing human activities at very little cost, especially for those nations with abundant domestic reserves, and those with the early lead technologically were best positioned to take advantage of these energy sources. First coal and later petroleum increased the speed and efficiency with which new lands and peoples could be colonized, in turn yielding access to an increasing variety and quantity of natural and human resources for industrial development (Heinberg, 2005, chap. 2).

Understanding that late capitalism is a *world-system* helps one to recognize the motivating forces behind the capitalist drive to enforce dependency. International political and economic equity coupled with localized gains in self-sufficiency would fundamentally alter the character of the system and its ability to serve dominant people and institutions by concentrating wealth and power in their hands. Socio-economic equity and sustainable self-sufficiency would virtually eliminate the dependency enforcing aspects of the world-system and unchain the masses from service to a globalized economy that simultaneously entraps and depletes them.

We will now explore important aspects of enforced dependency that constitute late capitalist globalization. Each aspect of globalization addressed below articulates in numerous ways with other aspects of the overarching system. Specific aspects of the system addressed therefore serve as lenses through which we can gain perspective on the whole of late capitalism as a dependency-enforcing system. The dependency-enforcing aspects of late capitalism discussed below create far-reaching effects, both geographically and in the sense that they serve deeply structural roles within neoliberal globalization.

How Money and Debt Enforce Dependency

According to economist Richard Douthwaite (2004), the modern money system creates enforced dependency because currencies used are created outside the community. In order to assemble the means to trade, communities must sell to outside markets at prices determined externally (p. 116). The money system also enforces dependency because it is based on debt (Douthwaite, 1999a, chap. 1, 2004, p. 118; Rowbotham, 2000, pp. 90–102). This debt-and-interest-based system requires infinite economic growth in order to avoid currency collapse so that economic growth has become a requirement for nations and for the global economy as a whole. Infinite economic growth within a limited system (the earth) is a defining contradiction of the capitalist world-system (Daly, 1999; Douthwaite, 1999a, pp. 25–26, 1999b; Kovel, 2002, chap. 3–4, 6). Therefore, it is important to understand how the money system perpetuates unsustainability. We will explore how debt-and-interest-based money fosters unsustainable dependence on a self-destructive global economy that is consuming its own base (Kovel, 2002, chap. 3–4, 6). It is also importtnat to understand potential currency reforms and innovations that could encourage and support sustainable living. In his *Ecology of Money* (1999a), Douthwaite explores how the way modern money is created influences the social and ecological effects of its use. He calls for (re)localization of some forms of currency as a strategy for promoting sustainability. Much of our discussion in this section draws upon Douthwaite's work.

In our efforts to comprehend how the money system contributes to enforced dependency, we begin with an explanation of the functions of money. Money serves three key purposes: it serves as a medium of exchange, as a store of value, and as a unit of account. These functions can conflict with each other. In an inflationary environment when prices are rising, the store of value function of money decreases, and people tend to spend money rather than saving it since they assume prices will be even higher later. In a deflationary environment when prices are falling, the medium of exchange role of money is inhibited. People see money as a good store of value and tend to save it since they assume they will be able to buy more with it later. In both cases, reactions of people to price changes tend to reinforce the cycle of inflation or deflation and further inhibit one of the functions of money. Modern money does not function optimally and simultaneously as *both* a medium of exchange *and* a store of value (Douthwaite, 1999a, pp. 11–12, 27–28).

Money functions no better as a unit of account. An analysis of the processes of cost-benefit analysis illustrates this point. Since money has no fixed value with regard to anything tangible, its use as the basis for quantification in cost-benefit analyses opens the door for manipulation of these analyses to suit desired outcomes. Since cost-benefit analysis relies on conversion of future costs and benefits to net present values (a process of "discounting" estimated cash flows related to future costs or benefits by a chosen rate of compounded interest), present costs and monetary profits are emphasized over benefits and costs that could occur in the long term. The rate at which profits made today are assumed to be able to grow or at which future costs lose their economic significance (the discount rate) is a crucial factor in coloring these analyses to encourage or discourage a given project or extractive enterprise. In cost-benefit analysis, present profits are favored over similar profits in the future, a function of the time value of money estimated through use of the discount rate. The tendency to value present over future profits intensifies with increases in discount rate used in the analysis. Cost-benefit analysis can be used to justify projects that will yield profits now, even if they are projected to have significant costs in the distant future (Douthwaite, 1999a, pp. 29–30). In cost-benefit analysis, it is also difficult to include costs or benefits such as human and environmental health, aesthetics, ecosystem integrity, cultural survival, and long-term sustainability that can be difficult or impossible to quantify monetarily (Moore, 1995).

The time value of money, along with the dependency of modern people on money, create a framework for decision making in which people tend to exploit and consume each other and nature for short-term financial gain—especially when the costs of doing so are not quantifiable or are likely to occur in the distant future. Widespread exploitation for profit further distances people from each other and nature and thereby increases dependency on money as a means to satisfy material, social, and service needs, and this increased dependence in turn drives further exploitation for profit. In today's societies, most people need money because they do not own or have access to enough land or other resources to produce for themselves what they need to survive. By extension, they are also dependent upon the institutions that create money.

Commercial banks create most modern money by lending out more money than they have on deposit, a process known as fractional reserve banking. Money created in this way exists purely as a function of borrowing. When loans are repaid, the money created by them ceases to exist. People must continually borrow from banks in order to maintain money in circulation (Douthwaite, 1999a, pp. 16, 21–22; Heinberg, 2005, pp. 187–190; Rowbotham, 2000, p. 90). The process of fractionalization is repeated with borrowed money when money lent by one bank is deposited into another, in which case money that was created from nothing in the first place is treated as a basis for further fractionalization (Douthwaite, 1999a, p. 20; Rowbotham, 2000, p. 15). The assumption is that all depositors will not converge on a bank at once demanding the full value of their deposits (Douthwaite, 1999a, p. 17; Hudson, 2005, p. 18). The loan-to-deposit ratio through which commercial banks create money is typically governed by the central bank to which the commercial banks report. Central banks use this and other mechanisms to control the money supply and, by extension, to control inflation/deflation (Douthwaite, 1999a, pp. 18–20). The value of modern money is determined solely by its acceptability to others—it is backed by nothing but faith (Douthwaite, 1999a, p. 21; Heinberg, 2005, p. 188).

Because most money is created in this debt-and-interest-based system, we are directly or indirectly *dependent* upon the existence of debt in order to have money at all. When we realize that debt is actually the source of money in circulation, it comes as no surprise that borrowing is heavily encouraged in consumerist societies. As anyone who has been in debt knows, the need to service the debt and the consequences that would follow from not doing so constrain one's choices and actions. We are heavily dependent on money to obtain goods and services while, at the same time, the very need for this money and the socially entrapping way that it is created impinge upon our life choices and actions. Most of us have little choice but to participate in, and thereby support, the money system, and the participation of each of us strengthens both the overall system and its ability to imprison us.

We rely on a money system that encourages and reinforces our dependency, even while the mechanisms within the overall economic system that support the value of money and prevent currency collapse are themselves quite fragile. The potential for economic depression is ever present because many circumstances can create an atmosphere in which people are unwilling to borrow money. Unless effective measures are taken in time when a downturn threatens, layoffs typically result from the slowing of economic activity and the concomitant reduction in the money supply. Layoffs further deepen the economic downturn and further decrease the willingness of people to borrow money. In such an atmosphere, the velocity of money circulation drops since even the wealthy feel uncertain of their economic future and hold onto their money, thereby reinforcing the downward spiral. The system that results from creating money through debt is therefore inherently unstable and prone to boom and bust cycles (Douthwaite, 1999a, pp. 22–23).

One particular form of debt, national debt, also influences the money supply and creates its own forms of enforced dependency. National debt entails the government going into debt to cover the difference between taxes levied and expenditures made. Rowbotham describes the process of using government bonds to finance national debt:

> When government bonds are bought by the nonbanking sector, funds held in various savings institutions (pension, life assurance and trust funds etc.) are brought back into everyday circulation, the sums being re-distributed into the economy through public services and other spending. Thus, monies relied upon for future payments are recycled into the economy, in parallel with a debt undertaken by the government, and registered against the nation's assets. However, when government bonds are bought by the banking sector, additional money is created since the purchase is made against, or using, the bank's fractional reserve. Just as with private/commercial debt, additional bank credit is thus created and new deposits of bank credit result. (2000, p. 96)

The mechanism of national debt used to finance public sector activities creates interdependencies among the savings and banking institutions, the national government, and citizens for whom government services are performed. All of these entities are also dependent upon economic growth to generate needed tax revenues to keep the national debt from spiraling out of control and triggering a currency crisis.

Dependency of individuals, communities, and entire nations on debt-and-interest-based money is also enforced through trade. In order to assemble the means to participate in the global economy using commercial-bank-created money, one must borrow funds from a commercial bank and agree to repay the loan plus interest, or one must sell goods or services to others who pay with currency. Profits from lending activities of commercial banks flow to their central offices. Therefore, repayment of loans held by nonlocal banks creates a systematic drain on local economies through interest paid and through the resources of the local economy being traded away for money to service debts (Douthwaite, 1999a, p. 21; Shuman, 2000, pp. 107–120). A co-dependency is established through which individuals, communities, and even nations—through their borrowing activities—help produce profits for corporate banks upon which they depend for access to the money they need to provide for themselves and satisfy their economic aspirations. This co-dependency is deepened by modern living patterns in which people are removed from land and other ecosystemic sources of sustenance and must rely upon global supply chains through which they purchase the necessities of life.

To summarize, we have discussed several ways in which debt-and-interest-based money contributes to enforced dependency. The economic system, due in part to the way money is created, is inherently constraining to human choice and activity and is always vulnerable to self-reinforcing economic downturns. The money system also tends to concentrate wealth and power in large banks that continually siphon away the material wealth of communities to which they do not belong. The loans provided by these banks allow individuals and groups to engage in trade and other economic activity through which they might support themselves and benefit their communities, but the fruits of a significant portion of their efforts flow away from them and their local communities in the form of repayment of debt to non-local corporate banks (Shuman, 2000, pp. 107–120). The ability of banks to create money therefore represents an important form of power and control. Any person or entity willing to accept a given currency increases the power and control of entity that created it (Douthwaite, 1999a, 2004, pp. 115–116). The need to purchase necessities, in particular, creates important forms of social dependency on money since necessities such as food and water are *required* for life itself.

International finance and debt also contribute to a multivalent world-system of enforced dependency. International debt fosters and maintains an imperialist political economy which, for former colonies, creates relationships and experiences similar to colonialism but without the overt conquering of lands (Miller, 1999). We will explore below in a section on the Bretton Woods paradigm how structural debt in the Global South contributes to political and economic enforced dependency of peripheral nations on the world-system of globalization (Rowbotham, 2000). We will also consider how dollar hegemony, initially enforced through the Bretton Woods system, produces economic benefits for the United States at the expense of other nations and thereby contributes to unsustainable co-dependency in international finance and debt.

When we understand how money is created and used, we begin to see how banks operate under a rentier's regime made possible through government. By virtue of their ability to engage in mortgage lending, banks also hold title to vast amounts of real property which can be claimed in the event of borrower default (Rowbotham, 2000, p. 186).

We now turn our attention to economic growth as a dependency enforcing aspect of the capitalist world-system.

Economic Growth

The industrial capitalist mode of production and debt-and-interest-based money undergird the capitalist tendency toward exploitation and its requirement for growth, both of which are unsustainable. People who rely on money to satisfy their needs are by extension dependent upon the growth required to keep that money in circulation. Interest must be paid on bank-created money, which means that the value of sales in the economy must grow in order to avoid system collapse. The growth necessary for repayment of loan principal plus interest can be realized through inflation, through increased output, through capturing market share from others, or through a combination of these strategies. The money supply must also grow continually in order to accommodate economic growth within the world-system and to allow interest to be paid (Douthwaite, 1999b, chap. 2).

Since output cannot continue to grow within a finite earth, the requirement for economic growth that is built into the debt-and-interest-based money system is inherently unsustainable (Daly, 1999; Douthwaite, 1999a, p. 25, 1999b, chap. 15; Homer-Dixon, 2006, pp. 200–201; Kovel, 2002, part I; Meadows, Randers, & Meadows, 2004). Vandana Shiva claims that growth is "theft from nature and people" (Shiva, 2000, p. 1). Every job, every product—even in the "information economy"—relies on the extraction and use of natural resources, including human resources. Growth in the money supply, required by the system, occurs through continually expanding debt. Servicing the loans that are the source of money requires continually expanding production and, therefore, continually expanding use of nature and people in the process of production. The growth economy is currently bumping up against both natural limits and the limits of the capitalist system to contain its own contradictions.

We gain a clear picture of the exponential character of the growth built into the system through examining the rule of 72. This mathematical rule of exponential growth allows one to estimate the time to double of a compound-interest-earning investment. Seventy-two divided by the interest rate per period equals the number of periods to double. So, an investment accruing 8 percent interest annually would double in nine years, and if the doubled money were to be left invested at the same rate, it would double yet again in another nine years. The rule of 72 applies to debt as well. A debt on which interest is owed that is left unpaid will double in size in the same way as an investment earning interest. According to the International Labor Organization (2005), the global economic growth rate for 2004 was 5 percent. At that continued rate of growth, the global economy would double in slightly less than 14 and a half years. When one considers that all future doublings occur based upon an exponentially expanding base, the power of the required growth in the global economic system becomes evident.

The need for growth is also a primary driver behind the neoliberal obsession with opening markets in parts of the world not yet fully exposed to commercialized living. It is also a primary driver behind the systematic destruction of subsistence cultures wherein one could live—albeit very modestly in terms of material wealth—independently from the money economy. Destruction of subsistence increases the number of people forced into dependency upon the commercial distribution system. As Shiva (2005) claims, growth is theft from people of systems of self-determination and independence and the replacement of these systems with enforced dependency. Left unchecked, global capitalism will destroy the few remaining subsistence cultures in an effort to satisfy the global economy's appetite for new consumers and natural resource inputs.

The push for growth that drives the opening of new markets globally also requires that people everywhere be subjected to a barrage of advertising to convince them to purchase and consume ever more products. In the United States, increased consumption has been heavily encouraged in recent decades, though this spending spree has been fueled by crushing levels of consumer debt (Clark, 2005, p. 11). The easy credit provided to consumers in the United States temporarily ameliorated the effects of global overproduction while also perpetuating and deepening enforced dependency globally.

Technological advantages that have developed historically also mean that growth is not evenly distributed globally. Producers who lag behind competitors in use of new technologies tend to see their profit margins shrink when compared to early adopters who innovate successfully because the new production methods typically decrease production costs and place downward pressure on prices. In such an atmosphere, very large producers who are able to innovate early and capture profits before prices drop have the edge. Once prices fall, large producers capture market share from their smaller competitors who are driven out of business. These monopoly aspects of the economy further the tendency of globalization to force everyone everywhere to use many of the same materials and products (Douthwaite, 1999b, chap. 2, 2004, pp. 116–117). They also decrease the resilience of local communities to both economic shocks and ecological disturbances. In food production, for example, local plants and plant varieties and locally adapted breeds of animals disappear, and the diversity they represent is often lost forever. This lack of diversity in production creates dangerous vulnerabilities, as demonstrated by the Irish potato famine in the nineteenth century. It also places increasing pressure on a defined set of natural resources. In more localized economies that existed prior to globalization, local materials, flora, and fauna were used to satisfy local needs. Today, even for those local resources that remain, local production for local use is often uneconomic when compared to purchasing materials and products from large global suppliers.

This process of creating global monopolies and monocultures depends upon cheap fuel for transport, a precondition that is increasingly in doubt (Campbell & Strouts, 2007, part 1; Deffeyes, 2001; Douthwaite, 2004, p. 118; Heinberg, 2005; Kunstler, 2005; Roberts, 2004; Simmons, 2005). As noted above, fossil fuel dependency may yet prove to be the Achilles heel of globalization since production of petroleum worldwide has peaked or will do so within a few years. If global economic growth is to continue, it will require an ever-expanding supply of dense energy that is convenient to transport and store and suitable for fueling existing transportation infrastructure. It appears increasingly unlikely that we will discover a means to capture sufficient energy sources that possess these qualities.

The global growth economy concentrates wealth and power while creating dependence. It also creates and deepens dependency while reducing socio-ecological resilience. It manifests path dependence in that, as the system persists through time, changing course becomes increasingly difficult because the cultures, resources, flora, and fauna on which alternate, steady-state, localized living patterns might be (re)constructed are increasingly sacrificed in the process of achieving economic growth. All the while, the capitalist growth economy increases the vulnerability of communities and societies to economic shocks and ecological breakdown.

We will now examine in more detail a specific and important source of both loss of resiliency and increasing dependency within the global economy: technological displacement of labor. We will also examine how this displacement increases the need for economic growth.

Technological Displacement of Labor and Elimination of Subsistence in the Growth Economy

Technological displacement of labor provides an important rationale for the fixation of policy makers on economic growth. Homer-Dixon summarizes the logic of economic growth as a remedy for labor displacement:

> In essence…the logic underpinning our economies works like this: if we're discontented with what we have, we buy stuff; if we buy enough stuff, the economy grows; if the economy grows enough, technologically displaced workers can find new jobs; and if they find new jobs, there will be enough economic demand to keep the economy humming and to prevent wrenching political conflict. Modern capitalism's stability—and increasingly the global economy's stability—requires cultivation of

material discontent, endlessly rising personal consumption, and the steady economic growth this consumption generates. (2006, p. 196)

But, according to Homer-Dixon, "Conventional wisdom rarely acknowledges the scope and relentlessness of technological displacement of workers" (2006, p. 195). Displaced workers also do not necessarily earn the same pay in new jobs. In the United States, where manufacturing has largely been outsourced, former manufacturing workers have in many cases been forced to seek comparatively lower-paying service jobs (Clark, 2005, pp. 10–11), and high-paying jobs in new industries and using new technologies typically require skills and knowledge that displaced workers do not have (Homer-Dixon, 2006, p. 195). Technological displacement of workers may eventually prove unsustainable and ultimately destabilizing to the capitalist system. Displaced workers are mostly dependent upon money and the market for fulfilling basic needs, and their displacement typically constrains their freedom and self-sufficiency. Control is further concentrated in the hands of business owners and managers as a result of replacing workers with machines. Owners and managers are interested in increasing their shares of business profits, but in the long term, because widespread technological displacement of labor erodes the purchasing power of the masses, displacement of workers may harm business leaders as well.

Technological displacement of labor and displacement of subsistence cultures are important sources of enforced dependency globally. Since displacement places downward pressure on wages and causes unemployment, it fosters consumer debt as a means to maintain demand. As noted above, consumer debt is an important source of enforced dependency. By systematically undermining self-sufficiency and placing downward pressure on wages, technological displacement of labor and displacement of subsistence cultures concentrate wealth, power, and social control in the hands of those who enforce widespread dependency upon the global economy.

We will now turn our attention to some important aspects of neoclassical economic thought that contribute to enforcing the dependency of global populations.

The Hegemony of Neoclassical Economics

One could claim that mainstream economists are agents of dependency enforcement. In society at large, the powerful disproportionately control which ideas circulate widely and in high places and, to a large extent, control approval mechanisms for those ideas that get branded as truth. Neoclassical economists have mostly nodded in approval to a system that has created crippling and permanent debt in the Global South, debt that represents the failure of modern economics (see Rowbotham, 2000, and Stiglitz, 2002). They have also promulgated the myth of the self-regulating market. We see the folly of this myth amidst the fallout of the continuing economic downturn and turmoil that began in fall 2008.

Reliance on incomplete models and statistical instruments that do not approximate complex reality gives economics the appearance of an objective science when, in fact, the very assumptions and rules embedded in the economic system are themselves political. According to Stiglitz (2002), standard formulas used in macroeconomics do not take unemployment into account:

> In some of the universities from which the IMF [International Monetary Fund] hires regularly, the core curricula involve models in which there is never any unemployment. After all, in the standard competitive model—the model that underlies the IMF's market fundamentalism—demand always equals supply. If demand for labor equals supply, there is never any *involuntary* unemployment.... The problem cannot lie with markets. It must lie elsewhere—with greedy unions and politicians interfering with the workings of free markets, by demanding—and getting—excessively high wages. There is an obvious policy implication—if there is unemployment, wages should be reduced. (p. 35)

According to Rowbotham, most academic economic literature contains little to no analysis of debt or money (2000, p. 15). We see from the arguments presented in this chapter that unemployment, falling wages, and the monetary and debt systems are critical factors in enforcing dependency among populations worldwide. Because these receive little scrutiny from academic economists or economists working in government and banks, we might assume that markets *are* working perfectly in the view of capitalist elites, or at least that they have generally been doing so.

The IS-LM model of macroeconomics is a useful case in point. The IS-LM model is the standard formula of current macroeconomic analysis used to manage economies through fiscal and monetary policies (Daly & Farley, 2004, p. 278). It exemplifies that capital accumulation is *the* goal within the capitalist order in the way it renders issues of environmental and social health, unemployment, distribution of wealth, and dependency indirectly visible at best. It also indirectly enforces dependency.

The IS-LM model hinges on achieving economic equilibrium in two areas: 1) savings (S) and investment (I) (the "real sector") and 2) liquidity preference (L) and money supply (M) (the "monetary sector") (see Daly & Farley, 2004, chap. 16 for a detailed explanation of this model). All values considered in achieving this equilibrium are monetary values, meaning that considerations related to quality of life, ecosystemic health, and socio-ecological sustainability are accounted for indirectly at best (through whether and how they affect the money supply and the uses of money).

In the standard macroeconomic view encoded in the IS-LM model, human beings are regarded as rational economic beings—*homo economicus*—for whom the *central* questions of *value* in the choices made by individuals are *economic* choices. *Homo economicus* is driven by competition to maximize his/her share of goods and services at the lowest possible cost. Choices made by *homo economicus* are not rational in any overarching ethical sense; they are rational only within a narrowly economic view of the world in which capturing material wealth is the only goal that "counts" (see Spretnak, 1997, pp. 219–221). The system of values inherent in neoclassical economics was both historically and culturally specific to Western capitalist societies, but it now holds sway in the globalized world-system.

That the IS-LM model remains a tool of choice among macroeconomists demonstrates the hegemony of capital in management and policy making worldwide. Use of this model as the "workhorse" of macroeconomics (Daly & Farley, 2004, p. 279) means that alternate values to capital accumulation are virtually absent from economic management and policy. Capital is the focus in economic decision making while other interests—even those that, if ignored, ultimately threaten the long-term viability of the capitalist order—receive little to no attention.

Perhaps the most important point about the IS-LM model with regard to socio-ecological sustainability is that natural limits and social health and resilience are invisible to the model except as these limits translate into prices, availability and costs of labor, and other monetized transactions. When the model is unaffected by clear price signals—even in the face of impending socio-ecological disaster—economic equilibrium can be pursued and possibly achieved—if only temporarily. This is exactly the process that is occurring with regard to oil and gas. In an oil depletion scenario, price signals will occur far too late for societies to prepare for constrained supplies.

The tendency of mainstream economists to subscribe to the cornucopian argument with regard to resources also belies the hegemonic orientation of neoclassical economics. According to this argument, when shortages appear, price signals trigger innovation which, in turn, creates substitutes for whatever resource is scarce. Therefore, in the view of mainstream economists, the value of every production input and commodity can be monetized, and the growth economy can adapt to any resource shortages that may occur. As members of indigenous subsistence cultures learned long ago, one species of animal or plant or the presence of clean water can be irreplaceable within a culture and an ecosystem. In such a society, the notion of infinite substitutability would be preposterous, but this idea predominates within the capitalist world-system today, even in the face of impending and severe shortages of resources essential to modern economies. In modern capitalism, the idea of infinite substitutability perpetuates an unsustainable economic order that favors the interests of the powerful at grave risk to all.

The management of inflation and deflation by policy makers also serves hegemonic interests. Inflation is characterized by rising prices; disinflation is characterized by slower than expected inflation, and deflation is characterized by falling prices. Inflationary periods can benefit debtors with loans at fixed interest rates because the real value of their repayment drops. Conversely, inflation negatively impacts the real income of creditors. Disinflation tends to positively impact creditors because expected inflation has not occurred, making the value of repayments, in effect, higher than expected. Deflation can benefit creditors so long as it does not continue long enough to cause a depression that triggers increased withdrawals of bank deposits and loan defaults. Both disinflation and deflation typically cause unemployment and lead to a net transfer of national income to creditors (the wealthy and banking interests) (Daley & Farley, 2004, pp. 291–294). Given the constituency of the U.S. Federal Reserve (bankers and other wealthy interests) for example, it is easy to see why that institution demonstrates an anti-inflation policy orientation (Daley & Farley, 2004, p. 301).

Nearly always, it is capitalist elites (those who were able to attend elite universities and those with access to the capital necessary to distance themselves from the world's growing poor populations) who advise highly placed decision makers or who undertake important economic policy decisions. As members of a hegemonic class, they tend to serve the economic interests of themselves and their economic and social allies, and they employ tools such as the IS-LM model and economic "common sense" such as a strongly anti-inflation policy stance in their advisory roles. This is not to say that all economists and capitalists are part of a vast conspiracy to dispossess the middle and lower classes. Although many knowingly do seek economic advantage for themselves and their social allies through any means possible, including brutal exploitation of other people and the environment, others engage in the world from positions of power with the intent of maintaining their relative social position but without an explicit intent of inflicting harm. Others are intellectually blinded to the damage done through use of standard economic tools such as the IS-LM model. After all, they learned to use such tools when they were students in respected institutions of higher learning. Whatever the conscious intent behind their actions, elite policy makers tend to reproduce the cultural hegemony of the capitalist order.

In order to have a sustainable world, we must fundamentally change our culture and its systems of valuation and exchange. Formulas such as the IS-LM model that place distance between analyzing the health of the economy and the health of humans and nature are part and parcel of the cultural landscape of neoclassical economics and capitalist society as a whole. One must recognize that economists who use these formulas are embedded in the capitalist culture which, by the very rules which constitute the system, privileges the economic interests of international elites over the well-being of others and the health of the environment. This system also privileges mainstream economists who serve the entrenched interests of the economic/political elite over radical economists who advocate change.

We have seen how neoclassical economics is indeed political in that it supports the hegemony of capitalist elites through perpetuating a narrow and hegemonic worldview. We have seen also that the capitalist values undergirding neoclassical economics directly conflict with socio-ecological sustainability in that they promote exploitation and the drawdown of critical resources, and we have recognized the virtual blindness of macroeconomic modeling to issues of socio-ecological sustainability. We must conclude that neoclassical economics contributes to enforced dependency by privileging the worldview and interests of capitalist elites in the global economy. Below, we will discuss additional specific examples of how the worldview and interests of the privileged have been encoded in the structures and rules that govern the global economy and enforce dependency.

We now turn our attention to the post-war Bretton Woods institutions and monetary system, both of which are informed by neoclassical economics. We will see how the Bretton Woods paradigm contributes in important ways to enforcing dependency within the world-system.

The Bretton Woods Paradigm

In this section, we explore how the post–World War II economic paradigm established at the United Nations Monetary and Financial Conference, held at Bretton Woods, New Hampshire, in July 1944, undergirds a global system of enforced dependency. John Maynard Keynes—whose reflationary economics formed the foundation for the New Deal and whose policies are typically credited with ending the Great Depression—is often incorrectly credited as the architect of the Bretton Woods institutions. In fact, his proposal for an International Clearing Union (ICU) was rejected by the American delegation prior to the conference (Rowbotham, 2000, pp. 13, 37–45). Keynes' Clearing Union, had it been implemented, would likely have greatly reduced the dependency-enforcing aspects of post-war international finance and debt. In this section, we will examine how and why debt in the Global South is a structural feature of the Bretton Woods economic paradigm.

We began our exploration by contrasting what might have been with what in fact became the central organizing institutions and features of the Bretton Woods paradigm. The Bretton Woods conference served as a forum for the allied nations to devise strategies and institutions for financing the rebuilding of Europe after the war, to facilitate post-war international trade, and to prevent future economic depressions. At the time, most colonies of the industrialized West had not yet gained independence (Stiglitz, 2002, p. 11). The Bretton Woods delegates created the framework for the World Bank and the International Monetary Fund (IMF), two institutions that have dominated Global South development ever since. The IMF, which began operation in 1945, was created to provide an international pool of funds upon which member countries could draw to help resolve temporary balance-of-payments deficits that threatened the stability of their currencies (Rowbotham, 2000, p. 35). At Bretton Woods, the U.S. proposed creation of the International Bank for Reconstruction and Development that would become known as the World Bank. This institution would lend to developing nations and to countries working to rebuild their economies that had been shattered by the war (Rowbotham, 2000, p. 43; World Bank, 2010). The foundation for the General Agreement on Tariffs and Trade (GATT), which came into force in 1947, was also laid at Bretton Woods (Kaplinsky, 2005, p. 14).

In his proposal for an ICU, Keynes, head of the British delegation, focused on the need to avert macro-economic destabilizing effects of trade imbalances. Recognizing that balance-of-trade surpluses and deficits were self-reinforcing, he hoped to avoid some nations becoming permanent creditors and others permanent debtors. Keynes argued that creditor nations experience boosts in demand for production and an influx of money, which in turn spur further investment and the seeking of additional markets for exports, while debtor nations see their domestic industry and agriculture fall into a self-reinforcing cycle of recession as their home markets are eroded by imports and their currency is drained abroad (Keynes, 1941/1980, pp. 42–66; Rowbotham, 2000, p. 37; Vegh, 1943). These self-reinforcing cycles can eventually create a global-scale contradiction of oversupply in an atmosphere of constrained demand that cannot be easily resolved.

Keynes' proposed ICU was designed to promote greater equity among nations by virtually eliminating trade imbalances. He proposed an international currency, the bancor, to be used for all international trading. A nation would accrue bancors by exporting, and importing would result in debits to a nation's bancor holdings with the ICU. Nations would be encouraged to maintain a zero balance in their ICU bancor-denominated accounts (Keynes, 1941/1980; Vegh, 1943). Since bancors could only be redeemed through international trade and would otherwise be worthless and because both net creditor and net debtor nations would incur minor fines as a result of carrying positive or negative balances, nations would be encouraged to spend rather than save bancors. By reducing incentives to generate a surplus of trade, the ICU would promote equity in the distribution of the benefits of production within highly productive countries. Without such a system, a few people in surplus producing nations typically benefit greatly from capturing and investing foreign money earned through exports, and most people involved in surplus-producing economies, in effect, work to export their wealth to other nations. ICU benefits to nations running a temporary trade deficit were equally apparent: they would not experience a net outflow of their national currencies to purchase imports and, therefore, would avoid currency destabilization and recession that would result from a negative balance of trade. Through the ICU, both net creditors and net debtors would be encouraged to restore the balance of trade (Rowbotham, 2000, pp. 39–40).

The ICU proposed by Keynes would have promoted greater self-sufficiency and self-reliance within the word-system and would therefore have promoted system resiliency and sustainability as compared to the current system in which imbalances of trade are allowed to self-perpetuate. Had Keynes' proposal carried the day at Bretton Woods, we would be much less likely to see wealth so intensely concentrated in the hands of international elites and transnational corporations who benefit directly from the enforced dependency of imbalanced trade and Global South debt, both of which are used to leverage the capital and financial dominance of transnationals and wealthy capitalists within the world-system.

The very potential of Keynes' proposed ICU to promote equity, both within and among nations, was resisted by the United States and its capitalist leaders who saw America emerging as a superpower following World War II. Britain, represented by Keynes, was in decline at this point, and its productive capacity had been severely impacted by the war. The United States, on the other hand, had seen its productive capacity rise as a result of the war, and its infrastructure and factories remained intact. With the U.S. having recently emerged from the Great Depression, the American delegation was concerned with maintaining and expanding trade surpluses as an outlet for American productive capacity and a vehicle for avoiding a postwar recession (Rowbotham, 2000, p. 41; Vegh, 1943).

The U.S. proposal made at Bretton Woods centered on conducting international trade in a free market using national currencies. Under this proposal, nations running a trade surplus would be under no obligation to spend their surplus earnings by purchasing the exports of debtor nations. The U.S. proposed creating the IMF as a stabilization fund to which all nations would contribute according to the size and vigor of their economies. The Fund would hold reserves of all national currencies proportional to the relative strength of their economies. Any nation that experienced a negative balance of trade which threatened to upset its economy could borrow from the Fund on a short-term basis in order to avert an economic downturn or currency crisis (Rowbotham, 2000, p. 43). IMF stabilization activities would also prevent future global depressions because the bank's loans would provide the liquidity necessary to maintain aggregate demand in the global economy. IMF loans would also encourage countries to maintain employment during an economic downturn so as not to compound existing problems (Stiglitz, 2002, pp. 12, 196).

According to the American proposal, nations running trade surpluses would be allowed to accumulate these surpluses and would be able to exchange them for gold held by debtor nations. The American delegation claimed that, using gold as the international currency meant establishing a neutral currency, but since the U.S. held at least 70% of all world gold reserves at the time, the American delegation also insisted that gold should be valued in dollars and that all other currencies should be valued relative to the dollar. This currency policy became known as the gold exchange standard. The convention eventually adopted the U.S. proposal. In the case of Great Britain, a U.S. war loan was made conditional upon agreement to the American proposal. With the approval of the U.S. proposal at Bretton Woods, the dollar, in effect, became an international currency (Dormael, 1978; Rowbotham, 2000, pp. 43–44).

Securing the dollar as the sole international currency bestowed certain advantages upon the U.S. It could run trade deficits and still maintain the acceptability of its currency—as long as confidence in the dollar remained stable and national gold reserves remained high enough to satisfy demands by foreign countries to convert their dollar holdings to gold. America's negative balance of trade during the Vietnam War, eventually produced circumstances in which international confidence in the dollar waned and the countries of the world increasingly converted their dollar reserves to gold, to the point that it seemed the U.S. might not be able to honor its gold exchange commitments. At this point in 1971, the Nixon administration unilaterally ended exchanges of dollars for gold and thereby ended the Bretton Woods gold exchange standard, but the dollar has continued to serve as a world reserve currency for several reasons noted below in the section on dollar hegemony (Hudson, 2005, pp. 22–25).

At Bretton Woods, the stage was set for a post-war world-system which carried forward in time the inequities created during European colonization and which deepened the enforced dependency of Global South nations even as they gained independence. As we will discuss in more detail below, when the U.S. succeeded in creating an international economic system that would allow it to maintain its trade surpluses, it ushered in an era of increasing concentration of wealth and power and growing enforced dependency of nations that would become mired in unpayable debt. When the U.S. succeeded in making its national currency the de facto international currency, it also laid the groundwork that would allow it to manipulate other nations of the world into financing its debt. The Bretton Woods agreement, through its refusal to enforce a balance of trade among nations and its promotion of laissez-faire trade, ensured the economic supremacy of the U.S. and sealed the fate of Global South nations as perpetual debtors (Rowbotham, 2000, p. 46). These successes of the U.S., however, also set in motion processes that would eventually lead to a crisis of capitalism in the global economy: overproduction in an atmosphere of declining purchasing power in important centers of consumption such as the U.S. and widespread, crushing national and consumer debt in the U.S. and Europe. In fact, the political success of the U.S. in getting other people and nations to finance its standard of living and its defense spending have created a situation in which the world, faces a growing potential for currency crises in many nations, including the U.S., that could completely destabilize the global economy. We will explore these issues in more depth in the next chapter in the section on dollar hegemony.

Chapter 4:

Neoliberalism: Deepening Enforced Dependency

The global economy has an in-built tendency to increase inequality. It is also inherently unreliable and the monoculture it creates puts excessive pressures on the environment. We should therefore attempt to both change the way it works and to build local alternatives to it.

—Richard Douthwaite

In this section, we analyze how the neoliberal economic regime has intensified tendencies inherent in the Bretton Woods institutions to enforce the dependency of Global South nations within the world-system. We will see that the IMF, the World Bank, and post-war free trade regimes typically drive and support the enforced dependency of former colonies on the industrialized powers. We will see how debt enforces the political and economic subservience of Global South nations to the industrialized core and how loan conditions force debtor nations to offer up at bargain prices their natural resources, labor, and industrial and infrastructural assets. We will see how the advantages gleaned through this system by dominant powers and groups become self-reinforcing and self-perpetuating and how systemic inequalities among nations breed complex forms of dependency—not only for nations and citizens of the Global South, but also for those of the First World.

It is important to emphasize that enforced dependency did not begin with the neoliberal era, nor even with Bretton Woods. The relative positions of winners and losers in the current world-system were in many cases established during the colonial period, and one can reach even further back in time to locate people, circumstances, and resources that contributed to creating current inequities. The neoliberal era, however, provides a most glaring example of raw political force wielded to the advantage of the powerful and wealthy at the expense of others.

The neoliberal political wave has washed over the world energized by economic rhetoric and policies that, at best, thinly disguise a global, systematic process of consolidating economic and political power (see Harvey, 2005). Though it is difficult to document that particular people, institutions, and nations *intended* to create a world-system that looks and behaves like the one we have today, it is possible to show how international policies, practices, and institutions have, by and large, systematically advantaged the proponents and enforcers of the neoliberal regime at the ultimate expense of almost everyone else. Many significant advantages for proponents of neoliberalism have been secured through the actions and policies of the Bretton Woods institutions—the IMF and the World Bank—working in concert with GATT and the World Trade Organization (WTO). In this section, we will focus on the policy content and practices of the IMF, the World Bank, and GATT/WTO as prime examples of dependency enforcing agents of global capital. We will see how the forms of dependency enforced through these institutions systematically reduce the resilience, not only of debtor nations, but of nations and peoples everywhere, making neoliberal political economy among the biggest threats to local and global socio-ecological sustainability.

According to David Harvey (2005),

> Neoliberalism is in the first instance a theory of political economic practices that proposes that human well-being can best be advanced by liberating individual entrepreneurial freedoms and skills within an institutional framework characterized by strong private property rights, free markets, and free trade. (p. 2)

In the neoliberal era,[1] which began in the 1980s, three central factors have propelled globalization and furthered enforced dependency: conditions imposed by the IMF and World Bank through their lending processes, increasingly aggressive free trade agreements, and the increasingly blunt use of political and economic power wielded by nations issuing world reserve currencies (the U.S. first and foremost among these). The power flowing to the U.S. from international use of the hegemonic dollar proved particularly instrumental in terms of U.S. influence. However, in recent decades, the economic power of nations has been progressively eclipsed by that of transnational corporations that have captured the political leadership within nation-states and have therefore been able to use national and international politics as platforms for extending their global investments and increasing

---

[1] In 1979, Paul Volcker was named Chairman of the Federal Reserve in the U.S., and he ushered in a set of monetary policies designed to fight inflation no matter the costs in terms of unemployment or other forms of social dislocation. Margaret Thatcher was also elected prime minister of Great Britain in 1979, and Ronald Regan was elected president of the United States in 1980. Spreading from these focal points of power and policy, neoliberalism became an organizing framework for the global economy (Harvey, 2005, p. 1).

their profits (Rowbotham, 2000, p. 47). As Stiglitz notes, in the neoliberal era, "the West has driven the globalization agenda, ensuring that it garners a disproportionate share of the benefits, at the expense of the developing world" (Stiglitz, 2002, p. 7). Under neoliberalism, structural inequalities built into the Bretton Woods paradigm have given birth to a world-system in which the economic and political position of nations in the Global South approximates that of colonies with regard to First World nations and transnational corporate interests. Peripheral nations perpetually depend on forms of economic assistance that deepen and enforce their dependency upon the very institutions, corporations, and nations that strip them of the infrastructural and business assets, decision-making power, natural resources, and social support that could serve as bases for creating more free, self-sufficient, and sustainable societies (Greider, 1997; Kaplinsky, 2005; Manley, 1987; Perkins, 2004; Rowbotham, 2000; Stiglitz, 2002).

The lopsided world-system that derived from Bretton Woods gained further momentum during the neoliberal era under the political leadership of Great Britain and the U.S. We will now more closely examine the ideology undergirding the neoliberal agenda for globalization.

Neoliberalism is an extreme form of market fundamentalism that gained worldwide prominence in the 1980s when it became the economic platform of British Prime Minister Margaret Thatcher and U.S. President Ronald Reagan. Neoliberal ideologues believe in small government and in the ability of a self-regulating market to serve effectively as the ultimate arbiter of economies and of all social life. Neoliberals promote privatization of all or nearly all public industries, services, and agencies; capital market liberalization; and the removal of all barriers to trade. In promoting free trade, neoliberal policy makers have typically opposed government regulations of all kinds, including those that safeguard environmental and human health and that regulate working conditions and wages (Achbar et al., 2004; Black, 2001; Harvey, 2005; Moyers, 2002). Liberalization is supposed to stimulate the economy by moving resources from less to more productive uses, but it has often destroyed jobs as a result of international competition. In the manufacturing sector, for example, corporations seeking production cost advantages in terms of wages, regulations, productivity, and other factors may relocate their production facilities, thereby causing unemployment in communities left behind. This footloose behavior of capital in an era of free trade also places downward pressure on the sovereign rights of nations to create and enforce environmental and labor regulations (Achbar et al., 2005; Black, 2001; Shuman, 2000, chap. 2).

In the 1980s, with Margaret Thatcher at the helm in Britain and Ronald Reagan as president in the United States, the IMF and the World Bank became "missionary institutions" for neoliberalism (Stiglitz, 2002, p. 13). A "purge" occurred at the World Bank that redirected the Bank's efforts toward reducing the power and role of government while increasing privatization and free trade (Stiglitz, 2002, p. 13). Although the objectives of the IMF and the World Bank remained distinct, their activities in any given country became increasingly intertwined. The World Bank began to provide "broad-based support in the form of structural adjustment loans" but only did so following IMF approval, and the IMF, originally created to assist on a short-term basis during an economic crisis, became involved in long-term development policy in countries experiencing perpetual states of crisis (Stiglitz, 2002, pp. 13–14). According to Stiglitz (2002), the neoliberal policies of the IMF in particular became dogmatic and expressive of a naïve faith in markets to self-correct so that the institution promoted liberalization as an end in itself (pp. 31–32). Prime Minister Thatcher touted ideology as fact in her claim that "there is no alternative" to the neoliberal agenda (Douthwaite, 1999b, chap. 5).

According to Stiglitz (2002), the neoliberal ideology that permeated the policies of the IMF beginning in the 1980s contradicted both the ideas upon which that institution was founded and the stated goals of its programs:

> Over the years since its inception, the IMF has changed markedly. Founded on the belief that markets often worked badly, it now champions market supremacy with ideological fervor. Founded on the belief that there is a need for international pressure on countries to have more expansionary economic policies—such as increasing expenditures, reducing taxes, or lowering interest rates to stimulate the economy—today the IMF typically provides funds only if countries engage in policies like cutting deficits, raising taxes, or raising interest rates that lead to a contraction of the economy. (pp. 12–13)

Furthermore, it takes capital and entrepreneurship to create new firms and jobs needed to end an economic crisis, and the austerity programs and high interest rates imposed by the IMF in the neoliberal period have resulted in a lack of both (Stiglitz, 2002, p. 59). We will address below the specific content and processes of the structural adjustment programs referenced by Stiglitz. It is important to note that, historically, the U.S. government played a central role in developing its strong domestic economy, but Global South nations have been effectively denied the opportunity to do the same under the neoliberal paradigm, in many cases because they are subject to the economic and political influence of the IMF and World Bank (Stiglitz, 2002, p. 21).

The ideology of neoliberalism has also been unevenly applied by the IMF, thereby increasing the relative position of strength of the First World in relation to the Global South. The influence of the IMF extends well beyond those countries that have loan agreements with it. In accordance with Article four of its charter, The IMF generates annual reports for every nation in the world in order to verify that each is adhering to the agreement under which the IMF is organized. According to Stiglitz (2002, p. 48), because these reports are used as a means of grading a nation's economy, they serve as ideological vehicles for advancing the neoliberal agenda. Peripheral countries have to pay attention to this grading in order to avoid frightening away current and potential investors while core countries can ignore them (Stiglitz, 2002, p. 48).

Economic policy guided by neoliberal ideology is also highly exclusionary to vast numbers of people worldwide because policy maker adherence to neoliberal doctrine tends to increase the income gap between rich and poor within nations and among nations.[2] These gaps are also an indication that many Global South societies have become mired in perpetual states of debt, underdevelopment, and maldevelopment (Homer-Dixon, 2006, p. 192; Manley, 1987; Robotham, 2000). Furthermore, according to former World Bank economist Partha Dasgupta, ideological support for growth economics and trade liberalization (by the industrialized world, major international development banks, and national governments worldwide) has encouraged practices that overlook impacts on the environment and on intertemporal human well-being. Dasgupta identifies how neoliberal economic policies can produce situations where economic and social development occurs according to standard means of measurement—typically the gross domestic product (GDP)—at the cost of drawing down or damaging natural resources and human well-being. Long-term prospects for future generations are negatively impacted in exchange for economic growth *now* (Dasgupta, 2001). Thus,

---

[2] It is important to note that some countries such as China have both successfully captured market share globally and grown their economies during the neoliberal era. Kaplinsky's (2005) empirical analysis of global trade demonstrates that these gains create a de facto reduction of similar opportunities for other nations. Therefore, although some national economies have attained a relative level of success during the neoliberal era, rising economic activities for one nation or business have come at the cost of reduced opportunities for others due to systemic overcapacity in production and the resulting squeeze on prices (Kaplinsky, 2005). Neoliberalism, therefore, is not a tide that lifts all boats. It is also important to note that China's trade successes have been supported by keeping the Chinese currency at an artificially low value, thereby creating a positive atmosphere for exports. This policy has been practiced in a global atmosphere of neoliberalism, but it is quite contrary to neoliberal doctrine.

neoliberal policies enforced through the IMF and the World Bank often circumscribe the resilience and sustainability of communities and nations.

According to Homer-Dixon, neoliberal globalization also has failed to produce promised growth in middle-income countries, and some of the countries that are deemed to be economic growth success stories actually protected their economies from free trade:

> Middle-income countries hardly gained at all, including those in Latin America that aggressively privatized state-owned industries and opened their borders to trade and investment. Also, some of the countries that grew the fastest—…including China and India, but also Malaysia and Chile—actively protected their economies using capital controls and trade barriers. (2006, p. 192)

Economic globalization has typically benefited transnational corporations selling high-value-added goods and services globally (Homer-Dixon, 2006, p. 192). Benefits enjoyed by these entities may well derive from "excessive political lobbying and representation by powerful commercial interests" in global and regional free trade and regional economic block organizations such as the WTO and the European Union (Rowbotham, 2000, p. 3). Large corporate interests influence World Bank and IMF policies and loans as well (Rowbotham, 2000, p. 3). For example, global water corporations have been actively consulted by World Bank officials in the process of developing agreements for financing water projects (International Consortium of Investigative Journalists [ICIJ], 2003).

Stiglitz (2002) offers an insightful summary of his observations on IMF policies and procedures as central, epitomizing phenomena of neoliberal globalization:

> [At the IMF,] decisions were made on the basis of what seemed a curious blend of ideology and bad economics, dogma that sometimes seemed to be thinly veiling special interests. When crises hit, the IMF prescribed outmoded, inappropriate, if "standard" solutions, without considering the effects they would have on the people in the countries told to follow these policies. Rarely did I see forecasts about what the policies would do to poverty. Rarely did I see thoughtful discussions and analyses of the consequences of alternative policies. There was a single prescription. Alternative opinions were not sought.... Ideology guided policy prescription and countries were expected to follow the IMF guidelines without debate.... These attitudes…often produced poor results [and were] antidemocratic. (pp. xiv–xv)

Here, Stiglitz highlights the market fundamentalism of neoliberal economic policy and practice. He also points out an important aspect of neoliberalism: that it is entirely distinct from political democracy. During recent decades, neoliberal politicians have conflated individual freedom and democracy with free markets. Below, we will explore in more depth how neoliberal ideologues promote free market strategies that actively undermine democratic processes—placing freedom of the market over freedom of people to govern themselves.

We will now explore in some depth how neoliberal dogma was applied in practice in many countries as a condition of receiving economic development and economic stabilization loans.

Structural Adjustment Programs and Loan Conditionality

International finance and debt enforce the dependency of many debtor nations upon First World banking interests, thereby insuring that they remain in line economically and politically with neoliberal globalization led by the First World. During the neoliberal period, the IMF and the World Bank have evidenced a high degree of integration at the policy level and have often combined their efforts within given countries. World Bank development loans have often been approved on the condition that an IMF structural adjustment program was in place. At times, the World Bank has even sought IMF endorsement of its loan agreements (Rowbotham, 2000, p. 49). It is important to acknowledge that the bulk of foreign debt incurred by Global South nations is commercial and created through fractional reserve banking. These funds, therefore, are not lent from one *country* to another. Still, it is important to elucidate how loan policies and strategies of the IMF and World Bank embody neoliberal ideology, especially because these policies influence commercial lending to debtor nations. Commercial loans are not typically forthcoming to nations that are not also supported by the IMF and/or World Bank (Bradshaw & Huang, 1991, pp. 323-325; Rowbotham, 2000, pp. 49, 98).

We begin our exploration of IMF and World Bank lending as embodiments of dependency-enforcing neoliberal ideology by exploring how loan conditions and structural adjustment programs (SAPs) tend to create economic recessions while, at the same time, undermining debtor nations' ability to address the social and economic causes of recession. We will explore how and why debtor nations find themselves trapped in a perfect storm of needing funds from the IMF and World Bank while, as a condition of receiving these funds, they must continually erode the foundations from which a healthy domestic economy might be built. Borrower nations find it difficult to escape this trap, and they become increasingly vulnerable to economic and political exploitation—a classic case of enforced dependency.

Countries can experience economic crises for many reasons. During the post-war period, IMF stabilization and World Bank development loans were sought by many peripheral nations burdened by legacies of colonialism and enforced dependency. These countries have in recent decades become characterized by large poor populations concentrated in and around cities (see Davis, 2006, and Araghi, 1995). These populations have suffered the destabilizing effects of both depeasantization (which will be examined in some detail in chapter six) and the neoliberal drive to open markets. Difficulties faced by nations attempting to deal with these problems are compounded by downward pressure on profits of latecomers to technological advancement (Douthwaite, 1999b, chap. 2), downward pressure on the prices of commodities in a global market (see Kaplinsky, 2007), international harmonization and enforcement of intellectual property policies (Garcia, 2004; Shiva, 2000 & 2005; Stiglitz, 2002, pp. 7–8), cultural imperialism, Bretton-Woods-induced systemic trade imbalances, agricultural subsidies provided to U.S. farmers (Norberg-Hodge, et al., 2002; Shiva 2000, 2005), aggressive international free trade agreements, use of nonlocal currencies that systematically deplete communities (Douthwaite, 1999a, 2004), and ecological and health crises. These factors conjoin within neoliberal globalization to thwart social and economic development in the Global South.

Since the 1980s, the World Bank and the IMF have made loans contingent upon debtor nations deploying sweeping structural adjustment programs. These programs integrate debtor nations into the world-system in ways that advantage transnational corporate and financial interests in the First World at the expense of the economic and social interests of most citizens in the Global South. Although external conditions such as the virtual monopoly status of transnational corporations, global commodity price variations, aggressive exporting, and protectionism practiced by industrialized nations can destabilize a nation's recovery from economic crisis (Rowbotham, 2000, p. 58), debtor nations are held solely responsible for repayment of IMF and World Bank loans. This burden of debt reinforces their often repeated dependence on external loans to address economic problems that may be less national than global. According to Rowbotham (2000), the ideology of "structural adjustment is based on the assumption that the cause of each nation's debt crisis lies entirely within its own economy. The economy must therefore 'adjust' to the wiser world economy" (p. 55). The opposite is often true: many of the crises debtor nations face result from legacies of colonialism and from external pressures of core entities within the world-system seeking to secure for themselves positions of relative advantage.

Structural adjustment programs most often entail all or most of the following: fiscal austerity (including reductions in public services such as publicly supported education, job training, and healthcare as well as the cutting of subsidies provided to the poor for obtaining the basics of life such as food, water, and transportation), raising taxes in order to pay external debts, privatization of public industries and services, elimination of barriers to trade, liberalization of capital markets, and currency devaluation (making a country's products cheaper to the outside world but more expensive to their own citizens) (Black, 2001, Manley, 1987; Robbins, 1999, p. 106; Stiglitz, 2002, p. 53). These policies almost always lead to recession or worse (Stiglitz, 2002, p. 38).

Additionally, loan conditions can be political in nature. Jamaica provides an example of a nation whose economic policy was a casualty of debt crisis. Jamaica's leftist and Third-World-solidarity-oriented economic and political philosophy as well as its resource politics (modeled on OPEC as a resource cartel) with regard to bauxite (from which aluminum is extracted) were all but obliterated as a result of its near currency collapse. Jamaica sought a stabilization loan from the IMF, and the strings attached to this loan radically rerouted its political path toward closer integration with the neoliberal economic project while reinforcing its peripheral status within the world-system (Manley, 1987; Black, 2001).

Loan conditions imposed by the IMF and World Bank often undermine democratic decision making in Global South nations and thereby undermine political and social coherence that could improve a nation's ability to advance self-determined domestic and international social and economic policies. These conditions also typically prevent nations from securing the public funding necessary to carry through progressive domestic social and economic policies. According to Stiglitz, for countries in dire need of credit, "Unless the IMF approves the country's economic policy, there will be no debt relief. This gives the IMF enormous leverage...." (Stiglitz, 2002, p. 43). According to Rowbotham (2000) the IMF and the World Bank use their financial leverage to circumvent the policy-making roles of sovereign nations:

> Structural adjustment has seen teams of World Bank and IMF economists virtually taking over the economies of debtor nations in an attempt to "turn them around." Exchange rates, government spending, labour laws, domestic deficits, taxation, welfare programmes, land tenure, environmental regulations, wage cuts and public service cuts—all of these have been subject to detailed requirements and constant scrutiny. (p. 56)

The depth and breadth of prescriptive SAP policies alone reveal them as undemocratic. Furthermore, the IMF and World Bank reveal their willingness to engage in undemocratic processes when they bar citizens from participation in negotiations and refuse governments permission to reveal to their citizens what loan agreements entail (ICIJ, 2003; Stiglitz, 2002, p. 51). According to Stiglitz (2002), undemocratic loan conditions can apply to the national governance process itself. He notes that "in some cases, the agreements stipulated what laws the country's Parliament would have to pass to meet IMF requirements or 'targets'—and by when" (pp. 43–44).

To make matters worse, these undemocratic policies have not benefited debtor nations economically:

> The first World Bank structural adjustment programmes (SAPs) were in Kenya, Turkey and the Philippines in 1980. None is a success today and the United Nations Economic Commission for Africa in 1993 found fifteen African countries clearly worse off after structural adjustment than before. (Rowbotham, 2000, p. 56)

Structural adjustment, rather than improving conditions in debtor nations, enforces their dependency on the world-system by perpetuating and deepening economic crisis. In Africa, SAPs have nearly always resulted in widespread unemployment, declining real incomes, economically damaging levels of inflation, capital flight, persistent trade deficits, rising levels of external debt, the destruction of the social safety net, and de-industrialization (Rowbotham, 2000, p. 57; Stigltiz, 2002, p. 46).

These outcomes demonstrate the fallacy of the assumption that free markets and freedom go together. The IMF and the World Bank do not facilitate blossoming economies in free societies but serve to enforce the dependency of client governments who, albeit sometimes quite unwillingly, force the will of more powerful economic and political interests onto their nations. The fact that the IMF has used a "boilerplate," one-size-fits-all approach to developing its loan agreements and that it has sought little input from outside experts or from national officials familiar with the countries in question (Stiglitz, 2002, pp. 47–48) demonstrates that its political agenda is driven by *outside* interests. The IMF claims it does not dictate the terms of its loan agreements, but they wield the power in one-sided negotiations because a country seeking an IMF loan is facing an immediate crisis and is in desperate need of assistance (Stiglitz, 2002, p. 42).

Staged dispersal of loan funds serves as yet another tool for international lending agencies to dictate and enforce neoliberal SAPs. Under a staged dispersal arrangement, if a country does not meet specific economic and policy tests and targets, disbursement of loan installments will be halted midstream (Rowbotham, 2000, p. 57). Sometimes meeting these requirements actually reduces a country's ability to repay its IMF debt, so that the conditions imposed cannot be justified in terms of the Fund's banking objectives (see the example of Korea in Siglitz, 2002, pp. 44–45) but are revealed for what they are: global policy tools.

In the case of the World Bank, typically a large proportion of money lent to nations for infrastructure projects is used to hire foreign contractors with the expertise necessary to successfully design and carry out development projects. Since employment is created wherever the borrowed money is spent, jobs resulting from development projects in the Global South are often created in the First World. Some degree of domestic economic growth may or may not materialize from investment in these projects, but in any case where foreign contractors are hired or where long-term management contracts are awarded to transnationals, global corporate interests benefit significantly from these loans while debtor nations are stuck paying the bill (Robbins, 1999, pp. 101–107). The World Bank has also repeatedly required privatization of state industries and services as a condition for loan approvals, a process which creates profit making opportunities for transnational corporations (ICIJ, 2003).

According to Stiglitz, there are several reasons for the failure of conditions to stimulate development. One is that loans create fungibility that may be poorly utilized (funds designated for a given purpose free up funds that may be unwisely spent elsewhere). He also cites poorly conceived conditions that deepen economic crisis combined with political unsustainability of policy initiatives as possible reasons for the failure of loan conditions to promote development. The political unsustainability of imposed policy initiatives may derive from public perception of these conditions as political and economic intrusions by a colonial power (Stiglitz, 2002, p. 46).

There are, however, some economic growth success stories of IMF policies. Botswana, for example, averaged more than 7.5 percent growth in the period 1961 to 1997 (Stiglitz, 2002, p. 37). But one must question whether these success stories have produced socio-ecologically sustainable and economically resilient societies. In some cases, nations that secured loans to address an economic crisis or to undertake development succeeded mainly at enriching their ruling elite (Stiglitz, 2002, p. 52), thereby enforcing dependency internally. External debt also serves as an important and continuing source of the vulnerability of the Global South to exploitation by core entities within the world-system. Enforcing peripheral status may in fact have been a goal of IMF and World bank lending—at least it seems resilience and self-sufficiency were not the primary goals. According to Rowbotham (2000), "The persistent and cumulative failure of the theoretical model that encouraged developing nations to 'borrow/invest/export/repay' suggests the nature, terms, and context of loans to the Global South have been such as to render these [debts] inherently unpayable" (p. 31).

It is important also to recognize that monies lent by the World Bank and the IMF may be largely created for this purpose. The World Bank creates and sells bonds to commercial banks to generate funds for lending, similar to the way a nation generates spendable funds through creation of national debt. The IMF requires 25 percent of its quotas deposited by member countries to be in gold. The other 75 percent can be in the national currency. National governments regularly generate the funds for these deposits by selling bonds, thus increasing the national debt. The IMF can also administer loan packages offered through commercial banks (Rowbotham, 2000, pp. 100–101). Thus, Global South debt is much more an obligation to the financial sector within industrialized nations and less a debt owed to the *nations* of the First World. Indebted nations are therefore dependent upon the global financial sector, the segment of the economy that has come to dominate all others in terms of growth, making it a formidable political and economic force.

We will now explore the dependency-enforcing aspects of implementing specific loan conditions imposed by the IMF and World Bank. We will also examine how free trade and the requirement of debtor nations to earn foreign exchange to pay their debts contribute to enforcing the dependency of nations in the Global South.

Fiscal Austerity

Fiscal austerity as a condition for loan agreements can include all or some of the following: reduced government spending, increased taxation, require-ments to reduce national and/or international debt, and requirements to raise interest rates. According to Stiglitz, IMF austerity programs have included demands to raise interest rates to as high as 20, 30, 50, or even 100 percent, a practice that makes domestic business investment virtually impossible (2002, p. 59). Although these strategies may help to reverse a currency crisis in the short term, they create a poor foundation for social and economic develop-ment. Reduced government spending can result in a near-complete disap-pearance of the social safety net, negative impacts to the scope and quality of education, reduced ability to address environmental problems, and reductions in the government's ability to stimulate domestic small business develop-ment. These impacts reduce a nation's potential to prepare citizens for jobs other than those in the low-skill manufacturing and service sectors. This lack of social investment virtually condemns a nation to dependence on core entities within the world-system as suppliers of high-tech products and services, and debtor nations lose out on developing citizens' potential to earn wages paid to highly skilled workers. Furthermore, the government loses out on taxes that could be generated by a higher percentage of the population engaging in skilled work, and this low tax base reinforces the government's inability to support social programs. Increased taxation—especially when combined with reduced social services such as unemployment benefits and subsidies[3] that ensure access to the basics of life—stresses people already living on the margin, and it forces some into abject poverty, virtually eliminating their chances to contribute to economic development in the future. Paying down government debt diverts money toward creditors and away from potential social investment. Raising interest rates slows the economy and makes it more difficult for consumers to purchase durable goods and homes, to start or expand businesses, or even at times to continue business operation. The economic slowdown caused by higher interest rates reduces the tax base and increases the need for social services at the same time that these services become less available.

Fiscal austerity strategies applied simultaneously are more likely to cause economic recession or even depression than to stimulate development. A nation caught in a cycle of debt and forced to implement fiscal austerity as a condition for loan approval is in a difficult position indeed. Such a country is likely to have its ability to implement domestic policy objectives and its control over its domestic economy systematically eroded, especially if it is forced to enter many successive loan agreements or renegotiate its debt. Countries caught in such a trap clearly suffer from enforced dependency (see Black, 2001, and Stiglitz, 2002).

Export-Led Development

Loans made to Global South nations by the IMF, the World Bank, and commercial banks are denominated in world reserve currencies. Therefore, in order for a nation to pay these external debts, it must earn foreign exchange by trading with First World nations. Export-led development represents a means of earning foreign exchange, and these development strategies have been a foundation of World Bank and IMF thinking since the late 1950s (Rowbotham, 2000, p. 50).

Since former colonies that once served as the raw materials plundering grounds for their conquerors rarely have developed as producers of a wide range of advanced, technologically complex products, Global South nations typically export raw materials, commodities such as agricultural and mining

---

[3] In Botswana, IMF-prompted removal of subsidies for food and kerosene undertaken as part of a wide ranging austerity program triggered riots (Stiglitz, 2002, p. 77).

products, and relatively simple manufactured products on a large scale. The price of raw materials and other products sold in the competitive global market falls to that of the lowest cost provider (Douthwaite, 2004, pp. 114–115). Rising production of commodities by debtor nations pursuing export-led development strategies also places intense downward pressure on prices (see Avramovic, 1986). Kaplinsky notes that prices for exported commodities have declined relative to imports of manufactured goods (2005, pp. 57–60).[4] Kaplinsky also notes that, in a globally competitive market, the lower the technological intensity of a given product, the more likely its price will fall (2005, pp. 184-185). When one considers that typical Global South domestic commodities and products have relatively little value added compared to the complex products of the industrialized world, one realizes that the playing field of the global marketplace is not at all level.

One might argue that complex products are manufactured in many Global South countries, but as Kaplinsky demonstrates when exploring the reasons for the demise of a garment producer in the Dominican Republic, if the production processes are easily reproducible elsewhere, transnational producers often relocate to take advantage of production cost advantages—for example, a currency devaluation in another country (2005, pp. 60–65). Furthermore, debtor nations typically see less than the expected benefit from factory production taking place within their borders because profits generated are mostly repatriated by First World corporate interests. These corporations may contribute little to a peripheral nation's efforts to earn foreign exchange, especially because many of their factories reside in tax-free zones (Achbar, et al., 2005; see also the Dominican Republic example offered by Kaplinsky, 2005, p. 60).

A focus on exports also often interferes with the domestic economy of the exporting nation (Rowbotham, 2000, p. 152). The harnessing of domestic production as a means of earning foreign exchange diverts resources and production capacity that could otherwise be used to support the domestic population. The export focus enforced through IMF and World Bank loan agreements therefore tends to increase the dependency of domestic populations in the Global South on the world-system for the provision of basic

---

[4] Kaplinsky's (2005) study of trade between Global South countries and the U.S., however, demonstrates that, although the relative price of exported commodities to imported manufactured goods declined, increased levels of commodities exports resulted in absolute gains in income generated from export activities. This finding does not mean that alternate economic strategies would not have generated even higher levels of income, and it does not take into account that commodities exports can come with high environmental and social costs (pp. 204–205).

needs. This situation is especially egregious in the case of export-oriented agricultural production in nations whose people are hungry. Export crops are grown on land monopolized by agribusiness instead of growing crops that could feed the people (Rowbotham, 2000, p. 7).

According to Rowbotham, "The obligation on debtor nations to direct an increasing proportion of their resources and economic effort to the export market has long been recognised as one of the primary causes of poverty and lack of internal development in the emerging nations" (2000, p. 6). According to Korten (2001), in Brazil, between 1960 and 1980, the conversion of small land holdings used to grow food for domestic consumption to agribusiness production of export crops displaced 28.4 million people, and in India, 20 million people were displaced over a 40-year period due to large scale development projects (p. 55). The displacement of subsistence farmers and indigenous communities enforces the dependency of the displaced upon the wage labor system—and upon the economic resources of their government, especially in the event that they remain under- or unemployed. These new and often desperate dependencies of a nation's citizens may contribute to the urgency of the national government's efforts to secure loans, thereby further enforcing the nation's dependency within the world-system.

Global South debt implies an imbalance of trade. Repayment of loans through export-led development means generating a trade surplus, which also means that at least some of the nations globally that had previously enjoyed a trade surplus would have to purchase the increased exports from debtor nations and see their own surpluses reduced or converted to deficits (Rowbotham, 2000, pp. 36–37). This scenario is politically and economically unacceptable to core powers within the world-system. [5] Export-led development, rather than being a way out of debt, contributes in important ways to perpetuating the debts of many nations in the Global South.

Free Trade

Under the Bretton Woods agreement, free trade was given high priority. The requirement to promote free trade of goods and services worldwide through removal of restrictions to trade was written into the charters for both the IMF and World Bank. "Countries were...permitted to seek a persistent trade

---

[5] It is important to note that the U.S. is an exception. It runs an extremely negative balance of trade of a magnitude that would not be possible for other industrialized nation. The U.S. is also the nation with the world's largest debt. This situation is possible because of the status of the U.S. dollar as a world reserve currency—and the only currency with which oil can be purchased from OPEC countries. We will further explore below reasons why the U.S. is able to run up huge debts without (as yet) facing a currency crisis.

surplus. The balance of international trade was left to 'free market commercial forces'" (Rowbotham, 2000, p. 46). In the neoliberal era, removing
barriers to free trade has become a standard condition for obtaining a loan
from the IMF or World Bank. Free trade is purported to level the playing
field in the global market through the removal of protectionist policies such
as tariffs and subsidies, but as we shall see, free trade policies tend to favor
First World nations and transnational corporations at the expense of the
Global South.

As economic protections are removed, weak economies are exposed to
competitive forces they often cannot withstand, especially in market areas
where economies of scale are achieved through applying capital-intensive
production methods (Douthwaite, 2004, pp. 114–115). Western industrialized
nations have also pushed for trade liberalization for products they export
while resisting liberalization in areas where it would negatively impact their
own economies (Stiglitz, 2002, p. 60). For example, the U.S. has pushed
Global South nations to eliminate barriers to trade whilst maintaining its own
subsidies in agriculture, thereby preventing Global South producers from
exporting some of their most abundant products and depriving them of export
income (Stiglitz, 2002, p. 6). These trade arrangements systematically
disadvantage Global South businesses which typically have neither the
capital nor the domestically produced technology to get ahead.

Free trade regimes also undermine the ability of national and local powers to regulate industries in the areas of social justice and environmental
protection. Intense competition among large corporations combined with
competition among debt-ridden countries seeking the opportunity to earn
foreign exchange constrain possibilities for regulation in areas such as
minimum wage, working conditions, and environmental protection (Moyers,
2002).

Free trade also paves the way for further concentration of wealth in the
hands of transnational corporate entities that can more easily achieve capital-
intensive economies of scale and undercut competitors with low prices,
thereby driving many smaller producers out of business (Douthwaite, 2004,
p. 117). In obtaining financing for capital investment, transnationals also
receive favored treatment that intensifies the concentration of wealth and
power:

> Size helps multinationals access capital, since they are generally able to obtain credit
> more easily, at lower rates of interest and on a more advantageous terms. Size also
> grants multinationals an advantage over smaller businesses when it comes to with-
> standing the pressure of debt. Banks and other lending institutions are less likely to

foreclose on large debts to Big Business than they are on small business debts. (Rowbotham, 2000, p. 156)

The purported mutual gains to be achieved among nations from specialization and trade in the globalized world-system are only possible in an atmosphere of full employment. In an atmosphere of significant structural unemployment, some producers are unlikely to find markets for their products. At the same time, the fact that transnational corporations can move their manufacturing base to take advantage of production cost savings reinforces the system of globalized production and trade by large-scale producers, making it nearly impossible for smaller producers to capture market share. Structural excess in production capacity places downward pressure on prices at the same time that labor saving technologies contribute to unemployment. Those countries that succeed in export-oriented development do so at the expense of less efficient producers elsewhere. Globalization creates winners and losers, and the poverty and inequity that characterize enforced dependency are integral to the process of globalization itself (Kaplinsky, 2005, pp. 230–231, 235).

Because it places downward pressure on prices, free trade also advantages wealthy consumers over poor producers (Douthwaite, 2004, p. 114), but the advantages of globalization reaped in core nations are likely to be temporary: "As corporations seek low-cost opportunities in the debtor nations, the wealthy nations export jobs abroad and suffer an influx of cheap products that destroy home markets" (Rowbotham, 2000, p. 6). The United States has witnessed the dissolution of its manufacturing sector as a result of these economic forces so that the foreign exchange earned by debtor nation sales to U.S. customers has been made possible only by American consumers slipping ever deeper into debt (Clark, 2005, pp. 10–11).

Free trade creates enforced dependency both within and among nations as it intensifies the concentration of wealth and power within the world-system, thereby depleting increasing numbers of individuals, families, communities, and nations who have, at the same time and paradoxically, become dependent on the global economy.

Privatization

IMF and World Bank loan conditions prompt nations to privatize public utilities, services, and infrastructure, thereby creating opportunities for which transnational corporations may be uniquely positioned due to their ability to obtain credit, their technological advancement, and their prior experience running large-scale, technically complex industries and services. Privatization of basic services such as water reinforces the power and reach of global corporate interests by handing them both new business opportunities and captive markets. The continual privatization of the commons, particularly commons that serve basic needs, raises the question of how those without money will survive. It also raises the question of who will speak on behalf of nature. Privatization tends to further concentrate wealth and power in the hands of a few while enforcing the dependency of the economically weak who are forced to obtain basic necessities from transnationals that bear no political or moral responsibility to them (Achebar, et al., 2005; ICIJ, 2003; Kovel, 2002, pp. 73–74). In more than a few cases deriving from loan conditions, the positions of transnationals have been advanced through political corruption, subterfuge, and conflicts of interest. IMF and World Bank officials have sometimes had ties to industries and companies that have directly benefited from privatization schemes they recommended (ICIJ, 2003). In some cases, domestic elites have also participated in buying up national assets at bargain prices (ICIJ, 2003; Stiglitz, 2002, p. 58)

These privatization strategies do not stimulate domestic capital development but transfer resources and enterprises to foreign investors at bargain basement prices (Ludwig, Blum, & Opitz, 2006), and profits earned by foreign investors in the Global South are mostly repatriated to the First World. Since government debts are paid with funds generated by various forms of taxation, the repatriation of profits from privatized assets purchased by foreign interests contributes to continued government indebtedness (Rowbotham, 2000, p. 125). Nations that are forced to privatize public assets forfeit the long-term potential for domestic economic benefits from enterprises sold under these conditions (Rowbotham, 2000, p. 63), thereby increasing their dependency on external loans and capital investment. Privatization most often functions to strip a country of assets rather than serving as a basis for economic expansion (ICIJ, 2003; Ludwig, et al., 2006; Stiglitz, 2002, p. 58) and may negatively impact the economy by encouraging unemployment (Stiglitz, 2002, p. 57).

In recent decades, the IMF has urged immediate privatization rather than waiting for proper regulation or competition to be in place. This urgency to privatize has created entrenched monopolies that need not heed the public interest (ICIJ, 2003; Ludwig, et al., 2006; Stiglitz, 2002, p. 56). According to Stiglitz, "Whether the privatized monopolies were more efficient in production than government, they were often more efficient in exploiting their monopoly position; consumers suffered as a result" (Stiglitz, 2002, p. 56). This urgency for privatization driven by the IMF and World Bank reveals the primary focus of these institutions on serving the interests of core entities within the world-system.

Capital Market Liberalization

Capital market liberalization opens a country's banking and currency systems to outside interests. According to Stiglitz (2002), liberalization of capital markets can be intensely destabilizing to a nation's economy, can negatively impact investment and growth, and can ironically involve a nation in purchasing U.S. debt:

> As bad…as trade liberalization was for developing countries…capital market liberalization was even worse. Capital market liberalization entails stripping away the regulations intended to control the flow of hot money in and out of the country—short-term loans and contracts that are usually no more than bets on exchange rate movements. This speculative money cannot be used to build factories or create jobs—companies don't make long-term investments using money that can be pulled out on a moment's notice—and indeed, the risk that such hot money brings with it makes long-term investments in a developing country even less attractive. The adverse effects on growth are even greater. To manage the risks associated with these volatile capital flows, countries are routinely advised to set aside in their reserves an amount equal to their short-term foreign-denominated loans…. Typically, reserves are held in U.S. Treasury bills, which today pay around 4 percent. In effect, the country is simultaneously borrowing from the United States at 18 percent and lending to the United States at 4 percent. (Stiglitz, 2002, p. 66)

Under conditions of financial market liberalization, global banks can attract depositors away from domestic banks while simultaneously encouraging loans to transnationals over those to local businesses (Stiglitz, 2002, p. 31). This process results in the export of capital and in jobs created outside rather than inside the country. Profits earned by transnational banks are largely repatriated to core nations and regions rather than reinvested in the host nation. In Ethiopia, this process meant that farmers were denied credit to purchase seeds and fertilizer. Domestic small farmers, unable to access agricultural inputs, were driven out of business, furthering their and the country's dependence on global agribusiness (Stiglitz, 2002, p. 31). The case of Argentina also demonstrates the dangers of capital market liberalization that competitively eliminates local banks:

> Before the collapse in 2001, the domestic banking industry had become dominated by foreign-owned banks, and while the banks easily provide funds to transnationals, and even large domestic firms, small and medium-size firms complained of lack of access to capital…. The challenge is not just to create sound banks but to also create sound banks that provide credit for growth (Stiglitz, 2002, p. 69).

The Community Reinvestment Act of 1977 in the U.S. was passed precisely to counteract the tendency for banks to lend only to privileged classes and big businesses rather than to diverse people and businesses in their local communities, including historically underserved groups (Shuman, 2000, chap. 4; Stiglitz, 2002, p. 70). Widespread access to capital provided to diverse groups, communities, and neighborhoods can stimulate economic development, but making what banks may perceive as riskier loans does not necessarily serve the interests of the financial sector, at least in the short term. The IMF and World Bank practice of forcing capital market liberalization supports the interests of the global financial sector at the expense of communities and nations.

Foreign Direct Investment

Foreign direct investment is the process of businesses building production facilities in foreign nations. It is associated with economic growth in cases such as Singapore and Malaysia (Stiglitz, 2002, p. 67). It is also associated with Ireland's Celtic Tiger economy (Kitchen & Bartley, 2007, chaps. 1 and 22). However, foreign direct investment may also destroy local competition as transnational firms that enjoy advantages of size and efficiency drive out local businesses, rendering them free to raise prices by virtue of their monopoly status (Stiglitz, 2002, p. 68). Economic growth fueled by foreign direct investment can also make the domestic currency appreciate, thereby making imports cheap and exports expensive and making external debts more difficult to pay. Meanwhile, relatively cheap imports can undermine domestically owned businesses that sell to domestic markets, a process which enforces dependency on outside producers and reduces the resiliency of the domestic economy (Stiglitz, 2002, p. 72).

Nations are often encouraged by the IMF and the World Bank to seek foreign direct investment as a means to develop. In their efforts to attract foreign investment, debtor nations usually succumb to pressure to roll back, refuse to enact, or refuse to implement environmental protections (Moyers, 2002; Rowbotham, 2000, p. 65). Environmental as well as worker protections are often treated as barriers to trade by the WTO, the IMF, and the World Bank. Foreign investors also repatriate large proportions of their profits. Therefore, these investments pay off first and foremost to foreign interests rather than for the domestic economy. Furthermore, countries attempting to attract foreign investors typically reduce corporate taxes, thereby constraining the ability of government to provide social support, invest in infrastructure repair and development, and undertake environmental protection and restoration.

Foreign direct investment may stimulate economic growth, but usually only temporarily. When opportunities arise to realize improved cost-of-production advantages elsewhere, foreign investors may pull up stakes and relocate, usually leaving the economy in worse condition than it was prior to experiencing external investment (Achebar et al., 2005; Black, 2001; Shuman, 2000). Through creating an atmosphere conducive to foreign direct investment, nations may reduce their resiliency to economic shock and their potential for self-reliance because they typically damage the domestic productive base and under-invest in social development and environmental protection and restoration in their efforts to attract foreign investors. Virtually all people who succumb to economic pressure to over-exploit their homelands and their people for profits, in a perverse twist, become increasingly dependent upon the very economic system that drives them to engage in such socio-ecologically damaging economic activity.

Loan Conditions and Enforced Dependency

IMF and World Bank loan conditions enforce dependency of the periphery on the core within the world-system. They also create or stimulate market forces that enforce the dependency of the middle and working classes and the poor in all nations upon those who continue to capture and concentrate wealth and power globally. We now turn our attention to the oil price shocks of the 1970s in order to examine how their economic impacts exacerbated the dependency of peripheral nations.

The 1970s Oil Shocks and the Global South Debt Crisis

The Arab embargo of oil exports to the United States in 1973 resulted from the U.S. support of Israel in the 1973 Arab-Israeli Yom Kippur or Ramadan War (Heinberg, 2005, p. 212). This embargo triggered a fourfold increase in global oil prices (see Manley, 1987, pp. 62–64). A second oil price shock in 1979 resulted from the Iranian Revolution which badly damaged that nation's oil sector. Global South nations that were pursuing industrialization faced new and unexpected costs for a commodity that had become essential to development and to producing and distributing essential products and services (Manley, 1987, chap. 3). Hunger and poverty increased sharply, and some regions even experienced an absolute decline in food grain consumption between 1976 and 1979 (Manley, 1987, p. 66). High oil prices triggered a global economic recession. The balance of trade for low-income countries turned sharply negative, and the debt burden for developing countries increased from $67 billion in 1970 to $438 billion by 1980. Making a living in practically any business became much more difficult. It also became more difficult for Global South nations to earn foreign exchange (Manley, 1987, chap. 3).

During this period, many countries capable of exporting large amounts of oil accumulated immense surpluses of dollars because OPEC oil sales were (and continue to be) transacted solely in U.S. dollars. Low absorption capacity for these funds in the domestic economies of OPEC countries encouraged investment of these dollars in the U.S. A sizable proportion of this flood of dollars generated by high oil prices was lent by First World banking interests to developing countries in desperate need of dollars to finance oil purchases and other imports. This process of lending excess petrodollars, known as petrodollar recycling (Clark, 2005, pp. 21–23), greatly exacerbated the debt situation of developing nations.

The value of the dollar also declined in the 1970s, in part due to the negative balance of trade maintained by the U.S. during the Vietnam War, and the U.S. economy experienced stagflation. Prices of goods rose in dollar terms at the same time that domestic economic stagnation reduced employment and aggregate purchasing power. Interest rates were increased sharply in the U.S. in 1979 in an effort to prop up the value of the dollar (Clark, 2005, p. 22), and debtor nations saw interest rates on their dollar-denominated, adjustable rate loans rise as a result (Manley, 1987, p. 70). This rise in interest rates, combined with deteriorating conditions for earning foreign exchange through exporting to the U.S., caused nations to default on their external debts.

This debt crisis in the Global South in the 1970s and 1980s resulted from the interrelated complex of dependency enforcing systems and strategies discussed above combined with historically specific circumstances deriving from fossil-fuel-dependent development. The oil shocks of the 1970s occurred for historically specific reasons but served to entrench global systems of power and exploitation as loan defaults paved the way for SAPs.

Enforcing the Dependency of Global South Nations

The growth economics and trade liberalization of the past fifty-plus years have resulted in further impoverishment and negative social and environmental impacts in many nations (Dasgupta, 2001; Stiglitz, 2002). Since the Global South debt crisis of the 1970s and 1980s, new loans by the IMF and World Bank have sometimes been made solely to allow a debtor nation to consolidate old loans with a new one (Rowbotham, 2000, p. 65). As anyone knows who has used one credit card to pay another credit card bill, the cycle of deepening debt imprisons the debtor, making him/her vulnerable to the demands of creditors. Many peripheral countries have found their entire export earnings insufficient to repay the interest alone owed on their external debt (Rowbotham, 2000, p. 51).

International finance has played a powerful role in enforcing the dependency of the periphery while privileging the core and transnational interests within the world-system. According to Rowbotham, "Debt...represents a powerful political instrument for subjecting debtor countries to international economic control and making them specialise at the level of production" (2000, p. 67), a process that reinforces their colony-like status and function within the world-system. To make matters worse, according to Kaplinsky, the terms of trade have steadily deteriorated for the Global South as the prices of their exports have fallen relative to industrial nation exports of agricultural and manufactured goods and knowledge-intensive services (2005, p. 187).

Within the world-system, IMF and World Bank loan conditions and SAPs have systematically served the interests of foreign creditors and investors at the expense of the people, environments, and economies of debtor nations. Stiglitz cites a particular example of IMF programs functioning to bail out Western creditors. According to Stiglitz (2002), when foreign creditors anticipate an IMF loan agreement with a given nation, they have weakened incentives to make sure the nation will be able to repay—a problem known as moral hazard (pp. 201, 207–208). In these and other dependency enforcing situations, IMF loans create profit opportunities for creditors rather than stimulating the domestic economy. External debt has served to undermine the long-term stability of entire societies and economies (Stiglitz, 2002, p. 209). This lack of stability characterizes the world-system while also undermining its resiliency and sustainability.

The multiple reciprocally reinforcing aspects of enforced dependency built into the system of global finance create a rentier economy of "resource grabs and debt dependency" (Hudson, 2005, p. xxvii). Rowbotham sums up this situation: "Debtor nations remit a perpetual tribute to the wealthy nations and their corporate interests, and are kept in a state of permanent monetary bondage as interest payments, profit repatriation and dividend payments siphon money from their economies" (2000, p. 123).

We now turn our attention to examining how U.S. dollar hegemony and U.S. debt enforce dependency within the world-system.

Dollar Hegemony, U.S. Debt, and International
Economic Codependency

With the Bretton Woods agreement in 1944, the U.S. dollar became the de facto world currency. The value of the dollar was pegged to gold, and the values for all other currencies were pegged to the dollar. The gold exchange standard ended in 1971 when the Nixon administration unilaterally refused to allow further exchanges of dollars for U.S. gold reserves and floated the U.S. dollar on the world currency market. Even without the backing of gold, the U.S. dollar has managed to remain the premier world currency. We will now examine how and why dollar hegemony is maintained in the world-system. We will also examine the global effects of that hegemony.

As discussed above, the fact that many international loans are denominated in dollars provides the U.S. with political and economic leverage, but dollar hegemony creates additional advantages for the United States at the expense of the rest of the world. During the 1970s, worldwide speculative movement out of U.S. dollars occurred, in part, as a result of continued balance-of-payments deficits run by the United States (Hudson, 2005, pp. 94–95). The Federal Reserve reacted by raising interest rates, thereby constricting the money supply and propping up the value of the dollar. Since the early 1980s, the value of the dollar and its status as a world reserve currency (meaning that U.S. dollars and dollar-denominated securities are held by non-U.S. central banks as well as by businesses and individuals around the world) has allowed the U.S. to accumulate huge balance-of-payments deficits. For the U.S., the total economic gain from supplying a world reserve currency (also known as seignorage) is equal to its cumulative balance-of-payments deficit on its import-export account. This deficit represents unpaid-for goods and services supplied to the U.S. by the rest of the world. This negative balance of trade has allowed the U.S. to maintain a relatively high standard of living for its citizens even as its manufacturing sector has been outsourced. In 2004, the total accumulated amount of this deficit was around $3 trillion, and it was accumulating at around $1.3 billion per day (Feasta, 2004, p. 5). The U.S. has been creating money from nothing and has been using it to purchase half again more imports than it exports. According to the Foundation for the Economics of Sustainability (Feasta),

> We can get a good idea of how big the $3,000bn subsidy has been by recalling that in 1998, the United Nations Development Programme estimated that the expenditure of only $40bn a year for ten years would enable everyone in the world to be given access to an adequate diet, safe water, basic health care, adequate sanitation and pre- and post-natal attention. (2004, p. 6)

The enormous gains reaped by the U.S. from seignorage account for the economic and military power of the United States. The U.S. has been able to develop its unparalleled military might because other nations finance its growing debt that, in turn, enables heavy military spending (Feasta, 2004, p. 5).

Although a large proportion of the now $15 plus trillion public debt of the U.S. (United States Treasury Department, 2012) is held by creditors in the form of Treasury bills and bonds on which interest is paid, payment of interest is actually financed through creation of further debt. The U.S. is able to import vast amounts of goods and services while simultaneously amassing a huge debt simply because of its position as the creator of U.S. dollars (Feasta, 2004, p. 5). This arrangement depends upon continuing international faith in a growing U.S. economy. It also depends upon the reluctance of creditors to pull their investments out of the U.S. economy due to the likelihood that doing so would threaten their own economic stability.

Michael Hudson describes the process through which the U.S. effectively gets others to both pay its debt and maintain their investments in the U.S. economy:

> The United States...[draws] on world resources through a novel monetary process: by running balance-of-payments deficits that it refuses to settle in gold, it has obliged foreign governments to invest their surplus dollar holdings in Treasury bills, that is, to relend their dollar inflows to the U.S. Treasury. (2005, p. 17)

This process is further supported by the fact that, in domestic markets worldwide, dollars are redeemed for domestic currencies at central banks, and central banks purchase Treasury bills with these dollars. Investing in Treasury bills—though they typically provide lower rates of return than investing in the private sector—has to date been deemed a comparatively safe investment by foreign central banks (2005, p. 30).

Through the aggressive use of seignorage, by 1971, "the United States [had] succeeded in establishing its own government debt as the key international monetary standard" (Hudson, 2005, p. 25). According to Hudson, post-1971,

> expansionary monetary and fiscal policies were pursued irrespective of their balance-of-payments consequences. In the face of a growing payments deficit the U.S. Government accelerated federal spending and money creation, and watched foreigners bear the cost of financing this spending spree. (2005, p. 25)

Hudson further explains the economic costs of U.S. policies to foreign countries:

Foreign countries that run balance-of-payments surpluses presently are obliged to keep their central bank reserves in the form of loans to the U.S. Treasury ad infinitum. These savings become part of the U.S. financial system rather than building up their own productive capacity. There is no hard-currency guarantee for the value of these loans as the dollar falls against the euro, yen and other currencies of economies running trade and payments surpluses. In domestic-currency terms, the values of dollars held in central bank reserves declines. (2005, p. xxviii)

Hudson continues: "In the past, nations had sought to run payments surpluses in order to build up their gold reserves. But now all they [are] building up [is] a line of credit to the U.S. Government to finance its programs at home and abroad" (2005, p. 30). U.S. military spending and purchase of foreign imports, in effect, translate into savings in foreign countries through their purchase of U.S. Treasury bills and bonds. Through this purchase of the U.S. debt and through other avenues in the global economy, many dollars spent overseas make their way back into the U.S. economy (Hudson, 2005, p. 32; Rowbotham, 2000, p. 125). These processes of simultaneously generating savings in foreign countries and purchases and investments in the U.S. economy increase the purchasing power and economic vitality of the U.S. while, at the same time, applying the brakes to other nations' economies as the savings stimulated effectively remove from domestic circulation the money earned from exports sold to the U.S.

During the oil price shocks of the 1970s, the U.S. succeeded in convincing OPEC members to hold much of their excess dollar earnings in U.S. Treasury bills. Therefore, the oil price rises were less problematic than they might have been for the U.S. due to the recycling of U.S. dollars spent on OPEC oil back into the domestic economy (Hudson, 2005, pp. 108–110). Between 1999 and 2008, the U.S. Treasury bill standard faced competition from Euro denominated investment opportunities, but the recent European economic crisis has allowed the dollar to regain its comparative value against the Euro. If the U.S. is to continue to convince OPEC to help finance its deficits, it must thwart the emergence of competing currencies and investment opportunities (Clark, 2005, chaps. 1 and 5; Hudson, 2005, p. 258). In particular, the U.S. must succeed in maintaining the petrodollar system according to which OPEC oil sales are denominated in dollars (Clark, 2005, chaps. 1 and 5).

With the near complete demise of the U.S. manufacturing sector along with global reliance on the U.S. as the consumer of last resort, the *status quo* of the global economy depends upon the U.S. government and consumers taking on increasing amounts of debt. The U.S. has actively promoted this international co-dependency which has allowed it to become the world's sole superpower—but this situation cannot continue forever. American deficits may become so large that they scare off creditors, and the rest of the world may tire of the U.S. dominating the global economy while undermining the ability of other nations to complete. Still, for many countries, there are risks to dumping dollar-denominated investments on the world market and abandoning the petrodollar system—both of which would drastically reduce the value of the dollar—though for some nations and regions, the risks may eventually be outweighed by the growing instability of the current monetary and economic system. According to Rowbotham, "The aggregate of national debts coupled with the private/commercial debt directly associated with the money supply places...[wealthy] nations in a position of permanent financial exposure" (2000, p. 97), and those exposed include creditors of the U.S.

The hegemony of the U.S. dollar, however, is perhaps beginning to crack under pressure from various sources. One point of pressure is the rise of competing currencies. Prior to the global economic downturn that began in 2008, the Euro was rising in value compared to the dollar. Another source of pressure has been stable and profitable investment opportunities offered by rising and integrating economies outside the U.S., particularly in Asia and Europe. At the time of this writing, the Eurozone as a whole has been destabilized by overly-indebted member nations and the negative economic effects of the Great Recession. As a result, the value of the dollar has risen relative to the Euro. Time will tell how the relative strengths of the two currencies play out within the global economy, but the recent rise of the Euro demonstrates the possibility that other currencies may eventually compete with the dollar on the world stage.

Another threat to U.S. dollar hegemony is represented by the power of OPEC, power that will only increase as peak oil enforces the dependency of economies worldwide on OPEC oil supplies, which are by far the largest in the world. Using their unique endowments of oil, OPEC nations could choose to extend credit to foreign purchasers conditional upon these countries allowing them to purchase their domestic assets, a condition that would extend their political influence. Such a move would weaken the political and economic power of the U.S., thereby weakening its ability to entice other nations to accept the Treasury bill and petrodollar standards (Hudson, 2005, p. 267).

We now turn our attention to the hegemony of global capitalist elites and culture in an effort to understand how this hegemony enforces dependency within the world-system.

The Hegemony of Global Capitalist Elites and Culture

Late capitalist culture and economy inform and reinforce each other recipro-cally. Global elites maintain hegemony through wielding their power and wealth in ways that promote continued concentration of power and wealth in their own hands, thereby furthering their ability to reproduce and reinforce hegemonic economic and political relationships.

The concepts of private property and the enforcement of property rights are central components of capitalist cultural hegemony (Achebar et al., 2005; Polanyi 1944/1957; Proudhon, 1890/1966). Commodification, privatization of public assets and enterprises, the extension of the concept of private property into ever more areas of life, and the enforcement of intellectual property rights all represent modern versions of enclosure of the commons. Capitalism has been able expand production and consumption globally, in part, due to private takeover of resources and services that were once considered public commons. Increasing privatization and commodification of essential resources such as land, water, seeds, and other genetic material essential to food production increase dependency of the world's communities upon globalized capital even as they increasingly concentrate wealth and power within the world-system (Barlow & Clarke, 2002; Garcia, 2004; ICIJ, 2003; Ludwig, et al., 2006; Shiva, 2000). That this dependency extends into areas essential to life further promotes the socio-ecological control and the entrenchment of capital.

The global concentration of wealth and power in the hands of transna-tionals and corporate and banking elites furthers the ability of capital to advance its economic and cultural agenda. We examined above many policies and practices that serve to concentrate wealth and power in progres-sively fewer hands. Growing income gaps between the rich and poor provide evidence that this concentration is occurring, and these gaps have never been greater than at this point in time (Homer-Dixon, 2006, p. 186). During the last decade of the twentieth century, despite promises that globalization would reduce poverty if developing countries would stay the neoliberal course (Stiglitz, 2002, p. 213), the numbers of those in dire poverty grew by almost 100 million people at the same time that world income increased by an average of 2.5 percent annually (Stiglitz, 2002, p. 5). Development proponents often point to relative rates of economic growth as an indication of economic convergence, saying that those developing countries experienc-ing high growth rates will catch up with the industrial world in terms of living standards. Homer-Dixon explains why citing these figures as evidence of near term convergence is a fallacy:

This may look like a convergence because incomes in poor countries are predicted to grow faster than those in rich countries. But it's not. The gap between poor and rich average incomes will continue to widen: although the average income of rich countries is growing at a slower rate, this rate multiplies a vastly larger income base—$32,000 annually per person in 2006, according to the [World] Bank, compared with $1,500 in poor countries. So the absolute size of the gap between the average incomes of rich and poor countries steadily widens. And it widens not just for a few years or even a few decades but for *hundreds* of years to come. (2006, p. 190)

Furthermore, core producers exploit their advantage in the global economy at the expense of peripheral producers who find it all but impossible to compete (Homer-Dixon, 2006, p. 200).

The corporate legal structure also contributes to concentration of wealth and power that promotes capitalist elite hegemony. Unlike people, corporations are immortal. When a person dies, his/her resources are usually distributed among surviving relatives and friends and possibly to chosen charities, foundations, and causes. Since corporations tend to outlive particular CEOs and boards of directors, they can typically continue to concentrate wealth and power for long periods of time so that successful transnational corporations possess economic power unknown to previous generations.

The global movement toward finance and speculation as the central vehicle for profit generation also reinforces the concentration of power and wealth because it takes money to make money through investment and speculation in the global economy, and the more money one has available, the larger the potential for profit. Although the possibility for extensive financial loss also exists, one cannot even play this game without significant capital. This new focus of economic activity, therefore, advances the hegemony of capitalist elites.

The continual focus on lowering inflation in the U.S. and Britain reveals how the interests of the wealthy and powerful also influence fiscal and monetary policy. Since inflation tends to benefit debtors over creditors, it is to be avoided. According to Stiglitz, "For the financier who has lent his money out long term, the real danger is inflation. Inflation may mean that the dollars he gets repaid will be worth less than the dollars he lent" (2002, p. 217). While there is agreement that no economy can succeed under hyperinflation, "there is little evidence that pushing inflation to lower and lower levels yields widespread economic gains commensurate with the costs, and some economists even think that there are negative benefits from pushing inflation too low" (Stiglitz, 2002, p. 220). In the global context, the IMF insists that "countries have an independent central bank focusing on [reducing] inflation" (Stiglitz, 2002, p. 45). This policy represents a clear creditor bias. If the IMF were more concerned with diverse and widespread economic development, it might focus at least as intently on employment and growth, which we have seen it does not (Stiglitz, 2002, p. 45). We see that the fiscal and monetary policies within globally hegemonic economies and within international banks that serve the interests of hegemonic capital serve to advance the interests of capitalist elites within the world-system, thereby helping these elites to perpetuate global economic culture in their own image.

Policy makers also tend to view the world through the eyes of large corporate entities since they are the economic heavyweights:

> Multinational corporations…have been accused of a catalogue of crimes; blackmailing national governments to grant them subsidies; exerting pressure to change government economic policy; asset-stripping; exploitation of the developing world; transfer pricing to avoid taxation; acquiring by patent law rights that ought not to belong to any single private interest—the list is endless. The fact that these issues have not been addressed, indeed are not even on the agenda, lends support to the concern that the tier of international governance is pro-corporate. (Rowbotham, 2000, p. 4)

Furthermore, corporate elites and international power brokers often conflate freedom with capitalism—free markets with free people. To its proponents, globalization is synonymous with progress, and developing countries must accept neoliberal globalization as the route to becoming an American-style capitalist society (Stiglitz, 2002, p. 5).

Considering what is on and what is off the neoliberal agenda also reveals its hegemonic foundations. Core entities stifle discussion of alternate economic strategies and policies, and leaders within the dependent periphery generally avoid openly questioning neoliberal imperatives (see Stiglitz, 2002, p. 43). There may be money to bail out banks but not to pay for improved education and social services, nor to assist those who become unemployed as a direct result of neoliberal policies (see Stiglitz, 2002, pp. 43, 81).

We have seen that the IMF and World Bank do not represent the broadly based interests of developing countries. By tradition, the head of the IMF has always been a European, and the president of the World Bank has always been an American. These leaders are chosen behind closed doors, and they often have little experience in the Global South (Stiglitz, 2002, p. 19). Stiglitz asserts that the IMF and World Bank are "driven by the collective will of the G7" (2002, p. 14) so that these institutions advance elite capitalist hegemony.

Those who speak for given Global South countries tend to represent similar interests. At the IMF, finance ministers and central bank governors govern the institution; at the WTO, trade ministers represent countries.

Representatives sent to each of these institutions tend to represent select constituencies within countries they represent: trade ministers representing the interests of the business community and finance ministers and central bank governors representing the financial community. The interests of a small minority—and a minority whose interests closely approximate those of the business and financial communities in the developed world—are advanced over the interests of the vast majority within many countries. This situation is profoundly undemocratic. Concerns for the environment and social justice are virtually ignored (Stiglitz, 2002, pp. 19–20) while the interests of national elites and global capital are advanced.

Homer-Dixon summarizes important cultural and economic processes that reinforce the hegemony of global capitalists:

> There are the social causes of denial [of the social and environmental problems created by globalization]. Probably the most important is the self-interest of powerful groups—corporations, government, agencies, lobbyists, religious institutions, unions, non-governmental organizations, and the like—that have a vested interest in a particular way of doing things or viewing the world. If outside evidence doesn't fit their worldview, these groups can cajole, co-opt, or coerce other people to deny this evidence. Some groups...will be much more effective in the effort than others, owing to their enormous political and economic power.... Our economic elites don't just encourage consumerism. Through their influence on the media and on our society's political process, they create, reproduce, and justify a pervasive and interlocking system of rules and institutions—from property rights and capital markets to contract and labor laws—that promotes growth and that, in the process, buttresses their power and privilege. A particular language of capitalism—a "discourse" of economic rationality and competition that penetrates into every nook and cranny of our economies, societies, and lives—helps us understand and abide by these rules and institutions. This language says that people maximize their pleasure from consumption and that they make decisions as if they were calculating machines, constantly weighing costs and benefits to evaluate their choices. Capitalism's language also says that our labor is a commodity to be bought and sold in the competitive marketplace. And it equates our personal identities with our economic roles in that marketplace.... For the vast majority of us who sell our labor in the marketplace, our economic insecurity and relative powerlessness impel us to play by the rules. And in capitalist democracy, playing by the rules means not starting fights over big issues like our society's highly skewed distribution of wealth and power. Instead, it means focusing on achieving short-term material gains—such as bettering our contracts with our employers. Put simply, our economic elites have learned, largely through their struggles with workers in the first half of the twentieth century, to protect their status by creating a system of incentives, and a dynamic of economic growth, that diverts political conflict into manageable, largely nonpolitical channels. And as long as the system delivers the goods—defined by capitalist democracy itself as a rising material standard of living and enough new jobs to absorb displaced labor—no one is really motivated to challenge its foundation. (2006, pp. 215–217)

As noted in chapter one, Marcuse (1964) calls this complex of processes "repressive desublimation." Antonio Gramsci (1971/1999) labels it "passive revolution" that serves to contain the contradictions of the capitalist system. Like Marcuse and Gramsci, Homer-Dixon argues that challenges to enforced dependency are unlikely to come from the upper echelons of society saying that "members of our economic elite rarely have qualms about the prevailing economic worldview because it sustains their status, and because they generally believe that they've achieved that status through their superior intelligence, guts, and drive" (2006, p. 218).

As Homer-Dixon suggests, cultural hegemony also inhibits challenges from below, not only through enforced dependency, but through hegemonically pervasive capitalist culture. According to Homer-Dixon,

> The tacit arrangement among our elites, our experts, and the rest of us is essentially symbiotic—a mutually gratifying and self-sustaining cycle of denial and delusion. Through our acquiescence in and often active support of modern capitalism, we legitimize our elites' and experts' status and power, while those elites and experts give us an overarching ideology of permanence, order, and purpose that lends our lives a sense of place and meaning. According to this ideology, economic growth is a panacea for all our social and personal problems. Growth equals health. Unfortunately…when we're in denial, we can't think about the various paths that we might take into the future. Nor can we prepare to choose the best path when the opportunity arises. Radically different futures become literally inconceivable—they are "beyond imagining"…in the same way the heliocentric cosmos was inconceivable to many people prior to the Copernican revolution. (2006, p. 219)

Stiglitz describes the situation thusly:

> We have a system that might be called *global governance without global government*, in which a few institutions—the World Bank, the IMF, the WTO—and a few players—the finance, commerce, and trade ministries, closely linked to certain financial and commercial interests—dominate the scene, but in which many of those affected by their decisions are left almost voiceless. (2002, pp. 21–22)

Through examining the closely reciprocal relationship between cultural and economic hegemony, we see that profound changes in the material circumstances of life create profound changes in consciousness and vice versa in a cycle that continually deepens enforced dependency within the late capitalist world-system. This process that characterizes late capitalism explains what might at first glance appear to be the surprising degree of global acceptance of hegemonic ideas and actions among both elites and the oppressed. Counter hegemonic groups and movements do exist, such as the Zapatistas in Chiapas, Mexico (Marcos, 2001), and indigenous groups fighting to maintain their traditional economies and cultural/spiritual ties to place (Grossman, 2005; LaDuke, 1999). These groups raise important challenges to globalization, but globalization as an economy and a way of life continues to spread—new markets are opened and the money economy colonizes the few remaining locally based and subsistence economies (Berry, 1987; ISEC, 1993; Martinez, 1997; Shuman, 2000).

Now that generations all over the world have grown up and are growing up in a thoroughly globalized world, the lines of cause and effect that informed the development of our economic, cultural, and ecological landscape are further obscured by absence of social memory. Young adults and youths have no direct experience with a world before globalization. The world as it is now is an unquestioned reality to many young people, and given Western culture's ingrained notion of progress, the world that is now assumed to be the best of all possible worlds. Young people have been acculturated into the values system of modernity (Spretnak,1997) and its outgrowth, the globalized world.

Globalized Society as a World-System of Enforced Dependency Lacking in Resiliency

Once subsistence cultures of place were broken down and colonies were folded into the capitalist system, the dependency of colonized regions was enforced through brute force and later through the creation and enforcement of economic rules and practices. Within the world-system, dependency that was originally enforced by nation states has continued in the late capitalist era in spite of the erosion of nation states' influence by the forces of globalization. Once an individual, a community, or a nation has become dependent upon the capitalist system, there is virtually no escape, and the dominance of the capitalist elite relies upon this dependency. The system is virtually impossible to escape for debtor nations faced with the choice to comply or collapse economically, but it is also difficult to escape for individuals who lack access to the productive capacities of the land or who lack the knowledge and experience necessary to make use of these capacities.

We have focused on macro-level analysis here, but dependency is enforced even at the very personal level of individuals and families. The urban poor are typically entrapped in the day-to-day struggle for existence, and wealthy and middle-class individuals, too, are often heavily indebted to banks. Many of those who are systematically disadvantaged within the world-system suffer from poverty, poor health, and lack of education, and each of these problems compounds and reinforces the effects of all the others in a process that creates powerlessness (Stiglitz, 2002, p. 83). Escape from extreme dependency is possible, as we shall see in later chapters, but it is very challenging—and, at least initially, it can come at a high cost in terms of personal security and social inclusion, a cost some simply cannot afford to bear.

Complex global systems of enforced dependency are highly unstable and vulnerable to near-term collapse. Continuing inequities within the global economy promote conflict, both within nations and internationally as late capitalist cultural hegemony loses its grip on containing its social and economic contradictions. At the same time, socio-ecological resiliency is continually sacrificed to fuel the engines of economic growth. As global society realizes diminishing returns on increasing social and economic complexity, the world-system becomes increasingly vulnerable to collapse (see Tainter, 1988). Furthermore, the likelihood that collapse will be catastrophic increases as globalization encourages tight linkages among components and processes of the world-system, making the boom and bust phases of the business cycle increasingly likely to trigger global-scale economic breakdown (see Homer-Dixon, 2006, chap. 9). Environmental damage and stress, and loss of diversity in human and ecological systems compounds these problems, further increasing the potential for global social, economic, and ecological disaster. The monocultures promoted within the world-system in agriculture, popular culture, materialism, and employment are inherently unstable due to their lack of diversity and their heavy reliance upon petroleum-dependent transportation of goods. Simultaneously, global free market competition in the world-system reduces resiliency because globalization channels power and wealth into the hands of a few while disenfranchising many small-scale, local producers and decreasing diversity in economies, ecosystems, and communities. These tensions serve as sources of ever-present and ever-increasing economic and social instability.

Though development bank and corporate officials may see themselves as helping the poor by investing in the Global South, these institutions and the dependency they enforce have the effect of serving powerful interests at the expense of the oppressed. These institutions actively resist a redistribution of wealth and power that would materially benefit the victims of privilege. Such redistribution would require hegemonic groups to adopt a virtually opposite set of priorities and interests to the ones they have advocated for many years, perhaps for entire professional lifetimes. Even if global capital were to genuinely promote economic growth in the Global South, growth itself is neither a desirable nor sustainable end in socio-ecological terms. The fossil fuels that drive the global shipping and mass production requisite within the late capitalist global economy are depleting. We appear to be reaching the physical—if not also the moral—limits of late capitalist globalization.

After exploring in some depth late capitalism as a world-system of enforced dependency, the question we are left with is this: if the dreams of late capitalism are counterfactual to aspirations for a socio-ecologically sustainable society, what kind of dreams should we dream? If we begin with the goal of eliminating systems of enforced dependency in an effort to restore or create diverse and resilient societies, I believe we are on the right track. I will explore possibilities for social change along these lines in chapters five and seven. Chapter eight will build upon all previous chapters in advocating for educational praxis as an important vehicle for sustainability-oriented social change.

Part Two:

The Road Ahead:

Setting Guideposts for Living and Learning

Sustainability

Chapter 5:

(Re)inhabitation: Place as a Concept and Construct for Sustainable Living and Learning

It may be the supreme irony of our time that abundance has encouraged selfishness and that scarcity may become the catalyst for a new covenant of cooperation.

—Dennis Lum

Partisans of neoliberal globalization have attempted to make home and geography irrelevant. Amid the blackmail of SAPs and the economic and political dominance of footloose capital, the power of nation states to regulate business activities has declined while monopoly industries have come to dominate markets worldwide. Monocultures in agriculture and consumer culture worldwide systematically reduce the cultural diversity and adaptation to place that make communities resilient to economic shocks and disruptions in global supply lines. Meanwhile, peak production of the oil and gas that drive the globalized economy has arrived, or will arrive in the next few years.

In the face of these challenges, this chapter is an exercise in hope—hope for (re)creating sustainable and fulfilling human/nature lifeways. By hope, I do not mean faith in a predictable or successful outcome. I mean taking stock of where we stand now and working toward rectifying the damage inflicted on people and nature by industrial civilization, even in the face of uncertain and likely uneven results (see Havel, 1990, p. 180). I offer a theoretical framework for ways of being in the world that nurture humans and nature simultaneously—physically, emotionally, intellectually, and spiritually. The depletion of the natural world and the widespread lack of recognition of the self in others and in nature threaten the very survival of our species and the health of the biosphere. This chapter is an effort to identify paths for coun-tervailing action in the midst of the global sustainability crisis, paths that hold the potential, and therefore the hope, for (re)generating sustainable lifeways.

In this chapter, I argue for (re)localization of communities as central to living sustainably. I consider Western and non-Western theories and cultural traditions of localized living as expressions of the need for humans to live within nature and within natural limits. Drawing on the philosophy and living praxis of sustainable indigenous lifeways as well as the work of sustainability activists and theorists rooted in the Western tradition, I articulate and elaborate a complex of interwoven themes for collective sustainability praxis. Indigenous models of inhabitation and reciprocal relationship with nature provide examples of potentially sustainable and fulfilling lifeways that have embodied sustainability for millennia. Recent movements such as the Transition Movement initiated in Europe, the Sarvodaya movement in Sri Lanka, and the Navdanya movement in India as well as widespread efforts toward "repeasantization" worldwide provide additional examples of living sustainability. I propose that the revitalization of indigenous cultures and the (re)localization and sustainability-oriented (re)characterization of non-indigenous communities represent parallel and complementary efforts. I further propose that the socio-ecological themes that orient sustainable societies coalesce in the concept and construct of place—place itself being a dynamic interweaving of nature and society. Place, as a both a concept held in the minds of humans everywhere and as a physical embodiment of human/nature relationship, serves as the mental and physical container for realizing socio-ecological sustainability. Place as a mental construct and sustainability praxis is socio-ecologically diverse. At the same time, sustainable manifestations of place embody certain socio-ecological themes.

This chapter builds upon the critical social theory of sustainability articulated in chapters one and two and on the theory of enforced dependency articulated in chapters three and four. These bodies of theory provide a foundation for elaborating themes of sustainable, place-based living and learning and for theorizing sustainability praxis. I argue that place-based sustainability is about creating localized self-sufficiency and self-determination. I propose that this work can progressively reduce the momentum of the capitalist world-system of integration and dependency while, at the same time, progressively increasing community resiliency.[1]

---

[1] This socially transforming engagement embodies the absolute negative moment of praxis. Absolute negativity consists of two forms of negation. First, it is characterized by counterhegemonic thought and action—the negation of the current world order, a vital step toward sustainability. The second negation consists of negation of the definition of praxis in

As noted in previous chapters, I recognize the breadth and depth of the challenges to sustainability represented by neoliberal cultural, political, and economic hegemony. I also recognize that there is a significant possibility that place-based sustainability praxis will be unable to successfully reverse the tsunami of socio-ecological destruction unleashed by the exploitation of nature and people that characterize the industrial, free market world-system. Still, as an expression of our love for our children and for nature, I argue that we must do all we can to (re)inhabit our places in sustainable ways—it is never too late to do the right thing, and the time has perhaps never been more ripe for (re)situating our cultures and economies within the bounds and bounty of place.

(Re)localization and Conceptions of Efficiency and Progress

Dismantling enforced dependency in the globalized world hinges upon creating alternatives through praxis. Many alternatives have been well developed theoretically and/or are in the process of being implemented in diverse locations and communities. A brief introduction to some key theories and projects will help us envision how (re)localization can work in practice. All of the works addressed below challenge neoliberal concepts of efficiency and progress by asking such questions as: Progress for whom? Efficiency for whom? Who or what pays for efficiency and progress? What are and what should be the economic and socio-ecological goals of efficiency and progress?

The work of Jane Jacobs (1969, 2000) represents a pathbreaking articulation of a rationale and a program for (re)localization. In *The Economy of Cities* (1969), she offers an important and detailed analysis of how and why an economy characterized by localized business ownership benefits local residents more than does an economy based on local jobs provided by nonlocal capital investment. Jacobs argues that communities should engage in import substitution development and should invest capital locally rather than exporting it through loans made outside the community. She also explains how multiplier effects that result from money spent locally recirculating within the local economy increases local economic wealth, diversity, and resiliency.

---

oppositional terms alone so that praxis can manifest as a creative force free from self-definition solely in relationship to the neoliberal order. Absolute negativity is a self-reflexive, historically situated process that translates critique into constructive praxis (see McLaren & Kumar, 2009).

Her insights on the economic and social value of localizing the economy represent a radical departure from dominant economic theory which touts the benefits of geographic specialization based on comparative advantage. Jacobs' place-centered, diversity-generating approach to economic development promotes increased resiliency when compared to more commonly used approaches to economic development such as export-led development at the national level and relocation incentives offered to transnational corporations by national, state, and/or city governments. Jacobs' work informs my conceptual framework regarding place-centered sustainability, and she has influenced a host of other theorists and practitioners who argue the importance of (re)localization of economic and social life including Shuman (2000), Kemmis (1990), and Calthorpe (1993). In her work, Jacobs (1969, 2000) proposes to refocus efficiency and progress toward benefiting diverse social needs within a broad community rather than benefiting monopoly capital, and she proposes to do so within a context of socio-ecological sustainability (2000).

In his book *Going Local*, Michael Shuman (2000) articulates a rationale and a program for local economic vibrancy and resiliency specifically targeted at reducing community vulnerability to footloose capital. In the age of globalization, footloose capital will often relocate production facilities in an effort to increase profits, even if doing so means sentencing an entire community to economic and social decay. Shuman advocates multiple strategies for reducing community vulnerability to both the vicissitudes of capital and environmental degradation. These strategies include import substitution, the creation and use of localized currencies, localized capture and production of renewable energies, localized ownership and control of businesses and banks, local investment programs for banks and pensions, widespread involvement in local politics, and devolving political control to local communities. Like Jacobs (1969, 2000), Shuman (2000) conceptualizes efficiency and progress within a socio-ecologically sustainable framework that would have the economy serve the community rather than the reverse.

In their book *Superbia!* Dan Chiras and David Wann (2003) construct similar arguments to those of Shuman (2000) with regard to (re)localization, but their focus is at the neighborhood level specifically within suburbia. In his book *The Next American Metropolis*, Peter Calthorpe (1993) also specifically addresses suburbia in his promotion of new urbanist design as a redevelopment strategy for creating a built environment conducive to socio-ecological sustainability and cultural vibrancy. The authors of these works advocate slowing the pace of community life by encouraging walking and biking and by taking the time to develop relationships with neighbors that can encourage sustainability-oriented community action. These authors question the idea that progress for people and communities translates to increasing the speed and isolation of community life.

Similarly to Shuman (2000) and Chiras and Wann (2003), economist Richard Douthwaite (2004) proposes that relocalization is essential for sustainability. He argues that global standardization of commodity culture places unsustainable pressures on people and ecosystems to produce uniform foods, housing materials, clothing, and other materials and products for people everywhere (2004, pp. 116–117). Douthwaite (2004) also argues that the global monetary system is inherently both unstable and unfair in the way it concentrates power and wealth in those nations that create world reserve currencies, foremost among these being the United States (pp. 115–116). He and his colleagues at The Foundation for the Economics of Sustainability (Feasta) argue for monetary reforms that address the looming economic crisis resulting from inevitable declines in global oil production.[2] As a means to increase local self-reliance and economic resiliency, Douthwaite (1999a) also advocates use of local currencies. These strategies confront the conceptions and measures of efficiency and progress proffered by monopoly capital in favor of measuring progress and efficiency in terms of resiliency and sustainability.

Richard Heinberg, author of *The Party's Over* (2005), and the makers of the film *The End of Suburbia* (Greene, 2004) call upon us to envision and create societies where people can live fulfilling lives without consuming unsustainably. Renowned petroleum geologist C. J. Campbell and permaculturist Graham Strouts call upon us to do the same in their book *Living through the Energy Crisis* (2007). In his books *The Solar Economy* (2002) and *Energy Autonomy* (2007), the late German parliamentarian and renewable energy activist Hermann Scheer argues for localized production and capture of renewable energy as a response to both climate change and peak oil. An important part of Scheer's argument entails the recognition that decentralized, renewable energy represents a significant opportunity to reverse the concentration of wealth and power in the global economy and in national and global politics. Rob Hopkins, founder of the Transition Movement that began in Europe and is spreading worldwide and author of *The Transition Handbook* (2008) offers a set of principles and processes for engaging in what he calls *transition initiatives*. These principles emphasize not only achieving material sufficiency within a mostly localized economy,

---

[2] The reforms Feasta promotes would also bring reductions in emissions of carbon dioxide while, at the same time, redistributing concentrated wealth to benefit the poor in the Global South (Feasta, 2008).

but also promoting socio-ecological sustainability. If sustainability were to become the measure of social success, as these authors and filmmakers argue it should be, neoliberal notions of efficiency and progress would be turned on their heads. Concentrations of wealth and power and the centralized, hierarchical decision-making structures that these concentrations engender would devolve into decentralized, locally-adapted governance structures and widely shared material wealth.

I have offered here only a sampling of the many cogent proposals for (re)localization as a sustainability strategy. In chapter seven, I will also explore how local food can serve as an appropriate nexus for sustainability praxis. Local activists, green politicians, educators, food producers, and other community members residing in villages, towns, and cities worldwide are currently articulating and implementing localized sustainability strategies tailored to their particular situations. In all of these efforts, we can recognize the value of making human to human, human to nature, and nature to nature ecological relationships visible and tangible within local processes of production and consumption. This immediacy can counteract the abstraction of modern, globalized living that obscures important relationships of cause and effect and, thereby, encourages collective violence against people and nature in the name of progress and efficiency.

My work of highlighting in this chapter the underlying and sometimes unstated themes that inform pace-centered sustainability praxis is not an idealist exercise. It is rooted in both the history and diversity of experience of human life on earth. The articulation of these themes is rooted in Gramscian (1971/1999) praxis. According to Gramsci (1971/1999), ideas are historically situated in parallel fashion to material circumstances. For Gramsci, ideas and materiality co-create each other within the broader process of history (p. 369). Community praxis of articulating themes for (re)localized, sustainable living derives from locally diverse, yet globally integrated, experiences with the palpable failures of neoliberal globalization. This praxis also embodies potential for an absolute negative break with the current paradigm and, therefore, the possibility for sustainability-oriented living outside and beyond the capitalist order.

Thematic threads of place-centered sustainability discussed below articulate with the overarching theme of (re)localization. These themes represent conceptual foundations for sustainable community praxis that can be adapted to specific natural and social contexts. Engaging in praxis guided by these themes will not, in itself, resolve the social contradictions of neoliberal globalization, but such action can reduce enforced dependency and increase community resiliency in diverse contexts worldwide. Communities engaged in place-centered community praxis can become less vulnerable in the event of a collapse of the late capitalist economic system and associated governance structures. In the event that nation states persist but become more or less paralyzed in their abilities to enforce rules and policies due to lack of funds and/or lack of energy resources, place-centered communities with strong local economies and with neighbors who know each other and share a history of collective decision making and mutual support will be better prepared for self-governance and self-sufficiency than communities that remain tightly tethered to globalization.

Ontological Foundations for (Re)Inhabitation

Place-centered sustainability praxis is rooted in an ontology in which humans, other life forms, and elements of life-sustaining systems exist—and can only exist and maintain their dynamic stability—through reciprocating relationships with each other. In many indigenous incarnations of such an ontology, human life is not seen as radically different, separate from, and superior to other creatures (Salmon, 2000, p. 1331)—a hierarchical relationship that would seem to justify exploitive relationships to nature. Place-centered sustainability recognizes that humans are embedded within nature and are subject to nature's laws and limits, though humans change nature and, through their cultural engagement in the economy of life, can also contribute to the diversity, resiliency, and vitality of natural systems (Grim, 2001; LaDuke, 1999; Mann, 2002; Martinez, 1997, 2010; Salmon, 2000, p. 1331; Sveiby & Skuthorpe, 2006).

Indigenous spirituality offers many examples of ontologies that recognize the holistic circle of life in which every thing and every being is related to every other thing and being. Many traditional indigenous cultures conceive of this relationship as familial so that the ethic of care one has for family extends to other people and the whole of nature (Grim, 2001; LaDuke, 1999; Norberg-Hodge, 1991/1992; Salmon, 2000; Sveiby & Skuthorpe, 2006). Some have argued that monotheistic religions such as Christianity desacralize nature and, through promoting the concept of human dominion over nature, encourage exploitation of other creatures and the environment (see for example Merchant, 1996). While the potential certainly exists for monotheism to encourage exploitation of nature, themes of stewardship of nature as a duty of righteous people also exist, for example, in Christianity (Barbour, 1991, pp. 74–77; Moyers & Casciato, 2006), demonstrating that monotheism can be reconciled with an ethic of care for nature and an understanding of the reciprocal roles of nature and humans in sustaining one another.

While I draw examples from indigenous cultural beliefs and practices in arguing for a holistic ontology of place and for other aspects of place-centered sustainability, I want to be clear that I do not advocate that non-indigenous people adopt indigenous beliefs and cultures. To do so is—besides being undesirable in terms of authentic praxis—impossible. One cannot and should not attempt to appropriate the history and culture of others whose relationships to place and concomitant spiritual beliefs represent authentic, multigenerational engagement with place through lived experience that constitutes the very being of individuals and cultures. Such appropriation represents a form of commodification and consumption of the other, even when well-intended. I also wish to avoid giving the impression that indigenous peoples have never made mistakes that caused damage to their places. According to Gonzales and Nelson (2001), "Some scholars have noted that Native nations have specific codes and instructions for taking care of the earth precisely because they did make ecologically harmful decisions in the past and consequently created specific stories and rules to learn from those experiences" (p. 498). I do argue that aspects of indigenous cultural practices and beliefs, together with certain cultural threads and practices within Western European and other traditions, can help everyone to identify themes for place-centered sustainability that can be adapted to their own places and cultures (see Armstrong, 1995; Berry, 1987; Bond, 2004; Grim, 2001; Jackson, 1996; Kemmis, 1990; LaDuke, 1999; Martinez, 1997; Nelson, 1983; Norberg-Hodge, 1991/1992; Ploeg, 2008; Salmon, 2000; Shiva, 2008; Sveiby & Skuthorpe, 2006). Engagement in the discursive and self-reflexive process of learning about sustainable systems of thought and action can spur the process of recharacterization of modern life, a process marked by a (re)-placement of individuals and communities within specific material and social contexts and within the rhythms of nature.

I would also like to be clear that, while I draw examples from diverse cultural traditions, I do not assert that all indigenous cultures form a homo-geneous grouping, nor do I claim that all industrial societies are the same. Some may find my mixing and matching of ideas across cultures disturbing, but I hope that it will be taken in the spirit in which it is offered: as a means to spark self-reflexive, critical, creative, and locally adapted thought and action toward sustainable inhabitation of place.

A holistic ontology of place is foundational to place-centered sustainabil-ity. In becoming more sustainable, we can learn from indigenous and other cultural traditions emphasizing the importance of reciprocating and sustain-ing human-to-human and human-to-nature relationships within the context of place. It is to these relationships that we now turn our attention as we explore the theme of the holistic circle of life and relationship as it is expressed in selected cultures and movements.

The Holistic Circle of Life and Relationship

Restoring a holistic ontology which recognizes the inseparable human/nature complex of life manifests as a process of sustainable human inhabitation of place. Wes Jackson notes that the Western European cultural tradition in the United States has yet to see itself as place-centered; most people in the U.S. have not sought to become native to place, to live from and within the bounds and bounty of a specific, tangible home on the land (1996, pp. 2–3). Instead, Americans have become increasingly mobile and have participated in extractive relationships with other people and places that have devastated communities and ecologies globally. As Jackson says, "Conquerors are seldom interested in a thoroughgoing discovery of where they really are" (1996, p. 15). I argue that the abuse of place by modern conquerors derives, in part, from perceptions of conquered spaces as *other*. Descendants of Western Europeans and other recent immigrants to the United States mostly lack an intergenerational relationship with place as a source of sustenance that must be cared for respectfully. Their relationship with place has been further stunted by economic globalization which distances people from the sources of their food, water, other necessities, and luxury goods, thereby obscuring their perception of the socio-ecological exploitation that occurs at extraction and production sites.

Depletion of oil, climate change, the recent near catastrophic collapse of the global economy, and the degradation and collapse of ecosystems globally demonstrate that we cannot continue to behave as conquerors, restlessly extracting desired wealth and moving on when resources run out. In order to make realistic assessments of lifestyles that are possible to sustain over the long haul, we need to (re)inhabit particular places, becoming intimate with local ecologies and possibilities for realizing new (old?) versions of the good life. As Jackson notes, "Human history forces upon us the terms of our coming nativeness as much as or more than does our freedom to choose" (1996, p. 18). (Re)inhabitation means developing an ontology of place as a holistic entity comprising intimate relationships among people, plants, animals, and natural systems such as rivers, prairies, oceans, and forests. Indigenous cultural traditions offer examples of what such an ontology might look like and the principles to which it must adhere.

Enrique Salmon (2000), an anthropologist and member of the Rarámuri (also known as the Tarahumara) tribe of the Sierra Madre Occidental in Mexico calls his culture's holistic ontology *kincentric ecology*. According to

Salmon, many indigenous peoples who have lived in and through long-term relationship to place have come to recognize that human relationship with nature is central to the meaning and the processes of life:

> Indigenous people in North America are aware that life in any environment is viable only when humans view their surroundings as kin; that their mutual roles are essential for their survival. To many traditional indigenous people, this awareness comes after years of listening to and recalling stories about the land. (Salmon, 2000, p. 1327)

Salmon recognizes the importance of reciprocal relationships between humans and nature in that, through mutual relationship, each provides and cares for the other as members of a family would do.

When he discusses the Rarámuri concept of *iwígara*, Salmon (2000) further elucidates how his people conceptualize their kincenetric place in the world in relation to other species and the whole of nature:

> *Iwígara* is the total interconnectedness and integration of all life in the Sierra Madres, physical and spiritual. To say *iwígara* to a Rarámuri calls on that person to realize life in all its forms. The person recalls the beginning of Rarámuri life, origins, and relationships to animals, plants, the place of nurturing [the Rarámuri homeland], and the entities to which the Rarámuri look for guidance....*Iwí* represents the fertility of the land.... It also means to unite, to join, to connect.... *Iwí* also makes reference to the Rarámuri concept of soul. It is understood that the soul, or *iwí*, sustains the body with the breath of life. Everything that breathes has a soul. Plants, animals, humans, stones, the land, all share the same breath.... *Iwí* is also the word used to identify a caterpillar that weaves its cocoons on the madrone tree.... The implication is that there is a whole morphophysiological process of change, death, birth, and rebirth associated with the concept of *iwí*. *Iwí* is the soul or essence of life everywhere. *Iwígara* then channels the idea that all life, spiritual and physical, is interconnected in a continual cycle. *Iwí* is the prefix to *iwígara*. *Iwígara* expresses the belief that all life shares the same breath. We are all related to, and play a role in, the complexity of life. *Iwígara* most closely resembles the concept of kincentric ecology. (p. 1328)

Given the intimacy ascribed by the Rarámuri to relationships among all beings and all things that comprise natural systems, it is not surprising that, for the Rarámuri, "the natural world...is not one of wonder, but of familiarity" (Salmon, 2000, p. 1329).

Furthermore, as Salmon (2000) illustrates, *iwígara* represents the interconnectedness of all life and all life systems, not only in the present, but intertemporally across generations. The centrality of the concept of *iwígara* to the traditional Rarámuri worldview implies a cultural focus on conscious participation in maintaining the health of the circle of life across the generations of the past, the present, and the future. In this way, the concept of *iwígara* resembles the seven-generations philosophy of the Iroquois Confederacy within which decisions were undertaken in an ethical context that required considering the effects of decisions on those who would live seven generations in the future.

Similarly, a holistic ontology of sustainability recognizes intergenerational equity, not only among humans, but among all beings and all living systems that comprise the circle of life, and this ontology represents the profound necessity of recognizing oneself in the other. Such an ontology parallels that of many indigenous worldviews by embodying socio-ecological living sustainability characterized by mutually sustaining relationships framed by concrete life experiences in specific places (see Salmon, 2000).

Many indigenous cultures that have lived sustainably in place or by moving seasonally and cyclically through traditional homelands share with the Rarámuri similar holistic, place-centered ontologies that emphasize intertemporal and interspecies relationships and an ethic of reciprocity. Jeannette Armstrong (1995), a native Okanagan and sustainability activist, writes of her people as embodiments of their homeland and keepers of the earth:

> We...refer to the land and our bodies with the same root syllable. This means that the flesh which is our body is pieces of the land come to us through the things which the land is. The soil, the water, the air, and all other life-forms contributed parts to be our flesh. *We are our land/place.* Not to know and to celebrate this is to be without language and without land. It is to be *dis-placed.*
>
> The Okanagan teaches that anything displaced from all that it requires to survive in health will eventually perish. Unless place can be relearned, it compels all other life forms to displacement and then ruin. This is what is referred to as "wildness": a thing that cannot survive without special protective measures and that requires other life forms to change behavior in its vicinity....
>
> The way we act in our human capacity has significant effects on the Earth because it is said that we are the hands of the spirit, in that we can fashion Earth pieces with that knowledge and therefore transform the Earth. It is our most powerful potential, and so we are told that we are responsible for the Earth. We are keepers of the Earth because we are Earth. We are old Earth. (1995, pp. 323–324; emphases in original)

Armstrong emphasizes the importance of responsible and respectful participation in the circle of life in recognition that humans have the capacity to change environments and that doing so in a way that disrupts natural cycles and the lifeways of other creatures is dangerous and can lead to decline of human/nature systems.

Like the Rarámuri and the Okanagan people, the Nhunggabarra aborigines of New South Wales, Australia, believe that all is connected: people, ancestors, animals, plants, and all parts of ecosystems. Therefore, a primary duty of people and community is "keeping all alive," practicing respectful care of ecosystems and others (Sveiby & Skuthorpe, 2006, pp. 170–172).

As noted above, indigenous cultures of peoples that have lived sustainably in place for long periods of time are by no means homogenous. In some cases, such as with the Koyukon culture, animals and other parts of nature, are not conceived of as family. They are still, however, treated with great respect since every entity in nature is seen to possess a spirit and intelligence and to watch over the behavior of humans who, if they behave disrespectfully toward nature, will reap retribution (Nelson, 1983, chaps. 2 and 12). These beliefs recognize human dependence upon the web of life and prescribe respectful behaviors toward nature. Respect often takes the form of using nature in ways that maintain the long-term productivity of natural systems and minimize waste, as in the case of using all parts of harvested animals and plants.

An ethic of respect often underpins indigenous cultural codes of conduct with regard to hunting methods and ceremonies as well (see Nelson, 1983). According to LaDuke, "When you take a buffalo, there is a Lakota ceremony, the Buffalo Kill ceremony. In that ceremony, the individual offers prayers and talks to the spirit of the animal. Then, and only then, will the buffalo surrender itself. That is when you can kill the buffalo" (1999, p. 148). Such ceremonies speak to the intelligence and spirit of other living creatures who, if not explicitly kin to humans, are nevertheless an integral part of the fabric of indigenous socio-ecology—animals are an integral part of the systems upon which humans must depend. The James Bay Cree ontology recognizes a hierarchy of beings that differs from some indigenous traditions but, nevertheless, maintains a holistic and reciprocating view of human/nature systems:

> This is a world in which there is a unified, but not rigid, hierarchy of beings descending from God to spirit beings to humans to animals. The metaphor and value of social reciprocity, and the moral responsibility that it highlights in social relations, permeate this social universe. When asking why an animal went in a trap, or allowed itself to be caught, the Cree answer with similar kinds of reasons to those they would offer for why a human gives food away to another person. That is, because it appreciates the need of the other. The implication is that it is a responsible thing to do as a moral social being. The separation between humans and animals is thus one of degree, and continuities of humans and animals, culture and nature are therefore assumed in the Cree symbolic universe. (Feit, 2001, p. 421)

Other cultural and spiritual traditions exist that present opportunities for developing and living place-centered ontologies. One of these traditions is Buddhism. Similar to indigenous, place-centered ontologies, Buddhist traditions emphasize the interconnectedness of all beings and respectful relationship to nature (Bond, 2004, p. 115; ISEC, 1993; Norberg-Hodge, 1991/1992). Belief in reincarnation and karma also encourage respectful treatment of others and consideration of long-term impacts of decisions and actions taken today. The continuity and unity of life is represented in the idea that people can be reincarnated as animals and vice versa. Buddhist spirituality, therefore, parallels important aspects of holistic indigenous ontologies. Unlike indigenous ontologies, Buddhist belief systems can be transferred from one place to another, yet Buddhist spirituality is highly compatible with sustainability due to its holistic ontology that encourages place-centered, sustainable living and respectful participation in the wholeness of life. As a spiritual tradition practiced worldwide, Buddhism has proven to be highly adaptable to diverse places where it serves as an effective spiritual and cultural form for socio-ecological living sustainability.

The case of Ladakh offers an example of Buddhist sustainability in action. Prior to the onslaught of modernity and global economic integration in Ladakh that began in the 1970s, Buddhist-inspired systems of subsistence and community decision making had served as foundations for sustainable living in a harsh Himalayan environment for centuries. Helena Norberg-Hodge (1991/1992) has lived for extended periods of time in the formerly independent kingdom of Ladakh, now part of India. Her writing about Buddhist society in Ladakh offers a counterpoint to the pervasive modern notion that place-centered, sustainable subsistence lifeways are mired in drudgery from which modern living promises to rescue all. Instead of liberation resulting from the opening of Ladakh to the global economy and Western culture, Norberg-Hodge has witnessed the breakdown of the holistic ontology and cultural practice that had informed traditional community life (1991/1992; see also ISEC, 1993). This breakdown has resulted in ecological degradation of the local environment and the loss of human-to-human and human-to-nature reciprocity that had been the foundation for Buddhist-inspired sustainable living. The breakdown of holistic ontology, the fragmentation of the community along generational and gender lines, and the concomitant abuse of the natural environment and other people are well documented by Norberg-Hodge as they play out in education, local markets, housing, agriculture, local governance, and other areas of Ladakhi life (1991/1992; ISEC, 1993). The disruption and damage done by the arrival of globalization in Ladakh and the resulting loss of local traditional culture exemplify the reasoning behind the warning articulated by Armstrong (1995) above regarding dis-placement: that when people live beyond the natural limits of place, they drive other people and nature to ruin. Traditional Ladakhi culture and spirituality, like indigenous ontologies discussed above, offers an example of a sustainable cultural and economic system, and like indigenous cultures and economies, Ladakhi lifeways have been colonized and fragmented by capitalist economy and culture.

Muted themes in Western culture also point to the potential for Western-European-based societies to engage in place-centered, sustainable living deriving from their own cultural traditions. These themes are muted because they have been almost entirely overwhelmed by capitalist consumer culture. The history of Western colonization worldwide also makes considering Western culture as a source for sustainable living problematic. After all, Westerners colonized and obliterated countless indigenous cultures that had developed sustainable, place-centered ontologies and lifeways. Still, even indigenous societies centered within the bounds of place often embody contradictions to the definition of socio-ecological living sustainability articulated in chapter one. War and raiding practiced against other tribes represent themes of unsustainability. Recognizing these contradictions emphasizes that sustainability is a process and not an end, and this recognition creates openings for widespread participation.

Political scientist Danniel Kemmis (1990) elucidates themes within mainstream U.S. cultural history that inhibit development of attachment to place as well as those themes that encourage attachment. He proposes that Americans should practice a politics and economics of place inspired by attachment to and care for place, including attachment to and care for other people who are part of place. Kemmis (1990) argues that the following cultural and political structures and phenomena have contributed to the dissolution of place as an organizing construct and context for American life:

- centralized, representative democracy (as opposed to more direct forms of community-based, participatory democracy through which people might work together face-to-face to solve their own problems),
- cultural and legal emphasis on individual rights (especially property rights) over community rights and welfare,
- complex government regulatory and procedural bureaucracies, and
- the highly abstract and complex modern economy operating at a global scale.

In his search for cultural, economic, and political foundations for healthy re-centering of human lifeways in place, Kemmis (1990) draws upon the agrarian community experience of small farmers (including that of his parents) and small farm communities in the not-so-distant past. He also draws upon Jeffersonian notions of agrarian republican governance as emblematic of highly participative, democratic self-determination within a context of place-based community development (chaps. 2–3).

According to Kemmis (1990), in agrarian communities, personal welfare depended on the general welfare of the community (p. 72). The fact that people had to depend upon one another in concrete ways, such as for assistance with building a new barn, meant that people's attachment to one another was more than psychological. If reciprocal interpersonal relationships were not developed and maintained, individual/community assets like barns might never be built, and individual families and the community would face tangible risks such as being unprepared for weathering winter storms with livestock and feed intact. In such communities, Kemmis claims that people learned to accept and rely on one another even if personal differences would have kept them apart if they had not actually *needed* one another (1990, chap. 6). This interdependence mirrors that observed by Norberg-Hodge in traditional Ladakhi society: "In the traditional economy, you knew that you had to depend on other people, and you took care of them" (1991/1992, p. 122).

Kemmis claims that lived experience with place as a teacher of practices of survival and fulfillment together with interdependence among community members in completing the tasks of life required in a particular place create *public values* that are *objective* in that they are based in commonly held knowledge and practices of successful inhabitation (1990, p. 75). Such values are not mere personal preferences but are rooted in community knowledge of place and in appropriate practices for living together well in place. Kemmis argues the importance of practices of inhabitation as concrete, specific, and tangible sources for the creation of public values: "It is precisely that element of concreteness which gives to practices their capacity to present values as something objective, and therefore as something public" (1990, p. 78). He argues that living within the bounds of place creates the necessary conditions for forming community relationships of respect and reciprocity in the form of public values:

> The shaping of their values was as much a communal response to their place as was the building of their barns.... The kinds of values which might form the basis for a genuinely public life, then, arise out of a context which is concrete in at least two

ways. It is concrete in the actual things or events.... [and] in the actual, specific places. (p. 79)

Public values can remain unwritten or be articulated as a declaration of shared values, a mission statement, or a community manifesto (see Ploeg, 2008, pp. 61, 190–191). For Kemmis, "to inhabit a place is to dwell there in a practiced way," and authentic public life requires the mutual inhabitation of place (1990, pp. 79–80).

Although Kemmis (1990) does not articulate a holistic ontology of place as fully developed as that of some indigenous cultures, he articulates appropriate processes for developing such an ontology within Western cultural traditions. Since he focuses on processes of place-centered cultural development rather than on relationships and practices relevant in only one or a limited number of places, his concept of inhabitation can be easily adapted to many places.

Kemmis' (1990) notion of inhabitation shares certain themes with other (re)inhabitation efforts based in the Western cultural tradition such as the Transition Movement discussed above (see Hopkins, 2008) and processes of *repeasantization* in Europe analyzed by Ploeg (2008, chap. 6). These themes include creative and sustainable adaptation to place that embodies the conscious development of public values, inclusive participation in decision making, and increased levels of community self-determination and resiliency.

And so, what specifically do holistic systems of belief and action have to tell us as we face the global sustainability crisis? They propose that nothing and no one exists to be exploited and that, when we use others in this way, we diminish ourselves by disconnecting from relationships vital to life (Salmon, 2000, p. 1331). As long as we continue to see nature as entirely other—as mere material available to satisfy our needs and desires—it remains unlikely that we will live sustainably in place—a way of living that is re-emerging as a survival imperative for humans and other life forms. This is not to say that the mere idea of living sustainably is enough to change the world. Rather, I argue that sustainable living is being thrust upon us by the damage globalized civilization has done to environments everywhere and by the economic and social crisis of capitalism as that system erodes its economic base. Within this context of concrete, material drivers for social change embodied in the sustainability crisis, recognizing the importance of a holistic ontology of place can blossom into new and revived cultural systems of sustainable, place-centered living. This recognition can contribute to Gramscian praxis (Gramsci, 1971/1999, pp. 333–334).

Sustainability education can play a role in building this awareness, in part through contextualizing content by engaging students in projects that offer direct experience in creating sustainable, place-centered community. Through a combination of responding to concrete ecosystemic constraints and opportunities and (re)conceptualizing our relationships to place, I argue that we can (re)develop place-centered ontologies that promote meaningful and sustainable participation in the community of all beings.

We will now explore resiliency as another constitutive theme for place-centered sustainable living.

Resiliency

According to Jacobs (2000), systems that are dynamically stable make effective use, through continual self correction, of information fed back to the system. Such systems resist collapse and disintegration (pp. 84–85) as do ecosystems that demonstrate resilience stability as defined by Odum and Barrett: "Resilience stability indicates the ability to recover when the system has been disrupted by a perturbation" (2005, p. 70). Jacobs also emphasizes how diverse uses and reuses of matter and energy in an economy are essential to its diversification and resultant resiliency, just as the diverse use and reuse of matter and energy flows is critical to development of resilient ecosystems (2000, pp. 43–63). Economic relationships of neoliberal globalization offer few prospects for communities and nations to develop economic resiliency. In the context of a capitalist economy, developing community resiliency would entail increasing economic diversity so that a given community is not entirely dependent upon profits generated by one factory, one mine, or one monocrop—a precarious situation indeed (Jacobs, 1969, 2000; Shuman, 2000). A resilient local economy would be composed of diverse businesses that produce goods and offer services for local consumption so that money spent within the community would produce extensive multiplier effects (Jacobs, 1969, chaps. 4–5; Shuman, 2000, chap. 2). Goods produced in a resilient local economy would rely mostly upon locally available resources and would use these resources in a sustainable manner (Douthwaite, 2004, pp. 119–121).

Such an economy would be well positioned to support a local community beyond the collapse of the globalized growth economy. In the long term, a resilient local economy would be based in a holistic ontology of place developed through praxis over many generations. Such an economy would make sustainable use of diverse local resources as well as wild foods and medicinal plants. A sustainable community could engage in food production activities that were themselves resilient and diverse by engaging in permaculture (see Hopkins, 2008, pp. 136–141) and other sustainable processes of agriculture, horticulture, and animal husbandry.

Sustainable communities globally would exhibit a high degree of diversity from community to community in utilizing place-appropriate technologies and production processes and in growing/hunting/gathering place-adapted food. As a result of creating diverse, place-centered communities globally, the species resilience of humans would rise. Similar to agricultural monocultures, homogenous human cultures are highly susceptible to disturbances both because everyone everywhere relies on many of the same inputs and products to fulfill basic necessities and because the provisioning of these inputs and products has become a unified and synchronized process globally (see Homer-Dixon, 2006, chap. 9). So, shortages of key goods grown, extracted, or produced in one region can quickly cascade into global shortages. If the corn crop in the U.S. fails, it is a problem for the world, not only the U.S. If global oil passes peak production so that supplies are constricted, it is a problem for everyone everywhere who relies on globalized transportation of goods and on petroleum products.

As long as people across the globe rely heavily upon more or less uniform foods and other necessities rather than upon a wide diversity of foods, other resources, and energy capture/production technologies (see Scheer, 2002, 2007) to fulfill their basic needs, the resiliency of the globalized provisioning system will remain very low indeed (see Douthwaite, 2004, and Homer-Dixon, 2006). What is needed to (re)vitalize systems of human/nature resiliency globally is (re)adaptation to place and (re)localization in providing for basic human needs. We must (re)adapt to place so that crises of basic needs, bound to occur from time to time, remain localized crises rather than cascading, global ones.

The diversity necessary for sustainable human inhabitation is reflected globally among indigenous societies, which is why Wes Jackson (1996) asserts that we must strive to become native to place in the modern era. Biologist and Native American activist Dennis Martinez observes that cultural diversity and biodiversity are inextricably interrelated (1997, p. 109). Anishinaabeg tribal member and activist Winona LaDuke (1999) concurs, stating that

> over 2,000 nations of Indigenous peoples have gone extinct in the western hemisphere, and one nation disappears from the Amazon rainforest every year.... There is a direct relationship between the loss of cultural diversity and the loss of biodiversity. Wherever Indigenous peoples still remain, there is a corresponding enclave of biodiversity. (p. 1)

Becoming native to place might be described as collectively engaging in long-term processes of learning to respond to and participate successfully with place in ways that maximize community health and resiliency as well as individual and community fulfillment. This engagement could be described as a process of building public values as product of living well in place and using these values to construct a set of ethical guidelines for governing life processes. Over the long term, these public values and codes of conduct would likely result in a collective ontology of place as a human/nature construct.

Processes of (re)inhabitation would produce more than human ideas, values, and codes of conduct. Concrete representations of diverse and holistic ontologies of place would include development and use of locally adapted seeds and animals owned/managed collectively as part of the cultural commons (see Shiva, 2008, p. 119). These tangible representations of successful human inhabitation have existed historically and still exist today, though many locally adapted plant and animal varieties have been lost in the globalization-driven push for efficient, standardized production. Vandana Shiva highlights the degree of standardization of current foodstuffs globally as compared to the historical diversity: "Humanity has eaten over 80,000 edible plants over the course of its evolution. More than 3,000 have been used consistently. However, we now rely on just eight crops to provide 75 percent of the world's food" (2008, p. 121).

In the Pueblo tradition, the holistic circle of life is both an idea and a tangible reality experienced as an integral part of daily life. It represents one of many diverse, place-centered lifeways capable of contributing in important ways to the resilience of human societies everywhere by resisting the standardization and synchronization of the globalized world-system and fostering localized resiliency. Pueblo Indian educator Gregory Cajete (2001) describes traditional Pueblo farming, not only as a process of diversification through adaptation to place, but also as the basis of a holistic ontology and spirituality of place in which humans and nature are conjoined: "The varied strains of developed corn were a direct result of our collective ecological understanding of the places in which we have lived through the generations. Corn became a sacrament and symbol of our life and relationship with the land" (p. 632).

The decentralization and diversification of energy production and capture would be another tangible representation of (re)inhabitation. As Scheer (2002, 2007) argues, because energy is central to economies and life, as long as we remain chained to global fossil and nuclear energy industries, we will be subject to overriding enforced dependency. By adapting our energy capture and production to place through decentralized and locally adapted use of renewable energy technologies and resources and through reducing the need to transport materials and products globally, we can increase both local economic resiliency and political self-determination, both of which are essential to successful and diverse (re)inhabitation.

(Re)inhabitation will not be easy. In the case of energy alone, we face a pervasive and deeply entrenched energy system that enforces dependency rather than creating self-sufficiency and resiliency: "the capitalist world-economy depends on...nonrenewable resources for nearly 90 percent of its total primary energy supply" (Li, 2008, p. 148). People engaged in processes of (re)inhabitation will also make mistakes and will continue to engage in forms of oppression of people and/or nature. The process of sustainable living represented by becoming native to place is historical in that we must start from the very difficult position of the current socio-ecological crisis in working toward sustainability. Furthermore, the process of becoming native to place is one that can never end because new circumstances will continually arise. Nevertheless, given the depth of the global sustainability crisis and the current and impending human and environmental catastrophes being driven by widespread human dis-placement, we must engage in action toward increasing the resiliency of our communities by wisely (re)inhabiting our places.

We now turn our attention to developing and enacting sustainable forms of leadership as another important theme in (re)inhabitation.

Authentic, Grassroots Servant Leadership

Processes and products of collective decision making are central to sustainable (re)inhabitation because these determine social structures, economic distribution, human uses of nature, and more. If hierarchical organization in societies promotes exploitation of both people and nature, then sustainable societies would engage in forms of decentralized, inclusive leadership in service to sustainable living. The concept and practice of servant leadership, articulated by Robert K. Greenleaf (1970/1991) serves as a model for sustainability-oriented, grassroots leadership that is adaptable to various cultural and spiritual traditions.

According to Greenleaf (1970/1991), a servant leader works to build the leadership capacities of others rather than to maintain his/her own position of power. The servant as leader sows the seeds of long-term change because the servant leader shares both the responsibilities and the fruits of the changes s/he leads with those who follow or who lead from their own positions. Greenleaf (1970/1991) contrasts the servant leader with the dominator leader who is "leader first, perhaps because of the need to assuage an unusual power drive or to acquire material possessions" (p. 7):

> The difference manifests itself in the care taken by the servant-first to make sure that other people's highest priority needs are being served. The best test, and difficult to administer, is: do those served grow as persons; do they, *while being served,* become healthier, wiser, freer, more autonomous, more likely themselves to become servants? *And,* what is the effect on the least privileged in society; will he benefit, or, at least, will he not be further deprived? (p. 7, emphasis in original)

Greenleaf's emphasis on increasing the autonomy, and therefore the leadership potential of those served corresponds well with fostering place-centered, sustainability-oriented community autonomy, and increased community self-sufficiency can serve as an effective vehicle for confronting and eliminating neoliberal globalization's ever-intensifying concentration of wealth and power. Servant leadership can therefore prove useful in combating enforced dependency and increasing individual and community resiliency.

In his discussion of authentic leadership, Terry (1993) offers theories that extend and sharpen Greenleaf's (1970/1991) concept of the servant leader. Like Greenleaf (1970/1991), Terry (1993) proposes that leaders must be concerned, not only with decisions made, but also explicitly with *how* they are made. According to Terry (1993), decision making must be transparent and open and must "acknowledge the significant features of the human condition" (p. 108). An authentic servant leadership of place would derive its authenticity from a holistic ontology of place rooted in public values that inform and are informed by processes and products of diverse, place-adapted (re)inhabitation.

Political leadership in such a context would entail much more than creating a political framework for voting and interest group competition. The leadership of (re)inhabitation would entail direct participation by community members in making decisions that affect their lives rather than providing only indirect avenues of influence aimed at appealing to elected or appointed decision makers (Kemmis, 1990). According to Kemmis, "This taking of responsibility is the precise opposite of the move toward the 'unencumbered self.' It is, quite simply, the development of citizenship" (1990, p. 113). It entails a "direct willing of the common good" (Kemmis, 1990, p. 85)—a process based in the development of public values and a holistic, intertemporal ontology and ethic of place. Such leadership and decision making represents a form of liberty realized through relationships to other people and to place rather than through radical individualism. Participatory servant leadership for (re)inhabitation is bottom-up leadership that is both figuratively and literally grounded.

As noted by Hopkins (2008) with regard to the Transition Movement, broad participation in community problem solving is required for successful (re)inhabitation. According to Hopkins, change needs to be fully embedded in community economic and political processes.[3] Inclusive leadership and governance for (re)inhabitation must encourage all community members to play roles in guiding community life. By contrast, placeless, globalized capitalism reduces opportunities for inclusive leadership by concentrating power, wealth, and decision making in fewer and fewer hands and in locations distant from affected communities. By increasing the capital to labor ratio over time, capitalism also eliminates the roles of many people in the productive forces of society (Li, 2008, pp. 142–143), thereby reducing opportunities for widespread leadership development. Shiva calls this a process of creating "disposable people" (2008, p. 2).

By contrast, traditional societies that embody a holistic ontology of place typically provide meaningful roles for all members of a given community, though leadership roles may be socially prescribed so that community members are not entirely free to choose the roles they play. Take for example the leadership model of the Nhunggabarra people of Australia:

> Nhunggabarra principles for organising society were *context-specific leadership and knowledge-based organizing.* Everyone in society had a leadership role in a specific area of knowledge, and the leader role shifted depending on the context and who, within that context, was the most knowledgeable. (Sveiby & Skuthorpe, 2006, p. xvii; emphasis in original)

Some traditional, place-based societies have engaged in unsustainable gender-based, caste-based, and other forms of oppression. Therefore, looking to just any traditional society for examples of community life on which to base sustainable (re)inhabitation would be problematic. Any community working toward socio-ecological living sustainability must carefully choose its examples and develop its own processes of leadership for (re)inhabitation

---

[3] Hopkins learned this lesson after spearheading the development of the Kinsale Energy Descent Action Plan through the Kinsale Further Education College in Ireland (Hopkins, 2008, p. 128; see also the action plan itself: Hopkins, 2005). Having visited the College with a colleague and a group of students in summer of 2008, where we discussed the plan and progress toward the goals articulated in it in 2005, I believe Hopkins' assessment is on target. The Action Plan (Hopkins, 2005) was developed by a small group, and even though it was endorsed by the town council, it was not fully integrated into the local economy and processes of town governance. Even though it is an insightful document that articulates many important ideas and goals for (re)inhabitation, once Hopkins left Kinsale, as did many of the students who helped to write the plan, community momentum for realizing its goals faltered, and progress has been slow.

that improves the health and contributes to the integrity of individual community members and subgroups. Still, as in the case of the Nhunggabarra, differential authority in leadership can be sustainable within a framework of servant leadership where building the leadership capabilities of others and engaging in inclusive, transparent, and participatory leadership are primary concerns.

The Sarvodaya Movement in Sri Lanka offers an example of community-based servant leadership that demonstrates the potential to serve as a platform for (re)inhabitation. Sarvodaya is a grassroots community-building and rural-development movement which began in 1958, eleven years after independence from British rule. Though it has no explicit connection to the work of Greenleaf (1970/1991), it offers an example of grassroots, decentralized servant leadership that has worked to promote community self-development and autonomy for decades. The movement promotes the development of collective leadership among those who might be considered the least likely to assume leadership roles: the poor, lower castes, women, and youth (Bond, 2004, chap. 1). Thereby, the movement embodies key aspects of servant leadership: the building of leadership capacity in others and benefitting the least advantaged (Greenleaf, 1970/1991, p. 7). Sarvodaya embodies the Gandhian ideals of truth, nonviolence, and "selfless service for humanity as the highest form of religious practice" (Bond, 2004, pp. 9–10). The movement adapts the Gandhian tradition to the predominantly Buddhist context of Sri Lanka by translating Gandhi's call for the uplift of all to "the awakening of all" while, at the same time, remaining a nonsectarian movement emphasizing "an underlying spiritual unity of all religions" (Bond, 2004, pp. 10, 13–14). The specific form of awakening embodied in the movement was inspired by Angārika Dharmapāla, who worked for restoration of Buddhism in Sri Lanka in the late nineteenth and early twentieth centuries and who emphasized the importance of moral worldly activity as spiritual practice (Bond, 2004, pp. 10–11). Sarvodaya seeks "the awakening of the individual with and through the awakening of society" (Bond, 2004, p. 14).

The Shramadana Movement, the earliest incarnation of Sarvodaya and an important part of the movement to this day, is a student work camp movement begun in 1958 by high school science teacher A. T. Ariyaratne and 40 of his students. These students and twelve teachers from their school conducted the first shramadana or "gift of labor" work camp (Bond, 2004, p. 7). Bond describes the motivation, purposes, and outcomes of this effort:

> Ariyaratne…wanted his students to gain some understanding of the condition in rural villages. When they went to Kanatoluwa, the students lived and worked with

the residents, who were considered outcastes by neighboring villages. The students helped the residents dig wells, build latrines, plant gardens, repair the school, and build a place for "religious worship." Building a place for worship was significant because even the clergy had previously shunned the people of Kanatoluwa.

The first shramadana camp was a great success. Students experienced a different aspect of their culture, and the project broke down barriers between the upper and lower castes. Doing manual labor alongside rural villagers changed the students' outlook, and associating, as equals, with "high-class gentlemen" from the city for the first time in their lives changed the villagers' perspectives. The neighboring villages were also affected: If the "upper-class people" from the city could work and eat with these so-called outcastes, then the outcastes could no longer be treated as inferior. (2004, p. 7)

In the earliest stages of a village's engagement in the Sarvodaya community development process, a village organizes various social support and networking groups. These typically include a preschool children's group and a school-age children's group as well as groups for youth, mothers, and farmers. These social support groups represent opportunities for building grassroots leadership capacity within the community, and the emphasis on youth leadership and involvement gives the movement long-term momentum (Bond, 2004, pp. 23–24). Through these groups, community development works from the bottom up, engaging people and resources that are available in every community.

Sarvodaya follows a practice of "project support project" in which villages that have successfully initiated community development and self-sufficiency programs support up to nine additional villages initiating such efforts (Bond, 2004, p. 58). This process, combined with the approach of the movement to generating leadership and development from within villages, helps to maintain the authentic grassroots character of Sarvodaya leadership that is an embodiment of the ideals of servant leadership.

Sarvodaya community development has achieved much success in its over fifty years of service, despite conflicts with donor foundations seeking to steer the movement toward integration with the global market rather than toward localized development for autonomy and self-sufficiency, and despite a smear campaign by the Sri Lankan government meant to discredit the movement (Bond, 2004, chaps. 3–4). In 2002, there were over 3,000 villages with Sarvodaya-supported economic development programs (Bond, 2004, p. 108), and the movement is spreading to other nations and cultures through "Sarvodaya USA, Savodaya Twente in the Netherlands, Savodaya Japan, and Savodaya UK" (Bond, 2004, p. 116).

The Sarvodaya Movement is not alone in its project support project approach to social change. The Transition Movement promotes similar processes, both for internal, localized community development and for inter-community support (Hopkins, 2008, pp. 142–143). It is important to recognize the value of communities networking with and learning from one another, useful processes that demonstrate localized resilient communities are not synonymous with isolated communities.

This last point also raises the question of appropriate scale in organizing for change. As Hopkins (2008) notes, issues of optimal community size and scope for grassroots organizing and action are difficult to define uniformly. I also prefer to leave issues of scale for effective change through servant leadership as an open question for individuals and communities. I concur with Hopkins who states:

> I have come to think that the ideal scale for a Transition Initiative is one over which you feel you can have an influence. A town of 5,000 people, for example, is one that you can relate to; it is one with which you can become familiar.... This concept of working at a neighbourhood scale is not a new one.... Ultimately, you will get a sense of what is the optimal scale for your initiative. (2008, pp. 143–144)

Because servant leadership is participatory, transparent, and broadly beneficial, social changes brought about through its processes are more likely to endure as compared to the philosophies and practices of single, dominator leaders. The Sarvodaya Movement is one case in point. Social changes undertaken though servant leadership are therefore more likely to contribute to the long-term development of public values that increase community sustainability.

In chapter seven, we will explore local food praxis as a particularly important vehicle for actualizing sustainable, participatory (re)inhabitation of place. I will develop a theory of sustainable food systems as embodiments of counterhegemonic social change that explicitly resist and create alternatives to enforced dependency while fostering place-centered sustainability and community resiliency.

Conclusions and the Role of Education in Place-Centered Sustainability

I have argued in this chapter that sustainable (re)inhabitation is built upon the following themes:

- a holistic ontology of place through which individuals and communities recognize and respect their relationships with and dependence upon nature as the source of all life,

- inclusive governance that is place-centered and that derives from public values (Kemmis, 1990, p. 75) developed through collective (re)inhabitation of place,
- (re)building community resiliency as a means to counteract systems of enforced dependency within globalized late capitalism, and
- the widespread practice of servant leadership.

Practices of (re)inhabitation can be perhaps as diverse as are the places in which people live, perhaps even more so.

Given the urgent need for (re)inhabitation as a fitting response to the sustainability crisis, and given the centrality of education in formative personal experience, it is important to ask what role(s) education might play in moving societies toward sustainable living. According to Cajete (2001), typical modes of education in the U.S. perpetuate dis-placement among students and the denial of human/nature holism: "Teaching about the reality of natural America as 'place' is not the intent of modern education, which is designed to condition students to view nature and places as objects that can be manipulated through science and technology for human economic ends" (p. 621).

In chapter eight, I will argue for re-placing both the content and the processes of modern education. Education should facilitate the conscious (re)embedding of humans within diverse, localized ecologies. Sustainable education should be education for healthy ecosystemic dependence rather than for domination and control of nature. Students should learn how to participate in (re)creating and maintaining healthy, resilient socio-ecological systems that provide a context for both people and nature to thrive.

Sustainability-oriented education today should be discursive, reflexive, and critical (see Freire, 1970/2000; McLaren, 2005; McLaren & Farahmandpur, 2005; McLaren & Jaramillo, 2007; McLaren & Kumar, 2009). Education appropriate for addressing the sustainability crisis would engage in the first negation of the capitalist order through critiquing the unsustainability and injustice of the capitalist world-system. The ultimate aim of sustainable forms of education would be absolute negation (see McLaren & Kumar, 2009) of the neoliberal order through engaging students and communities in transformative, place-centered praxis that embodies the themes of sustainable (re)inhabitation outlined here. Through engaging people in (re)inhabitation, sustainability education would have a decolonizing (see Grande, 2007) effect on communities. It would provide one context for educating the whole person within her/his ecosystem and within community. As Jackson says regarding higher education,

Our task is to build cultural fortresses to protect our emerging nativeness.... One of the most effective ways for this to come about would be for our universities to assume the awesome responsibility to both validate and educate those who want to be homecomers—not necessarily to go home but to go someplace and dig in and begin the long search and experiment to become native. (1996, p. 97)

In the long term, however, education as we have known it in the modern era might lose its relevance. Norberg-Hodge has the following to say about "education" in traditional Ladakh:

With the exception of religious training in the monasteries, the traditional culture had no separate process called "education." Education was the product of an intimate relationship with the community and its environment.... Education was location-specific and nurtured an intimate relationship with the living world. (1991/1992, pp. 120–121)

Industrialized education, by contrast, is abstract education. It has increasingly focused on training people to possess generic skills that can facilitate their serving as cogs in the vast industrial machine. It has increasingly *not* taught people to be self-sufficient but, instead, to become dependent upon the monetized economy. If education is to become truly place-centered and sustainable, everyone would become an educator (and a student at the same time), and the perceived need for specialized educators and for abstract forms of education would diminish.

It is my hope that educators today will direct their attention to serving societies in transition toward sustainability and that the time and resources afforded to education today will be used to help people learn what they will need to know for healthy, place-centered living. I hope that educators will, to the great benefit of people everywhere, gradually work themselves out of their jobs. We who are educators should realize that the education system is in many ways part and parcel of the unsustainable paradigm of globalized industrial capitalism. Knowing this, we should do what we can to make a contribution to the rebirth of education as an integral aspect of (re)inhabitation that permeates and sustains community life. We will further explore these possibilities in chapter eight.

Chapter 6:

Enforced Dependency in the Food System

Farmers and others in rural America aren't like those of us who make our way in the cities and suburbs. For these rural people, the loss of their land and their way of life creates incomprehensible despair, more severe than the death of a loved one. It's as if all the family members who had worked that soil before them and all the children and grandchildren who should one day inherit that opportunity had suddenly been murdered by an unseen assailant. You don't just lose a farm. You lose your identity, your history, and, in many ways, your life.

—Joel Dyer

As we have seen, the growth economy is bumping up against limits of all kinds: limits to available resources, to fresh water, to oil and other fossil fuels, to the ability of the planet to absorb pollution, and to the ability of global consumers to maintain purchasing power in an atmosphere of simultaneous overproduction and concentration of wealth. The contradictions of the system have intensified to the point that it can no longer contain them. A major source of strain is the need to feed a global population that has doubled since the 1960s and now numbers of seven billion. This challenge is, not only a question of the productive capacity of the agricultural sector, but also a question of the structure and character of the food system, which is the focus of this chapter.

Through food production and consumption, we write upon the world and express the story of our relationships to each other and to nature. Food provides a unique window for viewing and comprehending the current crisis in the world-system. In the face of seemingly overwhelming challenges to our quality of life, and perhaps our very survival, food also offers unique creative opportunities to resist enforced dependency and build community and ecosystem resiliency.

Although our choices may admittedly be limited, we each contribute to shaping the character and processes of the food system. Every food-related thing we do—all the way from purchasing fast food to growing our own food, from supermarket shopping to participating in community supported agriculture—is a political act, and this chapter focuses on the political implications of food. In this and the following chapter, I will examine the political economy of food and the opportunities to reshape that political economy in ways that build community and benefit nature. My central questions are these: What makes a food system sustainable? And what are the socio-ecological implications of creating such a system? As an educator, I am also concerned with how student engagement with sustainable food systems can enhance sustainability education and praxis.

In this chapter, I analyze the food system from community, national, and international perspectives. I elucidate the social and ecological problems caused by industrial food production. I then situate industrial agriculture within the broader context of globalized political economy, showing how the industrial food system is emblematic of broader trends in the crisis of capitalism: centralization of production, control, and profits in the hands of a few achieved at the expense of enforcing dependencies worldwide. In chapter seven, I will highlight the potential for the sustainable food movement to play a central role in the local/global struggle to address both the sustainability crisis and the crisis of capitalism. I will also draw useful connections between sustainable food and sustainability education.

Industrial Agriculture as Enforced Dependency

In pushing people off of the land, preventing them from growing their own food and making them dependent on wages in order to feed themselves and their families, the Green Revolution represents a particularly important aspect of enforced dependency. The Green Revolution began after WWII and continues today. It involves use of petroleum-based pesticides and herbicides, natural gas–based nitrogen fertilizers, hybrid seeds, and petroleum-fueled farm equipment to plant, care for, and harvest vast monocultures. In the Green Revolution era, farm size has increased dramatically as agribusiness entities have taken advantage of economies of scale made possible through industrial methods. Subsistence growers and small farmers who produced diverse crops and raised small numbers of animals were unable to compete with low-cost, often imported, industrial agricultural products. At the same time, a complex of forces in national and global political economy have driven large numbers of rural people from the land, many of them into the ever-expanding rings of squatter slums surrounding today's megacities. These forces include:

- national and international pressures to modernize,
- the need for nations to earn foreign exchange through growing export crops,
- costs of employing Green Revolution methods, prohibitive to all but very large agricultural interests,
- the ability of the rapidly growing international agribusiness sector to purchase land formerly occupied and worked by subsistence farmers, and
- post-war national and international policy. (Araghi, 1995; Homer-Dixon, 2006, chap. 1 and 3; Norberg-Hodge, et al., 2002, chap. 1)

The ability of the increasingly concentrated agribusiness sector to finance and purchase industrial agricultural inputs has been an important technological factor in displacement of the rural poor, but as we shall see in this chapter, concentrated, export-oriented agriculture is also a product of politics.

Agribusiness has profited immensely from industrial agriculture. Meanwhile, rural subsistence producers have been forced from self-sufficiency to enforced dependency. Global capital has benefitted as rural migrants have poured into cities to become exploitable workers and new customers. Global South nations have faced new challenges of engineering economic growth to provide employment for vastly expanding urban poor populations while also funding desperately needed social services and servicing external debt.

The social and ecological problems of globalized, industrial agriculture derive in large measure from the assumed separation of humans from the environment. I argued in chapter five that living sustainably requires a holistic ontology of place that (re)situates people within local ecologies and serves as a platform for sustainability-oriented praxis. I also argued that many indigenous cultures as well as certain cultural themes in Western thought offer examples and sources for (re)developing sustainability-oriented, holistic ontologies of place. Industrial agriculture negates such an ontology, and it destroys rather than supports diverse and reciprocating cultures. It is a one-size-fits-all, technologically and energy-intensive system that overruns nature as opposed to working with it.

If we hope to reduce enforced dependencies in the food system, we must understand how and why industrial agriculture functions as an agent of enforced dependency. Developing this understanding is my goal in this chapter. We begin our exploration of industrial agriculture as a dependency-enforcing system by exploring Marx's notion of metabolic rift. Marx's theory derives from his recognition of perhaps the central contradiction of modern industrial society: the illusory separation between humans and nature. We will explore how metabolic rift permeates Green Revolution agriculture and fosters destruction that far surpasses that caused by agriculture in Marx's time. We will then explore how and why both the ecological and social crises of agriculture enforce unsustainable dependencies. Finally, we will situate industrial agriculture within a world-historical analysis of global depeasantization and the dependencies it enforces.

Metabolic Rift and the Ecological Crisis of Agriculture

Marx's theory of metabolic rift is an important point of departure for analyzing how industrial agriculture embodies the rift between humans and nature in the modern Westernized worldview. According to Marx, industrial agriculture is robbery of both the soil and workers. The theft results in new dependencies on nonlocal and nonrenewable or exhaustible agricultural inputs used to maintain fertility in damaged soils. For Marx, industrial agriculture also robs workers everywhere of their self-sufficiency by alienating them from the land. Marx observes that modern agriculture embodies these contradictions because it fails to recognize and respect a metabolism in which nature and humans are mutually and reciprocally engaged. It also fails to recognize that the ability of people to create wealth depends upon the wealth-creating potential of nature (see Foster, 2009, pp. 175–187). Marx states:

> Capitalist production collects the population together in great centres, and causes the urban population to achieve an ever-growing preponderance. This has two results. On the one hand it concentrates the historical motive force of society; on the other hand, it disturbs the metabolic interaction between man and the earth, i.e., it prevents the return to the soil of its constituent elements consumed by man in the form of food and clothing; hence it hinders the operation of the eternal natural condition for the lasting fertility of the soil.... But by destroying the circumstances surrounding that metabolism ... it compels its systematic restoration as a regulative law of social production, and in a form adequate to the full development of the human race.... All progress in capitalist agriculture is a progress in the art, not only of robbing the worker, but of robbing the soil; all progress in increasing the fertility of the soil for a given time is a progress toward ruining the more long-lasting sources of that fertility.... Capitalist production, therefore, only develops the techniques and the degree of combination of the social process of production by simultaneously undermining the original sources of all wealth—the soil and the worker. (quoted in Foster, 2009, p. 176)

Marx noted that the importation of guano from Peru to fertilize English fields facilitated the continued industrialization of agriculture while creating a dependence on nonlocal inputs as a substitute for nutrient cycling within localized food systems (see Foster, 2009, p. 177). To Marx, emerging industrial agriculture was emblematic of a fundamental social contradiction: metabolic rift between humans and nature. In creating this rift, capitalists were sowing the seeds of destruction for modern societies.

For a time in the twentieth century, industrial agriculture proved spectacularly successful at increasing yields by making marginal lands viable for agricultural production and by intensifying the production processes themselves (Pretty, 2007, pp. 140–141). According to Pretty:

> In the last half century, agricultural area has increased from 4.5 to 5 billion hectares, increasing annual world food production from about 1.8 to 3.9 billion tonnes. Over the same period, the number of chickens has increased fourfold to 16.4 billion, pigs twofold to 950 million, cattle and buffaloes by about a half to 1.5 billion, and sheep and goats by forty percent to 1.8 billion. The key drivers of production intensity have grown rapidly too, with irrigation area doubling to 270 million hectares, tractors and other agricultural machinery more than doubling to 31 million in number, and fertilizer use up more than fourfold to 142 million tonnes per year. (2007, pp. 140–141)

As Marx's analysis predicted, however, industrial agriculture, reliant as it is on nonlocal inputs, has proven to be ecologically unsustainable as well as socially destructive. It has fostered and enforced new forms of dependency that have systematically depleted both people and place, making both increasingly vulnerable to exploitation. Contemporary authors have highlighted the many interwoven contradictions of metabolic rift that have cascaded throughout the food system and entire societies (see Bell, 2004, chaps. 1–3; Buttel, 1980; Norberg-Hodge, et al., 2002; Pfieffer, 2006; Pretty, 2002, 2005, 2007). Like Marx, these authors argue that industrial agriculture depletes the very systems upon which it depends: local ecologies and human communities and cultures.

Industrial agriculture erodes its own foundations in several interdependent and mutually reinforcing ways. In the nineteenth century, Marx remarked on the transfer of nutrients from one place to another as a result of intensive farming. He noted that farming was depleting the soil in some places while other places were becoming overwhelmed with problems of disposing of concentrated nutrients. Today, due to the concentration of animals in feed lots and other large scale production facilities, manure has become a problem waste (Norberg-Hodge, et al., 2002, p. 41; see also see Buttel, 1980, p. 46), and costs are high for transporting it to farms where it could serve as fertilizer for depleted soils. Absence of locally available manure for most farms results in increased use of inorganic fertilizers for which the major chemical feedstock is natural gas (Buttel, 1980, p. 46; Greene, 2004).

Industrial agriculture also wastes water and depletes limited sources of fresh water such as rivers and aquifers. In the U.S., industrial agriculture uses 85 percent of all freshwater used by people annually (Pfeiffer, 2006, p. 15), and much of this water comes from aquifers that are being drawn upon at well above their recharge rates. The Ogallala Aquifer that supplies water for much of America's Midwestern breadbasket is being drawn upon at 130 to 160 percent above its recharge rate (Pfeiffer, 2006, p. 17). Large amounts of grain grown in the U.S. Midwest to feed animals in feed lots further compounds problems of water overdraft in the industrial food system.

Industrial agriculture also decreases the water retentive capacity of the soil by systematically depleting it of organic humus. These reductions in the soil's ability to hold water could prove devastating in terms of food production in an era of climate change in which rainfall patterns are changing and glacial sources of water that serve millions are shrinking rapidly. Industrial agriculture also vastly exacerbates soil erosion, resulting in increased sediment loads in rivers and streams and massive losses of topsoil necessary for supporting plant life (Buttel, 1980, pp. 46–47). According to Pfeiffer, "It takes 500 years for nature to replace 1 inch of topsoil. Approximately 3,000 years are needed for natural reformation of topsoil to the depth needed for satisfactory crop production" (2006, p. 11).

Industrial agriculture also contributes to salinization of soils through waterlogging in the process of flood irrigation, and it is a major contributor to toxic contamination of water through pesticide, herbicide, and fertilizer runoff and seepage. Agricultural runoff transports inorganic pesticides and fertilizers into rivers, lakes, reservoirs, and oceans where they contaminate ecosystems virtually everywhere. Eutrophication of waterways carrying heavy nutrient loads from agricultural runoff leads to massive dead zones spreading offshore from major river drainages such as the Mississippi River (Norberg-Hodge, et al., 2002, p. 39; Pfeiffer, 2006, p. 18; Pretty, 2007, p. 141).

Industrial agriculture relies on monocultures and typically does not engage in crop rotations. These practices increase the severity of pest infestations and the consequent use of petroleum-derived pesticides (Buttel, 1980, p. 46). These chemicals threaten the dynamic stability of ecosystems by introducing toxins into food webs at all levels, thereby threatening the health of humans and other creatures (Buttel, 1980, pp. 46–47; Norberg-Hodge, et al., 2002, p. 39; Pretty, 2007, p. 141). Pesticides can contribute to aggressive behavior in exposed individuals, and since many are endocrine disrupters, they are likely contributors to the early onset of puberty observed in industrialized countries. These chemicals have not been tested for safety in combination with each other, which is how they occur in the environment, nor have they been tested over the long time spans necessary for identifying potential relationships to cancer (Norberg-Hodge, et al., 2002, pp. 54–55). Furthermore, pesticide usage appears to have reached a point of diminishing returns. According to Pfeiffer, "In the last two decades, the use of hydrocarbon-based pesticides in the US has increased thirty-three fold, yet each year we lose more crops to pests" (2006, p. 23).

In a world of depleting fossil fuels and climate change, food production must eventually rely entirely on natural processes for fixing nitrogen in the soil. Since pesticides damage the nitrogen-fixing symbiosis between bacteria in the soil and leguminous plants (Fox, Gulledge, Englehaupt, Burrow, & McLachlan, 2007, p. 10284), conventional agricultural methods are creating soil that is, not only depleted of plant nutrients and organic matter, but poisonous to the very soil processes upon which we will soon need to rely for nitrogen necessary for sufficient crop yields. To make matters worse, the damage we inflict on the soil now will negatively impact food production even after we cease applying chemical pesticides (see Fox, et al., 2007).

According to the International Commission on the Future of Food and Agriculture (ICFFA) (2008), industrial agriculture is also a major source of pollution-causing global warming—perhaps the most dramatic and threatening result of metabolic rift:

> According to the Stern Review Report on the Economics of Climate Change, agricultural activities directly contribute 14 percent of greenhouse gases. However, this is not the entire picture. Land use (largely referring to deforestation for globalized agriculture) accounts for 18 percent, and transport accounts for 14 percent.... Thus, a significant percentage of emissions from both the land use and transportation categories can also be attributed to industrial food and agriculture systems.... Some estimate that at least 25 percent of global emissions are related to non sustainable agriculture (p. 13).
>
> Industrial agriculture contributes directly to climate change through emissions of the major greenhouse gases—carbon dioxide..., methane..., and nitrous oxide.... Carbon dioxide emissions are largely caused by the loss of soil carbon to the atmosphere...and the energy intensive production of fertilizers.... According to the 2007 IPCC [Intergovernmental Panel on Climate Change] Report, nitrogen fertilizers account for 38 percent, the largest single source of emissions from agriculture. Chemically fertilized soils release high levels of nitrous oxide.... Ruminants [also] produce methane via enteric fermentation which increases when cattle are fed intensive feed. At 32 percent this is the second largest source of emissions. An additional 11 percent of agricultural emissions comes from intensive chemical cultivation of rice (p. 14).
>
> Conversion of natural ecosystems for industrial agriculture causes depletion of the soil carbon pool by 60–75% which is mostly emitted to the atmosphere as CO2. Some soils have lost as much as 20 to 80 tons of carbon per hectare.... (p. 15)

Industrial agriculture is also unsustainable in its reliance on nonrenew-able and rapidly depleting hydrocarbon energy and chemical inputs. Mecha-nization of agriculture, use of inorganic fertilizers and pesticides, and globalized transportation of foods translate to increased fossil energy inputs for industrial agriculture as compared to those for smaller, diverse farming operations focused on production for localized consumption (Norberg-Hodge, et al., 2002; *Power of Community*, 2006). Industrial agricultural practices enforce dependencies on fossil-fuel-based, toxic inputs that deplete energy resources and imperil the biosphere.

Dependencies on destructive practices are enforced through a vicious cycle of mutually- and self-reinforcing socio-ecological problems enumer-ated here: declining soil fertility, reductions in the soil's water retentive capacity, reduced symbiosis between nitrogen-fixing bacteria and legumi-nous plants, and declines in the organic matter content of soil. Freshwater depletion and increasing water contamination further exacerbate the effects of this vicious cycle. In the short term and in the eyes of industrial agricul-ture, the solution to the problem of diminishing returns is typically more of the same: intensification of the industrial processes of farming, the very processes which caused the problems in the first place. Dependency-enforcing aspects of industrial agriculture also include the social pressures of the agricultural treadmill to be discussed below. According to Pfeiffer (2006), the costs of industrial agriculture are beginning to outweigh its benefits:

> In the United States, the equivalent of 400 gallons of oil is expended annually to feed each US citizen (as of 1994).... Between 1945 and 1994 energy input to agri-culture increased fourfold while crop yields only increased threefold. Since then, energy input has continued to increase without a corresponding increase in crop yield. We have reached the point of marginal returns.... Modern agriculture must continue increasing its energy expenditures simply to maintain current crop yields. (pp. 7–9)

We will now explore the social aspects of metabolic rift in industrial ag-riculture. We will see why changing the trajectory of modern agriculture is both so difficult and so necessary for addressing enforced dependency and achieving sustainability.

Enforced Dependency and the Social Crisis of Agriculture

Industrial agriculture has its origins in the extractive colonial economies that fueled European industrialization (Miller, 1999; Norberg-Hodge, et al., 2002, p. 5). During the colonial era, subsistence producers in Europe were also dispossessed as agriculture was transformed into yet another activity from which surplus value could be extracted by the owners of capital and land.[1] But agriculture during the early colonial era was not yet mechanized, and increased production required increased numbers of laborers (Lyson, 2004, p. 14). Only with the advent of industrial-scale farm machinery and the increasingly widespread availability of chemical inputs in the 18th, 19th and 20th centuries could farming emulate the factory mode of production. Maximization of profits through economies of scale in industrial farming required removal of still more smallholders from the land (see Ploeg, 2008, p. 126).

They were removed by a variety of methods: though force, through policy, and through the continually increasing speed and force of the agricultural treadmill (a phenomenon we will examine below). In Europe, interest subsidies for capital investment by farmers, tax reforms, and state-financed stabilization of prices for agricultural commodities made capital investment in agriculture attractive as compared to labor costs. New technologies were also developed and supported by state extension services. These technologies fostered intensification of production without increases in labor. Use of new technologies in growing vast monocultures was therefore facilitated by government and resulted in production cost advantages (Ploeg, 2008, p. 127).

The often forcible removal worldwide of small, independent producers from the land in order to pave the way for the profits of moneyed interests marks the opening of a deep metabolic rift that would eventually result in the near complete alienation of modern people from the land and their concentration in urban centers. The transformation was radical. In the United States, "Less than 100 years ago, most rural households ... sustained themselves by farming" (Lyson, 2004, p. 8), and a far greater proportion of the nation's population resided in rural areas. Today, only 2.3% of the workforce in the U.S. is employed in farming (Bell, 2004, p. 9). Globally rural inhabitants as a percentage of the total population declined sharply from over 70% to approximately 50% between 1945 and 1990 (Araghi, 1995, p. 339). The removal of people from the land resulted in widespread enforced dependency as people became increasingly dependent on wage labor, and perhaps most importantly from a capitalist perspective, they became customers to whom basic necessities could now be *sold*.

---

[1] See Pretty on the Scottish clearances through which the wealthy gained access to large areas of land which they converted from subsistence production to raising sheep for wool, an activity that required only a small labor force and, at least initially, yielded large profits (Pretty, 2007, pp. 183–192).

The Agricultural Treadmill and Concentration in the Food Industry

For those who remained in agriculture, a central source of enforced dependencies is what has been called the squeeze on agriculture or the agricultural treadmill. The treadmill involves mutually reinforcing factors and processes that drive prices for farm products lower while simultaneously requiring increased investments by farmers in agricultural inputs and other farm technologies.

The story of the agricultural treadmill goes like this. Many small farmers produce the same product so that no single producer can affect the price. A new technology allows early adopters to capture additional profits from increased production until that technology is widely accepted, production rises across the board, and prices fall accordingly. Late adopters or those who fail to adopt the new methods and tools also see their profits drop when prices fall to new lows due to gains in overall production efficiency and increasingly globalized price competition. This process repeats itself as new technologies capable of increasing production and yielding economies of scale are continually invented and adopted by farmers (Douthwaite, 2004, pp. 114–115; Norberg-Hodge, et al., 2002, p. 7; Ploeg, 2008, pp. 129–130; Pretty, 2005, p. 108). The farms that survive grow very large in an effort to maintain profits in an atmosphere of declining marginal returns per unit of land.

Pretty (2002) describes the concentration of ownership resulting from the agricultural treadmill:

> The farm is swallowed up so another farm can compete better, until that, too, needs to get bigger again. During the past 50 years, 4 million farms have disappeared in the US. This is equivalent to 219 for every single one of those 18,000 days.... In France 9 million farms in 1880 became just 1.5 million by the 1990s. In Japan, 6 million farmers in 1950 became 4 million by 2000. (p. 107; see also Bell, 2004, p. 2; Norberg-Hodge, et al., 2002, p. 6–7)

Though their farms become larger, the farmers themselves do not become richer. In fact, a large proportion of them become increasingly indebted because of their continual need to invest in new, expensive technologies in order to capture the profits that go to early adopters. They find themselves on a treadmill that runs faster and faster as new technologies for agricultural production are introduced. Many eventually cannot keep up and succumb to foreclosures, making way for further increases in farm size among the farmers who remain.

All the while, an increasing proportion of agricultural profits is captured by the agribusiness inputs and food processing sectors which are themselves assuming larger and larger scale and monopoly or near-monopoly status (see Norberg-Hodge, et al., 2002). As a result of the global squeeze on agriculture, "economic insecurity and inequality among farmers provide conditions conducive for the penetration of an integrated corporate system of food processing and distribution" (Buttel, 1980, pp. 45–46). The corporations that dominate this system, in turn, can *require* that specific production methods are used by their farmer suppliers, thereby reinforcing the effects of the agricultural treadmill.

The structure of farm subsidies in the United States has also contributed to the agricultural treadmill because these subsidies tend to favor large farms over small ones. According to Norberg-Hodge, Merrifield, & Gorelick (2002), "In the United States alone, $27 billion in tax money was earmarked for farmers in 2000, most of which went to large industrial farms" (p. 71; see also Lyson, 2004, p. 34). The provision of these subsidies, not only disadvantages small farmers within the U.S., it artificially depresses prices for agricultural products globally so that small producers worldwide find it difficult or impossible to compete in the global marketplace. Global transportation infrastructures also represent a huge hidden subsidy to industrial agriculture that allows it to out-compete small producers (Norberg-Hodge, et al. 2002, p. 25). Pretty (2002) provides statistics for the U.S. that show the concentration of farm ownership, the increasing size of farms, and the disproportionate effects of farm subsidies on small versus large operations:

> In the US, the changing numbers of farmers and average farm size show an interesting pattern. Farm numbers increased steadily from 1.5 million to more than 6 million from 1860 to the 1920s…then fell rapidly since the 1950s to today's 2 million. Over the same period, average farm size remained remarkably stable for 100 years, around 60-80 hectares; but it climbed from the 1950s to today's average of 187 hectares.
>
> However, hidden in these averages are deeply worrying trends. Only 4 percent of all US farms are over 800 hectares in size, and 47 per cent are smaller than 40 hectares. Technically, 94 per cent of US farms are defined as small farms—but they receive only 41 per cent of all farm receipts. Thus, 120,000 farms out of the total 2 million receive 60 per cent of all income. The recent National Commission on Small Farms noted: "The pace of industrialization of agriculture has quickened. The dominant trend is a few large vertically integrated farms controlling the majority of food and fibre products in an increasingly global processing and distribution system." (pp. 107–108)

Industrial agriculture increasingly generates widespread, enforced dependencies, in part, because it concentrates wealth and power through globalized competition and due to the political favoritism of large corporate entities.

Furthermore, modern farmers have grown increasingly dependent upon agribusiness corporations whose technologies fuel the agricultural treadmill and who increasingly dominate the purchasing market for farm products. In seed provision, for example, farmers must rely increasingly on highly concentrated providers who maximize their profits through their virtual monopoly status.

But dependency is not enforced solely on the input side of the agricultural equation. Agribusiness giants use stipulations in farm product purchase contracts to demand particular inputs and production processes. Farmer dependency on large, vertically integrated food processing giants is enforced through the size and market penetration of corporate entities that can typically dictate the precise products they desire farmers to produce, the specific means of production, and the price they are willing to pay for end products. Producers must engage with agribusiness corporations in virtually all phases of production: purchase of seeds and inputs, distribution of products, storage, processing, marketing, and even banking (Dyer, 1998, p. 110).

In the U.S., it is difficult to raise livestock without a contract with a major corporate processor who will guarantee to buy one's animals because, in many parts of the country, there is only one buyer (Norberg-Hodge, et al., 2002, pp. 10–12). According to Thomas Lyson, (2004), the situation is similar for vegetable production:

> Today, in the United States, about 85 percent of processed vegetables are grown under contract and 15 percent are produced on large corporate farms. Contract farming allows food processors to exert significant control over their agricultural suppliers. While the processor benefits from these arrangements, the major disadvantage to the farmer is loss of independence. Many contracts specify quantity, quality, price, and delivery date, and in some instances the processor is completely involved in the management of the farm, including input provision. (p. 45)

Contract farming also contributes to increasing farm sizes as agribusiness corporations strive to capture economies of scale (Lyson, 2004, p. 46). Norberg-Hodge, Merrifield, and Gorelick describe the concentration of ownership and control in both the agricultural inputs and food processing arms of agribusiness:

> In the United States, the three largest beef processors control 74 percent of the nation's beef-packing capacity. Another four companies control 84 percent of American cereal, and just two companies control 70 to 80 percent of the world's grain trade. Five agribusinesses account for nearly two-thirds of the global pesticide market, almost one-quarter of the global seed market, and virtually 100 percent of the transgenic seed market.... Nine companies now dominate the global seed market. (2002, p. 89)

The concentration of power and control in the food processing industry is reinforced by concentration within the retail food industry. In the U.S., "average market concentration of the top four retailers in individual metropolitan areas stands at about 75 percent" (Lyson, 2004, p. 53). These retailers are centralizing their management at corporate headquarters and forging supply chains with global food processors (Lyson, 2004, p. 53). As a result of their ability to foster price competition among their suppliers and realize economies of scale, they enjoy lower per-unit costs. Obtaining foods from diverse, localized sources would mean foregoing a portion of their profits and reducing their control over the food system. According to Norberg-Hodge, Merrifield, and Gorelick, "For each supermarket to sell more than a token amount of foods produced nearby would jeopardize the structures and continual shareholder profits on which the entire global food system is based" (Norberg-Hodge, et al., 2002, p. 67).

The management of global agribusiness also reflects the interests of a small number of people with particular class interests who stand to profit a great deal from controlling the industry. According to Lyson (2004), "Only 138 men and women sit on the boards of directors of the ten firms that account for over half of all the food sold in America.... Individuals who are recruited to sit on the boards of large American corporations come from similar social and economic backgrounds and belong to the same social circles" (Lyson, 2004, p. 54).

Effects of input providers and food processing giants capturing ever-larger proportions of food sales profits are discussed by Pretty: "Fifty years ago, farmers in Europe and North America received as income between 45–60 per cent of the money consumers spent on food. Today, that proportion has dropped dramatically to just 7 per cent in the UK and 3–4 per cent in the US, though it remains at 18 per cent in France" (2002, p. 111). The ever increasing share of profit that goes to agribusiness produces diminishing returns for farmers who have increased the scale and capital intensity of their operations, a process that reinforces the agricultural treadmill.

For small producers globally, their state of dependence and subsequent insecurity is emblematic of what it means to have one's resources and the control over one's life steadily reduced through enforced dependency. As discussed in chapter four, neoliberal development policies encourage export-led development, thereby promoting the growing of large monocultures for export. Many small farmers who had been relatively self-sufficient have been bought out by large interests or thrust into the industrial model so that they produce and sell commodities while purchasing food for their own consumption.[2] Producers everywhere exist in the precarious position in which fluctuations in the relative values of currencies, economic recessions in distant parts of the world, larger than expected crop yields globally, and other macroeconomic factors well beyond their control can spell economic disaster for their families and communities (Dyer, 1998; Norberg-Hodge et al., 2002, p. 69).

Low-cost imports of basic food necessities further undermine local self-sufficiency (Black, 2001; Douthwaite, 2004; ISEC, 1993; Norberg-Hodge, 1991/1992). According to Araghi (1995), between 1951 and 1972, a time when food consumption was increasing along with population, export subsidies for U.S. grain producers were a major factor in depressing grain prices globally and discouraging production in formerly self-sufficient or surplus-producing areas of the Global South (p. 349). These pressures have intensified in the neoliberal era. The International Commission on the Future of Food and Agriculture cites Mexico as a case in point:

> According to the FAO [Food and Agriculture Organization of the United Nations], the liberalized economic globalization model has led to a 54 percent increase of food imports between 1990 and 2000 by least developed countries.... Mexico, which traditionally has grown enough maize to feed its populations for centuries, has become a net importer of maize due to dumping of artificially cheap corn flooding in from the U.S. (2008, p. 20)

The squeeze on agriculture is also self-reinforcing because it destroys relationships of trust and reciprocity in farming communities. It makes farmers "increasingly less likely to look to their neighbors for answers, lest either party gain an advantage over the other." Instead, farmers turn to answers provided by the agricultural treadmill itself (Bell, 2004, p. 236), reinforcing their dependency. Bell interviewed a conventional farmer who said that, if he had an idea that would improve his earnings per acre, he would be stupid to tell others about it. He would keep it to himself, rent extra land, and make more money for himself (2004, p. 182). This farmer, like many others, had internalized the logic of the industrial system.

In sum, widespread enforced dependency in the food system derives from the metabolic rift in agriculture, its ecological fallout, and its two complementary manifestations in the social sphere: 1) the situation of farmers running ever faster on the agricultural treadmill or dispossessed by globalized industrial agriculture and 2) the concentration of populations in urban areas where people typically cannot or do not produce food them-

---

[2] The U.S. is not immune to the social and economic forces that promote large-scale, commodity agriculture. See Pfieffer, 2006, pp. 25–26.

selves. Buttel summarizes the structural ramifications of the agricultural treadmill in the U.S.:

> U.S. agriculture and its food system have exhibited five principal structural changes during recent decades: (1) a trend toward large-scale, specialized farm production units, (2) increased mechanization, (3) increased use of purchased biochemical inputs (and corresponding transfer of the input-providing function of agriculture to the non-farm sector), (4) a trend toward regional specialization of production, and (5) an increased level of food processing and interregional marketing. Moreover, these trends are interrelated and mutually-reinforcing. (Buttel, 1980, p. 45)

These structural changes translate to a loss of economic and decision-making control by both farmers and consumers and a corresponding concentration of economic power and control within the non-farm agribusiness sector.

For those producers subjected to the pressures of industrial agriculture, stress can lead, not only to economic dispossession—it can become deadly. The sense of shame and anger farmers feel for losing the family farm to debt can be immense, leading many to take their own lives, as large numbers of farmers are doing in the United Kingdom (Norberg-Hodge, et al., 2002, p. 84). In the United States, the intense squeeze on agriculture during the 1980s led to crisis levels of farmer indebtedness and widespread suicides among farmers. In a study conducted in 1989, the suicide rate among U.S. farmers was three times the rate in the general population (Dyer, 1998, p. 33). In the late 1990s, Dyer noted that suicide had become "the number one cause of death on America's farms" (1998, p. 3). According to Dyer,

> Many of the suicides in rural America are a reflection of its unique culture and belief system. Psychologists believe that rural suicides often fall into the category they refer to as altruistic suicide, a suicide in which a society's customs or rules sanction or even require the death.... A farmer who kills himself to allow his family to collect insurance money and save the farm is often thought to be honorable in the subculture of rural America. (1998, p. 33)

For these farmers, slipping off the back of the agricultural treadmill means utter economic devastation and destruction of self-esteem. In their eyes, suicide can appear to be a solution to personal pain and a means to help one's family escape a devastating financial situation.

According to Vandana Shiva (2008) whose Navdanya movement in India promotes organic, biodiverse farming, the agricultural treadmill is not necessary to high crop yields nor to sufficient farmer incomes (p. 110). There are alternatives to the treadmill for making a living in farming:

Navdanya's work over the past 20 years has shown that we can grow more food and provide higher incomes to farmers without destroying the environment and killing peasants. We *can* lower the costs of production while increasing output. We have done this successfully on thousands of farms and have created a fair, just, and sustainable economy. The epidemic of farmer suicides in India is concentrated in regions where chemical intensification has increased costs of production.... Biodiverse organic farming creates a debt-free, suicide-free productive alternative to industrialized corporate agriculture.... (p. 111; for data on increased yields, see Shiva, 2008, pp. 113, 116)

Effects on Rural Communities

Through the industrial food system, dependency is also enforced at the community level because the squeeze on agriculture reduces community *potenia*[3] and thereby increases reliance on systems and entities outside the community and beyond community control. Buttel (1980) references a 1947 study comparing the effects of large-scale agriculture on communities in the U.S.:

A now classic study by the anthropologist Walter Goldschmidt (1947)...pointed to some profound and deleterious impacts that the emergence of large-scale agriculture was having on rural communities in California.... The most direct consequence...involves the reduction of the size of the farm population that results from mechanization and increasing farm size.... A declining population results in a decline in rural community population [and sets] in motion a downward community and regional multiplier effect which accentuates the economic consequences of the original decline of population. (p. 47; see also Lyson, 2004, pp. 66–68)

---

[3] Many authors use the term *social capital* to denote a community's capacity to engage in healthy social production and reproduction and to generate and protect material community wealth. For these authors, social capital is comprised of networks of reciprocating relationships built upon a framework of trust in individuals and the community. Community members trust each other to both continue and support reciprocating relationships that provide mutual aid and security while also encouraging satisfying relationships (see, for example, Pretty, 2002, pp. 152–153). I prefer to use the term *community potentia* in order to avoid conceptualizing the creative potential of social networks as merely instrumental to capital accumulation. I draw the term *potentia* from Dussel (2008) who argues that the sole foundation of social power is "the political community, or the *people*" (p. 18; emphasis in original). He defines *potentia* as the possible, but as yet unrealized, consensual will of the community that can be expressed through formation of socially agreed upon forms and institutions of all kinds and scales (pp. 18–23). The community bonds referred to as social capital represent constitutive factors in community *potentia*. Conceptualizing community *potentia* as independent from the capitalist order creates opportunities for recognizing it as a potential source of radical social and economic change.

Based on his reading of the Goldschmidt study and other evidence, Buttel (1980) concludes that "the trajectory of agricultural development in the U.S.A. has had decidedly adverse impacts on the socioeconomic fabric of rural communities and regions" (p. 48; see also Norberg-Hodge, et al., 2002, pp. 70–71).

The decline of rural communities in the U.S. is closely related to the economic decline of small family farms spurred by the increasing velocity and momentum of the agricultural treadmill. Additional social phenomena compound the effects of economic decline as rural communities fall into a downward spiral of social upheaval and community disintegration. These phenomena comprise an overall loss of community *potentia*. According to Pretty, central features of a community's social capital or *potentia* include "relations of trust; reciprocity and exchanges; common rules, norms and sanctions; and connectedness, networks and groups" (2002, p. 152). Communities with high levels of *potentia*, according to Pretty, engage in "specific reciprocity" and "diffuse reciprocity" (2002, p. 152). Specific reciprocity entails reciprocal exchanges of goods, services, or support between specific people who engage in exchanges with the expectation of receiving a specific return from a particular person or group. Diffuse reciprocity entails providing goods, services, or support to other community members with the idea that, when you or your family is in need of assistance, someone or some group from the community will offer the needed help. Diffuse reciprocity derives from community norms of mutuality and support.

We discussed above how the agricultural treadmill increases competition among farmers. This high-stakes competition is antithetical to community connectedness, trust, sharing, and the offering of reciprocal assistance in either specific or diffuse forms. The agricultural treadmill is therefore an important factor in not only the economic decline of farming communities, but also the loss of relationships that characterize communities with high *potentia*. A community that lacks *potentia* is less resilient in the face of external shocks and pressures, and its population is more vulnerable to the enforced dependency that permeates the monetized and globalized economic system.

We have discussed the social crisis in agriculture and its causes with an emphasis on revealing interlocking and mutually reinforcing systems of enforced dependency. This critique can inform counterhegemonic food praxis, the topic of the next chapter. At this point, we can surmise that (re)localizing the food system would result in increased community resiliency, but in order to understand the global implications of local food as counterhegemony, we must first examine the role of industrial agriculture in supporting industrialization and capitalist market hegemony in the broader economy. We begin this investigation by examining the structural role of industrial agriculture in the United States. We will follow this discussion with a discussion of how the U.S. model for development became an ideological tool for U.S. global hegemony and neoliberal reforms worldwide.

The Structural Role of Industrial Agriculture in U.S. Development

Buttel (1980) argues that overproduction in agriculture has made industrialization as a whole possible in the U.S. He claims that the structure of agriculture has resulted in rural underdevelopment and decline as well as reduced environmental quality in rural areas. Furthermore, according to Buttel, part of the enforcement of enforced dependency derives from the state's attempts to balance its roles of legitimation and capital accumulation (1980, pp. 49–53). Buttel's argument is important to our exploration of industrial agriculture as a system of enforced dependencies because it elucidates important structural, historical, and ideological foundations for the industrialization of agriculture in America and because it situates industrial agriculture within a large-scale, hegemonic political project.

Buttel argues that two central roles of the state in a capitalist society are to facilitate capital accumulation while also ameliorating discontent arising from the resulting socio-economic inequity (1980, p. 48). As noted in chapter three, the debt-and-interest-based system of money creation in the modern global economy makes growth required if the system is to avoid collapse (Daly, 1999; Douthwaite, 1999a, 1999b). At the same time, politicians in representative democracies must secure funding to run effective political campaigns. Therefore, elected officials, who depend heavily upon large corporations for campaign funding, are likely to advance the processes of the capital accumulation advantageous to their biggest supporters over serving the needs of small farmers (Buttel, 1980, p. 49). According to Buttel, the following set of policies that characterize agriculture in the U.S. support capital accumulation:

(1) Fostering a commercial (rather than a subsistence-oriented) agriculture that produces a surplus of relatively inexpensive food for consumption by the urban working class, (2) encouragement of exports of food and fiber commodities, (3) rationalization of the agricultural economy, particularly with respect to the endemic tendency toward overproduction and price instability, (4) the underwriting of capital accumulation in the production agriculture and non-farm agricultural industries through agricultural research, and (5) assisting in the international expansion of agribusiness firms that trade in agricultural commodities or market their products in foreign countries. (1980, p. 51)

All of these policies encourage and enforce dependencies, not only domestically, but globally, and they promote the concentration of wealth and power in the globalized world. According to Buttel, these policies are

> aimed at directing the agricultural sector toward a particular path in the overall trajectory of socioeconomic development. This path for agriculture essentially involves the creation of surpluses (food surpluses to sustain an emerging urban working class, as well as the extraction of financial surpluses from agriculture for investment in the urban-industrial sector) and markets for the products of industry. Thus, *while certain aspects of agricultural policy* (especially price supports and agricultural research) *have clearly encouraged capital accumulation within agriculture, such internal accumulation need not be and often is not essential for underwriting accumulation in the larger political economy....* The production of large food surpluses was instrumental in fostering American industrialization.... [and] exports of foodstuffs ... transformed the U.S. from a debtor-nation in the 19th century to a creditor nation shortly prior to the Great Depression. (1980, p. 51; emphasis added)

Buttel also argues that support for agricultural research has ironically played an important role in diffusing widespread social unrest. He suggests that unrest could result from government policies that propel the agricultural treadmill and contribute to the bankrupting of farm families and the disintegration of rural communities. According to Buttel, "Agricultural research became attractive to policy-makers because the development of improved farm practices could steer farmers toward *individualistic* rather than *collective* solutions to their problems" (1980, p. 52; emphasis in original). If farmers could be encouraged, through the promise of new scientific and technological developments, to respond to economic pressure by competing with one another, collective social unrest could be averted, and both the legitimation and capital accumulation goals of the capitalist state would be served. As we have seen in our discussion of the agricultural treadmill, this describes the overarching trajectory of farming in the U.S. during the twentieth century.

Buttel summarizes the character of agricultural policies in the U.S. and points to political change as necessary in order to reverse the decline of rural America:

> Agricultural policy has been formulated not so much according to desired goals for the agricultural economy and rural society, but rather according to how agriculture and the rural social structure could be made to serve the accumulation and legitimation roles of the state.... Because of the continued subordination of the agricultural sector to broader corporate and state interests, it becomes increasingly clearer that in order to effect progressive changes that will enhance rural development and rural environmental quality, these will require a qualitatively different agricultural politics than has prevailed in the past. (1980, p. 53)

In the U.S., a reordering of agricultural policy from above is unlikely. Change must therefore be initiated from the grassroots level.

We now turn our attention to a world-historical analysis of how the U.S. model of industrial development served as an ideological tool to facilitate capital accumulation and enforce dependencies in the global economy.

A World-Historical Perspective on
Dependencies Enforced through Agricultural Policies

National and international policies and the global struggle for political hegemony in the post-war world-system have created a framework for massive migration of people worldwide off of the land and into urban centers. New dependencies have been created and enforced among millions of people worldwide who must now depend on money and the global economy as means to satisfy basic needs. Massive global depeasantization translates to massive new dependencies—and concomitant opportunities for capitalists to profit and both extend and deepen their economic and social power. Sociologist Farshad Araghi's article "Global Depeasantization, 1945–1990" (1995) anchors our discussion on global rural-to-urban migration and its implications. Understanding the factors promoting massive depeasantization creates openings for envisioning and undertaking strategic social change that can reduce enforced dependencies, particularly within the food system, which is a critically important leverage point for sustainability praxis. These strategies and examples of their application will be the focus of chapter seven.

From 1945 to 1973, national and international policies affecting agriculture facilitated capital accumulation and supported national agendas for economic and political hegemony, particularly in the case of the United States (Araghi, 1995, pp. 344–354). In the neoliberal era, these policies have continued to facilitate capital accumulation, but this accumulation has increasingly benefitted transnational corporations at the expense of nation states (Araghi, 1995, pp. 354–358). During the entire post-war period, national and international policies have produced massive depeasantization globally (Araghi, 1995, p. 339).

The early post-war period from 1945 through 1973 was marked by Cold War competition for political hegemony between the pro-market United States and the socialist Soviet Union.[4] Though ideologically opposed to one

---

[4] After 1973, the political hegemony of the U.S. began to fray as the gold exchange standard was abandoned by the Nixon administration and as the U.S. economy became mired in stagflation.

another, the aspirations of both were nationalist in orientation. During this era, national liberation movements across the globe served as platforms for Cold War competition between the two superpowers. Socialist movements striving for national political hegemony in the Global South together with the Soviet Union that was seeking global political hegemony found it advantageous to meld the idea of peasant struggle with that of worker liberation (Araghi, 1995, p. 344).

For its part in the early post-war period, the U.S. supported development agendas in post-colonial nations that promoted balanced economic development and growth through import-substitution industrialization. The idea was that developing countries could follow the path of U.S. development that was based on domestic markets. These nations could become wealthier and more independent as they moved from widespread family farming to industrialized production. Because market competition was becoming increasingly global in scale and due to other historical factors that Araghi details, possibilities for people worldwide to achieve the American dream were mostly illusory (1995, p. 345). Still, through the mid-1970s, the idea that other countries could duplicate the American experience dominated national and international approaches to agriculture and to addressing peasant unrest. In non-socialist parts of the world, land reform was encouraged within a pro-market atmosphere in order to defuse peasant unrest and, purportedly, lay the groundwork for industrial development through stimulating widespread demand and purchasing power (Araghi, 1995, p. 346). Large tracts of land were, however, excluded from redistribution with the idea that they would be used for large-scale, mechanized agriculture. Paradoxically, both concentration and redistribution of land occurred simultaneously.

During this period, political leaders envisaged a structural role for industrial agricultural production (seen as primarily for domestic markets) in stimulating industrial development. Although under certain circumstances in particular countries, capitalized family farms indeed produced surpluses as a result of state policies and land reform, in most cases, rural allotments ended up being subsistence-sized, especially after further subdivision in later years. Most rural producers did not produce surpluses sufficient to underwrite industrialization. State policies also facilitated the replacement of subsistence regimes with commodity relations at the same time that small producers were increasingly exposed to international market competition. A new international division of labor in food production began to emerge. In combination with U.S. subsidies for domestic agriculture and the use of U.S. food aid to reduce competition globally, these developments advanced the hegemonic position of the U.S. within the world-system (Araghi, 1995, pp. 347–349).

These changes also resulted in an internationalization of the American diet, thereby reducing food system resiliency (Araghi, 1995, p. 349), and they powerfully enforced dependencies as food imports rose sharply in the Global South while possibilities for subsistence declined. According to Araghi, "In the underdeveloped world as a whole the ration of food imports to food exports increased from 50 percent in 1955–1960 to 80 percent in 1975," and "between 1960 and 1980 both the rural population as a percentage of total population and the agricultural labor force as a percentage of total labor force declined in all regions of the Third World" (1995, pp. 350–351). Western Europe, Canada, and the U.S. were not immune to deruralization during this same era (Araghi, 1995, p. 351).[5]

As noted in chapter four, the declining ability of small producers to compete in a global market that began prior to 1973 (Araghi, 1995, p. 352) intensified as neoliberalism ushered in falling trade barriers and falling prices, which resulted in further massive deruralization globally. From 1973 to 1990 (except in the case of Africa, which may be due to extremely low rates of economic growth), a higher proportion of urban population growth resulted from deruralization than in the period from 1945 to 1973 (Araghi, 1995, p. 357), demonstrating the negative effects of neoliberal policies on small producers. Depeasantization facilitated capital accumulation in urban centers (Araghi, 1995, p. 352) and refocused the attention of leaders in the Global South away from small holder production and toward industrialization as a means to increase wealth and power (Araghi, 1995, p. 354).

With U.S. hegemony beginning to slip after 1973, and with the assistance of political leaders worldwide (Araghi, 1995, p. 359), capital has been reorganizing itself on a global scale (Araghi, 1995, p. 355). Export-led development strategies pursued by debtor nations at the behest of international lenders contributed to dispossession of peasant producers in the Global South and to enforcing their dependency on the global economy (Araghi, 1995, p. 356). Through loan conditions that both exposed producers to increasing market pressures and undermined the sovereignty of debtor nations in the Global South, the IMF and the World Bank have made replicating the American experience virtually impossible. All the while, American material success served as the dream and the promise of development.

---

[5] See Araghi, 1995, p. 351 for rates of deruralization in a context of overall population growth for nations and regions worldwide. Although in Latin America, in absolute terms, the rural population grew, it declined substantially in relative terms as vast numbers of rural dwellers relocated to cities.

Conclusions on Industrial Agriculture as a Dependency Enforcing System

Industrial agriculture reduces farm biodiversity and community economic diversity, resulting in losses of resiliency. Working within the context of post-war national and international policies, industrial agriculture has resulted in widespread dependence of urban masses on monocultural production methods and the globalized shipping industry, neither of which can be sustained. It has also been instrumental in removing people from the land as a source of subsistence and autonomy. Industrial agriculture fractures community life by systematically drawing down the economic and social vitality of farming communities and fostering the concentration of wealth and power in the hands of agribusiness. In its wake, industrial agriculture leaves heightened insecurity and isolation where there once existed reciprocity, community autonomy, and resiliency. Industrial agriculture embodies the political hegemony of the powerful within late capitalist paradigm. On a global scale, it reduces the resiliency of communities and nature at a time when that resiliency is of critical importance to the health and survival of communities and ecosystems. During the post-war period, national and international policies facilitated the rise of industrial agriculture and the demise of small producers. These same policies facilitated the enforcing of perpetual dependence of nations in the Global South within the world-system. Reversing these trends will be a steep uphill battle.

In the next chapter, we will discuss examples of communities creating more sustainable food systems. I will explore a phenomenon that Ploeg (2008) calls *repeasantization,* and I will show how this phenomenon represents a set of strategies for engaging in creative destruction and transformation of the current food system. I will argue that localized production and consumption of food is central to (re)inhabitation and community autonomy. I will demonstrate how building sustainable food systems can help heal social ecology by mending the metabolic rift between people and nature and reducing enforced dependency.

Chapter 7:

Food as Sustainability Praxis

The time has come to reclaim the stolen harvest and celebrate the growing and giving of good food as the highest gift and the most revolutionary act.
—Vandana Shiva

In this chapter, I conceptualize sustainable food activism as resistance to enforced dependency. Although the neoclassical economic system sees food as just another commodity, food can be a platform for radical socio-ecological change, in part, because it defines in crucial ways the relationship between people and nature. Food is essential for life, and control over food translates to social power. Changing our food systems, therefore, implies extensive changes in relative social power. This chapter lays a foundation for chapters eight and nine in which I argue that study of sustainable food systems, combined with student participation in local food work, serves as an effective vehicle for counterhegemonic sustainability education and praxis.

Sustainable Food as Counterhegemonic Praxis

Sustainable food praxis is a wellspring of agency motivated by both ideas and real-world experience. We begin our analysis of sustainable food praxis as counterhegemony by highlighting intellectual and material factors that can motivate food system praxis.

Drivers for Sustainable Food Praxis

Bell's (2004) discussion of the relationship between ideal and material factors in the economy of agriculture helps us situate Gramscian praxis within the agricultural realm:

> Material factors depend upon ideal factors and ideal factors depend upon material factors to attain their persistence. The economy of agriculture embodies cultural values like the virtue of competitive individualism as much as it undermines cultural values like the importance of communal ties. The cultivation of knowledge depends in part upon our sense of the material implications of that knowledge as much as it

leads to the persistence of those material implications…. There is no first instance or second instance of either the material or the ideal. They are in endless conversation…. (p. 147)

And so, we might ask: what are the intellectual and material drivers of sustainable food praxis?

I suggest that people are beginning to recognize the political and economic limits to industrial agriculture. Buttel summarizes the social contradictions of industrial agriculture in the U.S.:

> Economic stagnation threatens to usher in a situation in which rising food prices can no longer be compensated for by rapid growth in disposable family income. Likewise, the fiscal crisis of the state makes it unlikely that it will take action to increase transfer payments to or reduce the taxes on the poor to meliorate discontent over food price inflation. At the same time, the "need" to expand food exports in order to pay for burgeoning imports of petroleum can only exacerbate food (and overall) inflation and thereby escalate the demands of the lower income classes…. The hegemony of prevailing trajectories of development in the agricultural sector…appears to be encountering political and ecological limits…. These limits or contradictions may allow certain groups to take steps to seek change in the structure of agriculture and rural society. However, these steps are *unlikely* to be taken by a state apparatus whose primary functions are rationalizing the [capital] accumulation process to the advantage of large-scale property owners, on the one hand, and regulating social conflict and discontent, on the other. Thus…most contemporary initiatives to effect change in rural society are not deriving from state policy, but rather are coming from *local efforts* on the part of persons whose needs are not met by the present food system and rural economy. (1980, p. 55; emphasis in original)

Buttel recognizes the implications of fossil fuel dependency in agriculture in an era of oil depletion and increasing petroleum imports. He also recognizes both the economic stagnation that precipitated the recent global economic crisis as well as the need for grassroots, counterhegemonic food praxis as a means to revitalize agriculture and society.

While it is unlikely that most Americans and people globally grasp the structural contradictions of industrial agriculture and globalization with the depth of Buttel (1980), a growing number of people worldwide recognize that the food system is not serving them well on one or more levels: economic, nutritional, or political. They also recognize that industrial agriculture is damaging the very socio-ecological systems on which it depends and producing toxic food. This recognition can stimulate sustainable food praxis.

According to Buttel, faced with the pressures of globalization and the momentum of the agricultural treadmill, "The petty commodity producer increasingly faces a… difficult choice, essentially whether to cast one's lot with large-scale capital or with the working class" (1980, p. 57). Some producers realize that the current system is not meeting their needs (Buttel, 1980, p. 57). This realization, derived primarily from material experience, motivates some producers to engage in sustainable food praxis.

Both the ideal and the material, separately or in combination, can serve as launching points for counterhegemonic food praxis. Ploeg (2008) describes the praxis process as it relates to local food producers, whom he collectively calls peasants:

> The peasant condition is composed of a set of dialectical relations between the environment in which peasants have to operate and their actively constructed responses aimed at creating degrees of autonomy in order to deal with the patterns of dependency, deprivation and marginalization entailed in this environment. Responses and environment mutually define each other.... The responses shape the environment as much as the environment generates the responses.... Typical of the peasant condition, is that the responses unfold by means of constructing a resource base that allows for the co-production of man and nature. (p. 261)

Ploeg (2008) describes counterhegemonic praxis that could heal metabolic rift.

Buttel (1980) also emphasizes a need for class-based alliances in counterhegemonic food praxis. He points to the radical potential of a class alliance "primarily between the working class, and petty bourgeois farmers and 'independent' business people," noting that such an alliance "has occurred infrequently during the course of economic development in the U.S." (Buttel, 1980, pp. 55–56). An alliance of this sort is emerging in my town of Durango, Colorado, where local business people, small producers, and consumers are organizing to promote food system (re)localization as a boon to community vitality and resiliency. This counterhegemonic praxis connects local food activism to an overarching community movement toward (re)localization and (re)inhabitation. Like Buttel (1980), I argue that addressing the problems of industrial agriculture requires deep social change and a transformed political economy of food. Creating such a political economy requires widespread sustainable food praxis. We will now explore defining aspects of this praxis.

Beyond Organic

Organic production alone is not counterhegemony in agriculture (Bell, 2004, pp. 245–246). Large-scale organic farms can take advantage of disproportionate subsidies available to large producers, allowing them to outcompete local producers. Enforced dependency characterizes large-scale organic production. Wealth is extracted from local communities whose residents lose out on economic multiplier effects, economic vibrancy, and self-determination associated with strong local economies. Large-scale organic farms also use nearly as much energy as do conventional industrial farms, meaning that they do little to reduce fossil fuel dependency (Norberg-Hodge, et al., 2002, pp. 44–45). Furthermore, large-scale organic farms cannot cycle nutrients like localized operations can. They, therefore, perpetuate metabolic rift.

Building Community Potentia

If the hegemonic food system breaks down community and atomizes personal relationships, food system counterhegemony entails resisting enforced dependency and building community *potentia* through (re)creating resilient, reciprocating social networks. To accomplish these purposes, Buttel advocates structural changes in agricultural systems:

> Significant structural changes (beginning along the lines of encouraging a small farm system, localism in the food system, and emergence of worker-controlled enterprises in rural areas) are required to redress the fundamental problems faced by rural people and communities. (1980, p. 57)

In his study of Practical Farmers of Iowa (PFI), Bell highlights personal relationships, community networks, and an atmosphere of cooperation as important aspects of sustainable farming:

> Sustainable agriculture is a different *social practice* of agriculture.... The relations of knowledge within PFI have a different feel to them, a different way of experiencing others and of experiencing one's own self. And that different way is to recognize difference and to encourage it as a source of learning, change, and vitality, rather than as a threat to self and knowledge. That different way is the way of *dialogue*, rather than monologue.... PFI framers seek...to create a *dialogic agriculture*, an agriculture that engages others—men, women, family members, other farmers, university researchers, government officials, and consumers alike—in a common conversation about what it might look like. (Bell, 2004, pp. 17–18; emphasis in original)

According to Bell, sustainable farming explicitly builds community *potentia* and counteracts the distrust and competition among farmers that is typical of the industrial model. Farmers share ideas and observations, successes and challenges with each other in order to catalyze movement toward sustainable food production (Bell, 2004, p. 185; see also Carolan, 2006).

Similarly, sustainably-oriented producers interviewed by Kneafsey, et al. (2008) emphasized cooperative rather than competitive relationships among fellow sustainable farmers who seemed to respect each other's work and see it as a complementary part of the food system. Some even expressed desire to avoid competition with counterparts in nearby locations (p. 87). These same producers expressed ethical commitments to consumers to provide them with authentic, quality food through honest means (p. 88), indicating that trust and healthy community relationships are important.

Lyson (2004) calls food system (re)localization *civic agriculture*, a term which highlights the community building aspects of its counterhegemonic processes and strategies. According to Lyson,

> civic agriculture is the embedding of local agricultural and food production in the community. Civic agriculture is not only a source of family income for the farmer and food processor; civic agricultural enterprises contribute to the health and vitality of communities in a variety of social, economic, political, and cultural ways.... Taken together, the enterprises that make up and support civic agriculture can be seen as part of a community's *problem-solving capacity*.... Civic agriculture...is a locally organized system of agriculture and food production characterized by networks of producers who are bound together by place. Civic agriculture embodies a commitment to developing and strengthening an economically, environmentally, and socially sustainable system of agriculture and food production that relies on local resources and serves local markets and consumers. The imperative to earn a profit is filtered through a set of cooperative and mutually supporting social relations. Community problem solving rather than individual competition is the foundation of civic agriculture. (2004, pp. 62–64; emphasis in original)

Lyson (2004) advocates building community *potentia* through praxis.

Some sustainable farmers recognize value in locally produced and rigorously tested scientific knowledge. According to Bell (2004), in the sustainable agriculture movement, there ought to be room for both locally produced knowledge and scientific inquiry aimed at broader dissemination of findings. What matters most, in Bell's view, is the difference in relationships: moving away from technology and knowledge transfer to relationships based on trust, mutual respect, and reciprocity. For PFI farmers, experiments allow them to test their own methods and move away from relying on the testimonials of agribusiness advertising (Bell, 2004, pp. 190–193). Experiments represent an avenue for building community knowledge and trust.

Multifunctionality

Sustainable food systems are also counterhegemonic in their multifunctional and holistic orientation. Multifunctional holism in sustainable agriculture entails maximizing the resilience of all system participants so as to enhance the long-term viability of the system itself. According to Ploeg, multifunctionality is the use of "one and the same set of resources…to generate an expanding range of products and services, thus reducing the cost of production of each single product…and simultaneously augmenting the value added realized on the farm" (2008, p. 151). Bell notes that raising livestock is one means to achieve multifunctionality by gaining fertilizer inputs, being able to make use of hay produced during crop rotations, and selling value added products like milk, cheese, and meat (2004, p. 6). Multifunctionality can also increase the diversity necessary to inhibit pest infestations while also contributing to increased total food yields (Pretty, 2007, p. 145).

Pretty offers examples of sustainability-oriented multifunctionality:

> Many of the individual technologies [used in sustainable farming] are multifunctional, and their adoption results, simultaneously, in favourable changes in several aspects of farm systems. For example, hedgerows encourage wildlife and predators and act as windbreaks, thereby reducing soil erosion. Legumes in rotations fix nitrogen and also act as a break crop to prevent carry-over of pests and diseases. Clovers in pastures reduce fertilizer bills and lift sward digestibility for cattle. Grass contour strips slow surface run-off of water, encourage percolation to groundwater, and are a source of fodder for livestock. Catch crops prevent soil erosion and leaching during critical periods, and can also be ploughed in as green manure. Green manures not only provide a readily available source of nutrients for the growing crop, but also increase soil organic matter and hence water retentive capacity, further reducing susceptibility to erosion. Low-lying grasslands that are managed as water meadows, and that provide habitats for wildlife, also provide an early-season yield of grass for lambs. (2002, pp. 113–114)

Multifunctional holism in farming directly confronts conventional farming's linearity and narrow focus on large-scale production of commodities. Multifunctionality also reflects the diversity present in healthy ecosystems and is compatible with a holistic ontology of place.

No GMOs

Because relying on genetically modified seeds and organisms creates dependencies on transnational corporations, sustainable farmers should avoid their use. Furthermore, transgenic seeds cultivated in the open environment have polluted plant gene pools of traditional varieties of plants, an irreversible ecological act with far-reaching potential consequences (ICFFA, 2008, p. 30). When engaging in place-centered sustainability praxis, farmers must rely on technologies they can manage independently. These technologies should not pose unknown risks to locally adapted plants and animals.

Autonomy

Family and community autonomy are distinguishing aspects of what Ploeg (2008) labels peasant agriculture and what I call food system (re)localization. Particularly regarding inputs, peasants seek to distance their agricultural practices from markets so as to increase their autonomy and resiliency (Ploeg, 2008, pp. 1, 23, 49, 66):

> Central to the peasant condition...is the *struggle for autonomy* that takes place in *a context characterized by dependency relations, marginalization and deprivation*. It aims at and materializes as the *creation and development of a self-controlled and self-managed resource base*, which in turn allows for *those forms of co-production of man and living nature that interact with the market, allow for survival and for further prospects* and *feed back into and strengthen the resource base, improve the process of co-production, enlarge autonomy and, thus, reduce dependency*.... Finally, *patterns of cooperation* are present which regulate and strengthen these interrelations (Ploeg, 2008, p. 23; emphasis in original).

Distancing from the market economy is also accomplished through community reciprocity, pooling of machinery among farmers, saving rather than buying seed, "regrounding agriculture on available ecological capital," emphasizing the use of craft- and skill-based technologies rather than capital intensive technologies, internalization of as many aspects and phases as possible of the production process, maintaining system health and productivity over the long haul, orienting production toward internal standards of use value rather than externally determined exchange value, and "intergenerational transfer of farm units" (Ploeg, 2008, pp. 50, 62, 114–117). Buttel advocates similar strategies for small farmers in the U.S. to achieve autonomy and resiliency in the face of ever-expanding agribusiness and food processing entities that enforce dependency:

> Decentralization of the food system would allow both farmers and consumers to benefit via circumventing the increasingly pervasive food marketing industries, as well as facilitate environmental benefits from the increased crop diversity required in a more regionally self-sufficient agriculture.... Worker-controlled enterprises appear to be pivotal in establishing a higher degree of local self-sufficiency since community-controlled firms are more likely to orient their activities toward the utilization of local resources to meet local needs (1980, pp. 55–56).

Buttel's arguments prefigure the recent rapid growth in community supported agriculture and the slow food, locavore, and farmers' market movements in the U.S.

Distancing from the market economy is counterhegemonic in its potential to combat enforced dependency. It not only increases farmer autonomy, it also means that labor becomes an important part of the agricultural process once again (Ploeg, 2008, p. 156). Distancing production from global markets amounts to counterhegemonic resistance that allows sustainable producers to escape the clutches of the hegemonic agribusiness sector while also constructing viable alternatives to it.

A Rising Tide

Counterhegemonic local food praxis is widespread and growing. Ploeg notes that there are 1.2 billion peasants worldwide, a figure that includes European small farms (2008, p. xiv). Some 80 percent of European farmers are engaged in some form of repeasantization (Ploeg, 2008, p. 157). According to Ploeg, "Over the last decade and a half, Europe has witnessed a widespread process of repeasantization. This process mainly expresses itself qualitatively. It involves enlarging autonomy and widening a resource base much narrowed by previous processes of specialization that followed the script of entrepreneurship" (2008, p. 151). Small producers who sell farm and garden products locally are, in effect, incrementally replacing conventional industrial agriculture with a sustainable food system that (re)integrates humans with their environment while also increasing community *potentia*.

The demand side of the food system is also contributing to local food counterhegemony. According to the Institute for Grocery Distribution, as of 2005, 70 percent of British consumers wanted to buy local food, and 49 percent wanted to buy more local food than they were doing. In a space of less than ten years, farmers' markets had also increased from one to 550 in the United Kingdom. Several hundred organic vegetable box schemes were also in operation. Food sales through independent stores, box schemes, and farmers'markets increased 32 percent in the UK between 2004 and 2005 (Kneafsey, et al., 2008, p. 2). In the United States, the number of farmers' markets had dwindled to fewer than 100 in the 1970s. In 2003, the U.S. Department of Agriculture counted over 3,000 (Lyson, 2004, pp. 91–92).

In terms of volume of food produced, localized production demonstrates some striking successes, including in urban areas. In 1996, 80 percent of the poultry and 25 percent of the vegetables consumed in Singapore were provided from the city's urban agriculture (Pfeiffer, 2006, p. 71). According to Murphy, in 1999, there were "over 1,000 gardens in New York City and over 30,000 gardens in Berlin.... In densely populated Hong Kong, 45 per cent of local vegetable needs [were being] met through intensive cultivation on only six per cent of the land area" (1999, p. 1).

Recent food system praxis demonstrates that producing the means of survival outside of the market economy can reduce dependency on the globalized economy (Ploeg, 2008, p. 31). Ploeg highlights the counterhege-monic promise of food system (re)localization:

> This new form of resistance...basically searches for, and constructs, local solutions to global problems. Blueprints are avoided. This results in a rich repertoire—the heterogeneity of many responses thus becoming one of the propelling forces that induces new learning processes. Resistance becomes a form of production embody-ing a radical break from neoliberal globalization. Repeasantization is a form of radi-cal agency of uncaptured people. (2008, pp. 271–272; p. 274)

In order to gain an understanding of what food system counterhegemony looks like on the ground, we will now explore counterhegemonic food system praxis in different parts of the world. We begin with the transfor-mation of Cuba's food system from one dependent on agribusiness and integrated with the global economy to one focused on sustainable, localized production for localized consumption.

The Cuban Response to Soviet Collapse: Food System Relocalization with Government Support

As noted above by Buttel (1980), since capitalist states tend to support capital accumulation within the national (and now global) economy, national level policy support for decentralization of agricultural production is gener-ally not forthcoming. Therefore, changing the food system in the U.S. and many countries of the world will require localized counterhegemonic resistance (Buttel, 1980, p. 55).

The Cuban experience is a unique example of food system relocalization *with* government support. This support was possible because of Cuba's avowedly socialist orientation. Since the revolutionary victory in 1959, the Cuban state has redistributed wealth to a large extent and focused much attention on improving the well-being of average Cuban citizens. In its response to the collapse the Soviet Union and the food shortages that ensued, the government strongly supported the food system relocalization of the Special Period.

During the Special Period, which began with the 1991 collapse of the Soviet Union, Cuba converted from a high-input, fossil fuel–based agricul-ture system to one emphasizing diverse and dispersed production and localized self-reliance. The crisis of the Special Period was both intense and abrupt. Practically overnight, Cuba lost 85 percent of its international trade; 80 percent of its fertilizer, pesticides, and animal feed imports; and half of its food imports. By 1994, food production stood at 55 percent of its 1990 levels, and per capita caloric intake had dropped by 36 percent. Cuba averted the worst of the potential effects of the crisis through food rationing and food programs targeted at highly vulnerable populations (Pfeiffer, 2006, pp. 56–57). Cuba also reformed agricultural production. According to Murphy,

> Cuba responded to the crisis with a national call to increase food production by restructuring agriculture. This transformation was based on a conversion from a conventional, large scale, high input…agricultural system to a smaller scale, organic and semi-organic farming system. It focused on utilizing local low cost and environmentally safe inputs, and relocating production closer to consumers in order to cut down on transportation costs. (1999, p. ii)

These changes were stimulated by radically reduced availability of fuel, tractor parts, and chemicals formerly supplied by the Soviet Union (*Power of Community*, 2006).

The U.S. intensified the crisis with the passage of the Toricelli Bill in 1992. This law tightened the U.S.'s economic blockade of Cuba by banning "foreign subsidiaries of U.S. companies from trading with Cuba. Seventy percent of this trade had been in food and medicines" (Murphy, 1999, p. 9).

Prior to the crisis, Cuban agriculture was typical of that in the Global South. It was heavily concentrated on export commodity production of sugar, tobacco, and citrus while 60 percent of staples consumed by citizens were imported (Pfeiffer, 2006, p. 56). Murphy (1999, pp. 7–8 notes that, Cuba's participation in the global economy of industrial agriculture meant that it "was not able to produce enough food to meet the needs of its people. In the 1980s, 57% of the calories consumed by the Cuban people were still imported…. The diversification of agriculture that the revolution had intended was frustrated by the continued dependence on sugar as a source of foreign exchange…. Cuba initiated the National Food Program in the 1980s…" in response to increasing criticism of its industrialized agriculture. In an effort to increase self-reliance, especially around Havana, this program moved land from sugar production to production of vegetables. The National Food Program also laid important groundwork for the agricultural programs undertaken during the Special Period.

The Cuban response in the Special Period was made possible partly by land reform that had occurred since the revolution. Murphy describes the concentration of land ownership in foreign hands in Cuba prior to the revolution: "By 1959, corporations and U.S. citizens owned 75 percent of arable land in Cuba" (1999, p. 7). Following the revolution's success in 1959, land was redistributed to

> squatters, sharecroppers, and landless farmers. Fifty percent of the land in Cuba was nationalized, and more than 100,000 landless peasants became landowners over-

night. A second agrarian reform in 1960 further limited landholdings and most ex-
propriated land was converted into state farms. (Murphy, 1999, p. 7)

These reforms redistributed land and wealth in a nation where 8 percent of
the farmers had controlled 70 percent of the farmland (Pfeiffer, 2006, pp. 54–
55). Cuban land reforms increased the number and distribution of farmers
and diffused agricultural knowledge. This diffusion proved to be an important asset during the Special Period.

In 1993, the Cuban government engaged in another land reform through
which large state farms were converted to private cooperatives called Basic
Units of Cooperative Production that were owned and run by farm workers.
Through this program, 41.2 percent of the nation's arable land was converted
into 2,007 new cooperatives with 122,000 total members. The government
retained ownership of cooperative lands, but members were granted free
leases for growing food. The government then contracted with the co-ops for
specific crops and amounts to be grown and sold to the co-ops some of the
necessary inputs (Pfieffer, 2006, pp. 58–60; Murphy, 1999, p. 10).

The policy focus on increasing localized food production also extended
to private lands. The Special Period triggered an increase in private farming.
The government turned over close to 170,000 acres to private farmers who
farm rent free on government-owned land (Pfieffer, 2006, p. 60). According
to Murphy, "Even privately owned land in...[Havana], if not in use, was
turned over to those who wished to cultivate it.... If the owners objected,
they would be allowed six months to put the land into production themselves.
If the owners never cultivated the lot, use rights would then go to the
soliciting gardener" (1999, p. 13). According to Murphy, "Land access poses
the largest constraint to producers around the world" (1999, p. 4). In this
respect, Cuban land reform before and during the Special Period is particularly important.

During the Special Period, urban agriculture was a spontaneous and successful response to food shortages that was backed by government policies.

By 1994 a spontaneous decentralized movement of urban residents joined a planned
government strategy to create over 8,000 city farms in Havana alone.... In 1998 an
estimated 541,000 tons of food were produced in Havana for local consumption....
The Ministry of Agriculture...created the world's first coordinated urban agriculture
program that integrated: 1) access to land, 2) extension services, 3) research, and
development, 4) new supply stores for small farmers, and 5) organized points of sale
for growers and new marketing schemes, all with a focus on urban needs. (Murphy,
1999, pp. ii, 11–12)

Now, over half the produce consumed in Havana comes from the city's
urban gardens, and urban gardens nationally produce 60 percent of the

produce consumed. Significant amounts of produce from urban gardens is also donated to schools, clinics, and senior centers, and laws require urban gardens to use only organic methods (Murphy, 1999, pp. ii, 29; Pfeiffer, 2006, pp. 60–61).

The government's agricultural extension program in Cuba was also modified during the Special Period. Rather than focusing solely on information transfer from agricultural research stations and universities to producers, extension agents also focus on increasing community-based social learning processes. Extension agents in Cuba are based in communities long term. They assist producers with advice, inform them about workshops, and help them acquire inputs. These agents work closely with seed houses that "sell garden inputs, including seeds and tools, locally produced biological control products, biofertilizers, packaged compost, [and] worm humus." They also work with state research centers (Murphy, 1999, pp. 29–31).

Before the Special Period, Cuban scientists had already developed many organic methods of farming. During the crisis, the government embraced and supported these methods through a variety of programs (Pfeiffer, 2006, p. 58). Cuba strongly supported biological control of pests. According to Pretty,

> Key components of the strategy were the Centres for the Production of Ento-mophages and Entomopathogens (DREEs), where the artisanal production of bio-control agents takes place. By 1994, 222 DREEs had been built throughout Cuba and were providing services to cooperatives and individual farmers. (2005, p. 79)

Cuban farmers and gardeners also used intercropping, manuring, biopesti-cides, biofertilizers such as worm compost, pest- and disease-resistant plant varieties, crop rotations, cover cropping for weed suppression, green ma-nures, integration of grazing animals for nutrient cycling and meat produc-tion, and animal-based traction for tilling fields (Pfeiffer, 2006, p. 58). To increase organic urban fruit production, the Cuban government promoted planting and care of fruit trees in Havana (Murphy, 1999, p. 39).

The success of Cuba's rapid food system relocalization has been remark-able. In Havana, "there are now more than 7,000 urban gardens, and produc-tivity has increased from 1.5 kilogrammes per square metre to nearly 20 kilogrammes per square metre" (Pretty, 2002, p. 74). Furthermore, "sustain-able agriculture is encouraged in rural areas, where the impact of the new policy has already been remarkable" (Pretty, 2002, p. 74). Although animal protein production remains similar to the low 1994 levels in Cuba (Pfeiffer, 2006, p. 63), since 1995, average caloric intake per person has increased 33 percent from 1994 levels (Pfeiffer, 2006, p. 57). The transition away from industrial animal production toward a distributed production system has encountered obstacles such as lack of suitable animal feed, difficulties in waste disposal, and lack of refrigeration at local markets (Pfeiffer, 2006, p. 63). Still, Cuba has shown that, with a change in diet, grassroots organic agriculture can feed the people in a modern nation (Pfeiffer, 2006, pp. 57–58).

The Cuban example calls upon us to envision counterhegemonic food system activity as an important *urban as well as rural* response to fossil fuel depletion, climate change, and the socio-ecological damages of conventional farming. Prior to the crisis, urban Cubans, like many residents of cities worldwide, had come to see food production as backward and associated with disadvantaged and enslaved people (Murphy, 1999, pp. 3, 43). According to Murphy,

> In Cuba, as in many underdeveloped countries, gardening was never seen as a form of leisure.... Urban gardening was popularly associated with poverty and underdevelopment. [Prior to the Special Period,] Havana even had city laws prohibiting the cultivation of agricultural crops in the front yards of city homes. (1999, p. 12)

Since the Special Period, the importance of urban food production in a sustainable food system has become clear.

Cuba's response to the Special Period offers an example of successful government support for building a sustainable food system, but Cuba's situation is unique. It is literally an island of socialism in a largely capitalist world order. Those of us who live in capitalist nations can learn a great deal from the Cuban example regarding what works well in stimulating localized, sustainable food systems, but we should not rely upon the state to initiate change. In order to create sustainable food systems we must initiate change from below.

In order to provide additional examples of sustainable food system praxis, we will now explore food system (re)localization efforts in other parts of the world.

Food System (Re)localization in Other Parts of the World

The sustainable food movement is both extensive and global in nature. The European sustainable food movement presents a counterexample to Cuban governmental support of sustainable food because national government support for the movement has not been forthcoming. Ploeg notes that repeasantization runs against the political grain in Europe where it is characterized, not as a counterhegemonic movement, but as an addition to current methods of conventional farming:

The paradigm shift entailed in the process of European repeasantization has never been clearly articulated at institutional levels. This is because it runs counter to too many institutional interests associated with previous modernization processes. Admitting that such a far-reaching shift is occurring would imply that vested positions, scripts and routines need reconsidering.... Hence, shifts...resulting in multifunctionality are represented as something additional to farming, while the agricultural sector as a whole is conceptualized in terms of co-existence, meaning by this that alongside "productive farming" there are other "rural development" types of farming.... But earnings from "old" and "new" activities cannot be separated in order to compare them; it is their *unity* that matters. (2008, p. 155)

But European repeasantization, according to Ploeg (2008), does represent counterhegemony in that its processes are oriented toward increasing individual and community autonomy and decreasing dependence on the agribusiness and global food processing sectors. Repeasantization in Europe is characterized by diversification of outputs, on-farm processing of food products, shortening distribution lines to consumers, cutting out middlemen in distribution, increasing emphasis on craftsmanship and artisanship, shifting away from purchasing of inputs, reducing reliance on financial and industrial capital, and increased autonomy that "materializes in reconstitution of the resource base of the farm." European farmers engaged in sustainable food praxis are engaging in a *"regrounding of agriculture upon nature"* (Ploeg, 2008, pp. 153–156; emphasis in original) and a healing of the metabolic rift between humans and nature. European sustainable food praxis provides an important example for the U.S., not only because of policy similarities, but because of similarities in culture and levels of industrialization.

Sustainable food praxis is happening in Latin America as well, where it is producing some striking successes. In southern regions of Brazil, per capita economic output is higher than the average for the nation. The area is populated by small farms whose owners have increasingly adopted conservation tillage and no-till methods of farming. They have also diversified their crops and gardens; incorporated animals such as pigs, chickens, dairy and beef cattle, and fish in ways that promote the ecological health of their farms; and used green manure and cover cropping that is resulting in reduced problems with pests and soil erosion.

Perhaps the most important development in Brazil, however, is social. Farmers have formed associations to work and learn together and have increased their direct connections with their consumers (Pretty, 2007, pp. 116–119). Their sustainable food praxis is producing, not only local food for local consumption, but the community autonomy and *potentia* necessary for continuing and deepening the process of sustainable food praxis itself. The

Lovera family claims that "the biggest change over six years since conversion [to more sustainable practices] has been to their self-esteem—instead of being controlled by the agro-industries, they can choose what to raise and grow, when and how to market, and are fully linked to the internet and the outside world" (Pretty, 2007, p. 119). Pretty highlights the importance of community support networks:

> Working together is critical when large changes in our lives are to be made. Having a friend or neighbor to share the experience, to provide moral support, to succeed and to fail as you do, is an essential prerequisite for most landscape transformations. The fear of failure is enormous. It keeps people awake at night, it gnaws and worries at the edge of consciousness. Crossing the mental frontier is not easy, and most of us need some help, even if the local associations do not last forever. (2007, p. 117)

Pretty concludes that "cooperation was…fundamental to all agricultural and resource management systems throughout our early [human] history" (Pretty, 2007, p. 120). We can presume that cooperation is important to these systems today.

In India, Navdanya, a nongovernmental organization and network of seed savers and organic producers (Navdanya, 2009), focuses on provision of basic human needs from local resources: "food for the soil and her millions of microorganisms…food and nutrition for the farming family…food for local communities…and [only] unique products [such as spices] for long-distance trade and export" (Shiva, 2008, pp. 127–128). Navdanya has helped to set up 54 community seed banks throughout India where farmers share and save seeds. The organization is also "actively involved in the rejuvenation of indigenous knowledge and culture," and it has generated awareness of the hazards of genetically engineered seeds (Navdanya, 2009). Navdanya's efforts represent sustainable food praxis in that they promote individual and community autonomy, the building of community *potentia*, and socio-ecological sustainability.

People living on the land of the former Dravidian kingdoms, some of the driest lands outside of formal deserts in India, are also among the successful practitioners of sustainable food praxis in that country. In their region, land that had been historically farmed had been abandoned and invaded by Prosopis, a thorny tree. John Devavaram, Erskine Arunothayam, and Nirmal Raja spearheaded a new organization called the Society for People's Education and Change (SPEECH) through which they sought to "encourage rural people to increase their literacy by learning about their own places and what they can do to change them" (Pretty, 2007, pp. 126–129). SPEECH organizers called animators, "trained in participatory learning methods, conflict resolution, songwriting and storytelling," catalyze change in the villages by visiting every home to raise awareness of common community problems and the need for organizing and participation to address them.

As the people begin to work together, they form village committees to work on issues of their own choosing. Once the people begin to cooperate and develop trust among the group, they can begin to address problems of the land. Over a period of years, as trust and cooperation have grown, fields and rice paddies have been returned to productivity through the clearing of invasives, water harvesting, and the digging of wells for irrigation. Communities have also addressed issues of sanitation, education, and community economy (Pretty, 2007, pp. 126–131). Pretty concludes that "it is clear that new configurations of social and human relationships were a prerequisite for land improvements. Without such changes in thinking, and the appropriate trust in others to act differently too, there is little hope for long-term sustainability" (Pretty, 2007, p. 129).

In this case, as in many others, sustainable food praxis includes and requires the building of community *potentia* (Ploeg, 2008, p. 34). While value added is continually decreased in industrial, entrepreneurial farming, it is increased with "peasant driven rural development" at both the level of the farm and the level of the small farm sector (Ploeg, 2008, p. 156).

In Thailand, 100,000 farmers are organized into networks. This transformation was a direct response to the utter failure of industrial agriculture in that country in the 1990s as a result of the Asian economic crisis. This crisis left farmers with unpayable debts incurred in the purchasing of inputs at the same time that money earned from off-farm work of family members also dried up. As a result of this economic crash, farmers began to think differently. They have diversified their former monocultures, reduced or ended their use of inorganic fertilizers and pesticides, engaged in agroforestry, and focused their efforts on developing polycultures of intense production. Farmers continue to recruit new participants in their networks and to develop new farmer's cooperative groups called "local wisdom networks." The average farm size is less than two hectares (Pretty, 2007, pp. 120–121).

International organizations also situate localized efforts within a global counterhegemonic struggle. One such organization is La Vía Campesina which fosters international dialogue among those who make their living from the land outside the realm of industrial agriculture.

> By "building unity within diversity," the movement creates political spaces in which men, women, and youth from the Global North and South consolidate a shared identity as "people of the land," develop collective analyses, and struggle against the violence and disempowerment they experience daily as the dominant model's pro-

cesses of accumulation are unleashed in the countryside everywhere. (Desmarais, 2009)

La Vía Campesina and similar organizations serve as venues for sharing ideas and worldviews among sustainable food practitioners, thereby catalyzing praxis.

Examples of sustainable food praxis included here demonstrate its adaptability to diverse cultures and ecologies. Taken as a group, these examples and the example of Cuba's response during the Special Period demonstrate that sustainable food praxis is possible with or without government support, in widely diverse geographies, among people of diverse cultural heritage, and in socialist- or capitalist-oriented economies. The counterhegemonic thrust of sustainable food praxis does confront neoliberal globalization in that it seeks to distribute wealth and control within the food system rather than foster capital accumulation and monopolies within global markets. Still, citizens of capitalist-oriented nations can begin to build sustainable food systems within the capitalist order.

We now explore sustainable food praxis in the United States, arguably foremost among nations in its policy support for globalized capitalism and industrial agriculture.

Food System Counterhegemony in the U.S.

The Practical Farmers of Iowa (PFI) experiment with and practice many of the field techniques of sustainable farming practiced in other parts of the world (Bell, 2004). They also conduct agricultural research, both independently and in collaboration with university researchers. PFI member farmers are oriented toward smaller-scale, diverse, multifunctional practices that minimize or eliminate the need for off-farm (particularly chemical) inputs and that maximize total food yields, and they seek to improve the socio-ecology of the farm, the community, and surrounding nature. PFI farmers establish direct relationships with consumers (Bell, 2004, p. 205), a counterhegemonic market activity that minimizes or eliminates food processing industry control over their farms and farm practices. Understanding social forces that seem to promote or impede U.S. farmers from turning toward sustainable practices is important to understanding PFI activity as counterhegemonic. This exploration can help us identify challenges to increasing sustainable food counterhegemony in the U.S.

Bell argues that the social construction of farming and identity in the U.S. impede movement toward sustainable agriculture.

I argue that the reasons why most farmers...don't change to sustainable agriculture lie in matters of knowledge and its relationship to identity.... Farming requires the acquisition of a vast array of tricks of the trade—some ticks you buy...some tricks you learn...and some tricks you both buy and learn.... Once acquired, you can't take the time to continually question the stock of tricks you have at hand.... What you know is who you are.... Farmers are types of farmers...because of what they know, therefore do, and therefore identify with.... Knowledge has a history, a social history, and we connect ourselves to that social history.... Knowledge is a social relation.... And with identification with knowledge comes a sense of trust in it and those we received the knowledge from.... To give up a field of knowing and relating, is to give up both a field of self and its social affiliations and a field of trust in the secure workings of the world. (2004, pp. 14–15)

Once the soil has been damaged by years of monocultural production and once a farmer has invested borrowed money and cultivated his/her identity as an expert conventional farmer, there is a great deal at stake in changing course toward sustainable food production. Changing course is not only a question of adopting techniques that promote sustainability, but a matter of deep personal and social change. Ploeg makes a related point when he notes that "expressions of repeasantization are experienced as 'betrayals' [among conventional, entrepreneurial farmers], as forms of inappropriate behavior, and as blocking the free flow of resources badly needed for further expansion of entrepreneurial farming" (2008, p. 155).

Bell emphasizes that the process of making a change from conventional to sustainable farming often represents a phenomenological break in one's worldview and perception of self:

I think it significant that more than half the farmers we were able to speak to in detail about how they came to sustainable agriculture reported a similar experience: a sudden, disorienting change, in most cases during a period of severe economic stress, in which they had to rethink not only their farming practices but their practices of self. (2004, p. 154; see also Kneafsey, et al., 2008, p. 84)

Bell concludes that "identification with sustainable agriculture is commonly experienced as an intense, rapid, holistic crossing over. So much is at stake. A self. A farm. A way of knowing and doing them both" (Bell, 2004, p. 158; see also p. 236). According to Bell, for some farmers, this break includes an element of resistance to the manipulative hegemonic powers of industrial agriculture. He notes that some farmers speak of a new faith in sustainable agriculture that is rooted in conscience and truth free of power manipulations (Bell, 2004, p. 158). Similarly, in their study of sustainable producers in the United Kingdom and Italy, Kneafsey, et al., found that, "the motivations associated with their projects went beyond a response to difficulties experienced with conventional food production businesses, and were related to ethical positions on how food production should be practiced. In some instances, producers talked in visionary terms about the objectives that underpinned their involvement" (2008, p. 85). The findings of these researchers suggest a praxis in which the ideal and the material reinforce each other.

In his exploration of social drivers for sustainable agriculture, Bell (2004) cautions that there is no particular recipe of circumstances and/or personality traits that can determine who will and who will not move from conventional to sustainable farming:

> A number of surveys on sustainable farmers have been conducted, looking for the statistical factors that might help predict why one person turns to sustainable agriculture and another does not. They haven't found much…. Sustainable farmers are unremarkable in their age, their educational attainment, their political affiliations, their ethnicity, and their household sizes and structures. Their farms are noticeably smaller…. There is some…evidence to suggest that farmers are usually in their younger years, typically under fifty, when they make a commitment to sustainable agriculture…. it is probably an economic matter. Older farmers are more likely to be financially secure…. Economic stress is, of course, part of the standard crystallography of theories of social change…. Yet even here there is need for much analytic caution. For many of the sustainable farmers we spoke with, the mid-1980s farm crisis seems to have been the economic straw that broke the phenomenological back of their previous style of farming. It is probably no accident that PFI itself was founded in 1985, at the height of the 1980s farm crisis. (p. 159)

For some PFI farmers, however, the crisis came much earlier, two decades earlier in one case, and for some it came much later (Bell, 2004, p. 159). Still, Bell notes that most farmers who also suffered through the 1980s crisis are not sustainable farmers (p. 160). According to Bell (2004), some farmers who change course from conventional to sustainable farming do so because of "dialogic providence"—chance encounters with alternate knowledge systems and people involved in sustainable farming (pp. 162–163). Bell's (2004) notion that contact with ideas, combined with material experience, drives for sustainable food praxis among PFI farmers is an essentially Gramscian formulation counterhegemony. Health issues, such as personal experience with unintentional livestock poisoning with agricultural chemicals, play a role in some farmers' break with conventional agriculture, but health issues are still not an accurate predictor of a switch to sustainable methods (Bell, 2004, pp. 163–167).

Bell also found that sustainable farmers commonly related their interest in sustainable practices to an extended period of overseas living in their twenties or thirties, usually the Global South, sometimes as a teacher, a

Peace Corps worker, or a missionary (2004, p. 167). One farmer's Peace Corps experience led him to see U.S. culture from alternate perspectives and to develop a deep sense of mistrust regarding dominant domestic narratives of U.S. history (Bell, 2004, pp. 167–170). It seems that the combination of being exposed to new ideas and perspectives, combined with experiencing challenging material circumstances, can generate counterhegemonic sustainable food praxis among U.S. farmers.

Though issues of identity and knowledge construction may play important roles in keeping farmers on the conventional track, Pretty (2007) reminds us that losses accruing to farmers who convert to sustainable agriculture are not only psychological. Although industrial farming has its costs, transitioning to sustainable farming is also a costly and risky process: "During the transition period, farmers must experiment more and so incur the costs of mistakes as well as of acquiring new knowledge and information" (Pretty, 2007, p. 144). As noted above, sustainable food systems can be comparatively productive (see also a study by Pretty, 2002, pp. 82–83). PFI farmers have their own proof of success. In over 29 PFI trials, ridge tilling without herbicides resulted in the same yields as ridge tilling with herbicides (Bell, 2004, p. 222). But sustainable farming can be challenging as well. In the U.S., sustainable farmers typically farm fewer acres and, therefore, reap fewer benefits from agricultural subsidies. Furthermore, they often start from a financially weak position, which reduces their chances of getting loans and increases the likelihood that banks will call in their debts (Bell, 2004, pp. 237–238). Due to farmer indebtedness resulting from the agricultural treadmill, even temporary losses incurred in a transition toward sustainable farming may mean loss of the farm itself.

In a study of Iowa farmers, Carolan (2006) found that, among conventional farmers, there are also epistemic barriers to change. Carolan (2006) noted that many conventional farmers had difficulty seeing the benefits of sustainable agriculture, which often extend beyond the scale of the farm, but they could easily perceive the benefits of conventional methods, such as weed-free rows.

Conventional farmers in the U.S. may be among those who identify most strongly with industrial agriculture, but the increasing presence of sustainable producers, combined with rapidly growing consumer interest in sustainably produced food (Lyson, 2004, pp. 91–92), indicates that sustainable food counterhegemony is possible, even at the center of Empire.

Conclusions on Sustainable Food Praxis and Education

From local to the international levels, from Latin America to the U.S., from Europe to Africa and Asia, we see counterhegemonic praxis around sustainable food. I concur with Ploeg (2008) who characterizes this praxis as counterhegemonic resistance to a global capitalist order built upon enforcing dependencies. Ploeg states,

> The peasantry increasingly represents resistance…. The resistance of the peasantry resides, first and foremost, in the *multitude of responses* continued and/or created anew in order to confront Empire as the principal mode of ordering. Through, and with, the help of such responses, they are able to go against the tide. (2008, p. 265; emphasis in original)

We have analyzed examples of sustainable food action as counterhegemonic praxis. Praxis is about one's agency; it is about refusing to remain a mere object and victim of a hegemonic order. Ploeg draws connections between agency and local food activism saying, "The peasant principle is about facing and surmounting difficulties in order to construct the conditions that allow for agency…. It is also about subjectivity—the peasant principle implies that particular worldviews and associated courses of action matter" (2008, p. 274). He concludes that "the peasant principle is an emancipatory notion" (p. 262).

Local food counterhegemony, then, is a form of democracy made possible through the dispersal of food system control and the resulting openings for choice among both producers and consumers. Hamilton defines food democracy, a concept and practice that would be an ultimate outcome of food system (re)localization:

> The word "democracy" comes from Greek words meaning "people" and "rule"…. There are four essential pieces to the creation of a food democracy. The first is citizen participation; all actors in the food system must have a voice, and the contributions and concerns of each group must be considered. Second, informed choices are necessary. Questions, information, and knowledge about how food is produced are key. Third, a number of choices must be available to citizens. Although there are currently many types of food to choose from, most of the food is produced in the same faceless, industrial manner. Fourth, participation in food democracy must happen at the local as well as the national levels. One's food choices should be geared toward protection and development of the community, whether this means buying from farmer's markets or eating at locally owned restaurants. (quoted in Nabhan 2009, p. 193)

The sustainability crisis has put food in the spotlight. Community resilience requires that local communities and towns encourage sustainable food.

In light of this imperative, it is ironic that many colleges and universities with missions rooted in advancing the success of their communities and regions have either ignored or abandoned agriculture as part of their curricula. Higher education programs should teach both how to produce food and why doing so locally and sustainably is important. They should teach an integrated view of the food system as embedded in wider socio-ecological processes, and approach that, according to Buttel, has been rare in the U.S.:

> There are very important connections between the arenas of agricultural structure, rural environmental problems, and rural community and regional development. Unfortunately, these areas are almost always conceptualized and researched in isolation from each other in North American rural sociology. (1980, pp. 57–58)

Ploeg also notes that local food activism is hardly visible to professionals and leaders, many of whom have earned undergraduate and graduate degrees. He states that "current forms of repeasantization are barely understood by most scientists and politicians" (2008, p. 152). We might ask why this is. Perhaps it is because we only see what we have been trained and encouraged to see. Higher education for sustainability must teach people to recognize enforced dependency as a threat to socio-ecological health and survival. It must also teach them that there are viable, even vibrant, alternatives.

Part Three:

The Critical Role of Sustainability Education

Chapter 8:

The Critical Pedagogy of Sustainability:

A Call for Higher Education Praxis

Mainstream educational institutions are heavily invested in the maintenance and perpetuation of the old cosmology. Education is what molds and conditions people to "fit" into a society. Essentially, modern education conditions a person to be oriented to consumerism, competition, rationalism, detachment, individualism, and narcissism. Education supports the "consciousness" that has led to the ecological crisis and dilemma we face today.

—Gregory Cajete

To this point in this book, I have developed a critical social theory of sustainability as a normative conceptual framework for sustainability praxis, and I have developed a theory of enforced dependency as an unsustainable and pervasive phenomenon in the globalized world. I have also advocated sustainability praxis that engages both ideas and action in counterhegemonic dialogue, and I have argued that sustainable food system praxis should be a central focus for individuals and communities. I have also advocated place as an important context and construct for sustainability praxis that makes communities more resilient and less dependent on unsustainable social and economic systems.

In this chapter, I focus on higher education as a vehicle for addressing the sustainability crisis. I propose that college educators should practice a critical pedagogy of sustainability that includes involving students in service learning projects. College courses and programs should combine teaching and learning of a structuralized and critical view of the world-system (see Wallerstein 1974, 1976, 2008) with participation in transformative and transdisciplinary community action. Using this pedagogy, higher education could help move society toward sustainability.

I do not hold any illusions about the difficulties inherent in reorienting the pedagogies of U.S. colleges and universities toward sustainability. Not only do dominant educational priorities and practices demonstrate a glaring lack of concern with sustainability, higher education frequently perpetuates destructive worldviews and practices. If higher education is to effectively serve society, it must move quickly to critically examine the causes of current and emerging economic, social, and environmental crises. Higher education institutions must contextualize their conventional efforts within the context of sustainability, and professors and students must engage in praxis toward mitigating these crises whenever and wherever possible.

Even amidst the growing recognition of the seriousness of our situation, those of us who work to redirect higher education toward sustainable ends are often misunderstood within our departments, programs, and institutions. The hegemony of the status quo may act with blunt force to inhibit our efforts and silence our voices. In the face of these evident risks, I offer my critical pedagogy of sustainability as a contribution to sustainability praxis. This pedagogy addresses the sustainability crisis in its full complexity and can therefore serve as a platform for both individual educators and entire institutions to recontextualize and reorient their work.

The sustainability education I advocate in this chapter builds upon the working definition of sustainability and the critical social theory of sustainability articulated in chapters one and two. I argue that the critical pedagogy of sustainability must confront and disrupt patterns of enforced dependency elucidated in chapters three and four, and this pedagogy is place-centered for reasons articulated in chapter five. It engages students, faculty members, and the broader public in transdisciplinary praxis aimed at sustainable (re)inhabitation of place. My arguments are addressed primarily toward higher education in the United States, though they may also be relevant in other settings, including sustainability-oriented community activism. Below, I develop a set of claims that form the central tenets of the critical pedagogy of sustainability.

Enforced Dependency: A Critical Frame for Sustainability Education

As a central component of the critical pedagogy of sustainability, enforced dependency should be studied from a world-historical perspective.

In order to reveal enforced dependency as a pervasive and constitutive force within late capitalist globalization, the critical pedagogy of sustainability must engage with political, economic, social, cultural, and ecological facets of this global/local phenomenon. I believe it is especially important for sustainability educators to address the historical development of enforced dependency since European colonization. Sustainability educators should especially emphasize the post–World War II period and the Bretton Woods institutions as foundations for neoliberal capitalism. Doing so encourages students to appropriately situate critiques of late capitalist globalization in their historical and geographical contexts. Such study encourages students to conceptualize globalization as the current embodiment of a lengthy historical trajectory of colonization. This reading of history highlights relationships of domination and oppression of both people and nature as two sides of the same coin of modern capitalism and its commodifying and alienating economy and culture. It encourages students to make connections between such abstract and seemingly disparate phenomena as international finance and debt, the global oil industry, and the Green Revolution, on the one hand, and the growth of megacity slums and perpetual underdevelopment in the Global South on the other. A study of enforced dependency since the colonial period also emphasizes that economy is indeed political—that economic structures and practices represent means of producing and repro-ducing social relationships that reinforce the hold of the dominant over the oppressed and nature (see McLaren & Houston, 2005, pp. 176–177).

Because the U.S. is the premier global hegemon and enforcer of late cap-italism, the critical pedagogy of sustainability explicitly engages in critique of the United States as a dependency-enforcing empire (see McLaren and Farahmandpur, 2005, pp. 250–251). Although regulation of footloose global capital eludes nation states within the regime of neoliberal globalization, the U.S. still functions as an empire. It maintains a military force equipped with weaponry that dwarfs that of any other nation, and it has used and continues to use its military force—both directly and as a threat—in an effort to maintain the political and economic hegemony of the Washington consensus. The petrodollar system of conducting OPEC oil sales solely in U.S. dollars and dollar hegemony in international lending also promote export-oriented development in the Global South, the flip side of which is extreme overcon-sumption in the U.S., the source of the dollars much needed by debtor nations worldwide. Global price competition driven by both export-oriented development and free trade regimes, depletion of U.S. domestic oil reserves, and domestic deindustrialization have resulted in the outsourcing of the U.S. manufacturing sector, an extremely negative balance of trade for the U.S., and skyrocketing U.S. debt. At this point in time, the debts of the world's only remaining superpower threaten to destabilize the entire world-system.

Given the breadth and depth of U.S. complicity in perpetrating the sus-tainability crisis and the crisis of capitalism itself, I propose that U.S. citizens should engage in a form of self-reflexivity that critiques their collective agency as a nation. We in the U.S. have contributed in many ways—

willingly and knowingly or not—to widespread collective violence and ecocide ranging from war making to consumerism, and our nation has been the prime mover in creating systems of enforced dependency worldwide. The critical pedagogy of sustainability must engage students in critique of the U.S. as an empire in order to elucidate the mechanisms of enforced dependency in the world-system. Students should be encouraged to question the sustainability of this dependency and to envision and begin to create alternatives.

As William Clark notes in *Petrodollar Warfare* (2005), engaging in such critique is not inherently anti-American. A healthy democracy must have room for critical perspectives, especially those aimed at improving the long-term prospects for the democratic project itself. The critique I advocate is not snide America bashing, but a careful assessment of the effects of U.S. policies on the long-term health and sustainability of the nation and the world-system. Only when we see how we arrived at this precarious point in history can we begin to envision alternative, sustainable options.

Conjoined Decolonization and (Re)inhabitation

The critical pedagogy of sustainability focuses on decolonization and (re)inhabitation as two aspects of a unified sustainability praxis. The theory of enforced dependency unites the two like two sides of a coin.

The critical pedagogy of sustainability represents the synthesis between critical pedagogy and place-based education proposed in Gruenwald's (2003) article "The Best of Both Worlds: A Critical Pedagogy of Place." Gruenwald recognizes the close relationships between the exploitation of people and nature and notes that, while the critical tradition has emphasized the need for social transformation, it has often neglected to situate its analysis within a broader ecological context. He argues also that the pedagogy of place typically neglects to address the political context within which places evolve as human/nature constructs marked by social contradictions.

Gruenwald's work represents an important launching point for developing the critical pedagogy of sustainability. He proposes using decolonization and reinhabitation as frames for critical praxis. According to Gruenwald,

> Being in a situation has a spatial, geographical, contextual dimension. Reflecting on one's situation corresponds to reflecting on the space(s) one inhabits; acting on one's situation often corresponds to changing one's relationship to a place. Freire asserts that acting on one's situationality, what I will call decolonization and reinhabitation, makes one more human. It is this spatial dimension of situationality, and

its attention to social transformation, that connects critical pedagogy with a pedagogy of place. (2003, p. 4, emphasis in original)

Decolonization and (re)inhabitation are central themes in the critical pedagogy of sustainability.

Decolonization is a counterhegemonic activity that involves critically
examining the colonizing aspects of hegemony as these manifest in both the
inner landscape of the person and the landscape of the globalized world.
Hegemony can take the form of overt dispossession, oppression, and control,
such as that embodied in colonial conquest, imperialism, and externally
imposed neoliberal economic and social policies. Hegemony also manifests
in the continual colonization of the lifeworld of people who internalize the
value system and the logic of neoliberal globalization, with the devastating
result that they collude in their own oppression (Gramsci, 1971/1999, pp.
57–58).

In the Global South, cultural hegemony insinuates itself in people who
develop tastes for products of imperialistic cultures and concomitantly come
to value money over self-sufficiency, thereby enforcing their own dependency within the world-system. The colonized admire the "lifestyles of the
rich and famous," prefer modern fast foods and soda over traditional local
foods and beverages, and come to see non-modern lifeways as inferior to
industrialized living. The colonization of the lifeworld enforces the dependency of the colonized upon the unsustainable global economy and stimulates
unsustainable and oppressive consumerism.

In industrial societies, cultural hegemony often takes the form of repressive desublimation, a situation in which the material and sensuous desires of
people are met (desublimated) while their freedom to break from the status
quo is progressively circumscribed. According to Marcuse (1964), repressive
desublimation masks deep social contradictions by appearing to fuse interests
and goals across classes and other social divisions. Repressive desublimation
also drives unsustainable consumption in core countries and regions.

Both domination by force and cultural hegemony foster enforced dependency, a social power dynamic akin to colonization. The critical pedagogy of sustainability confronts hegemony through a process of critical self-
reflexivity in which students confront their own internalized assumptions
while also confronting parallel assumptions that support the late capitalist
system. According to McLaren and Farahmandpur, "Self-reflexivity is a
process that identifies the source of oppression, both from the outside and
from within, through participation in a dialectical critique of one's own
positionality in the larger totalizing system of oppression and the silencing of
others" (2005, p. 110). The theory of enforced dependency provides an

excellent critical lens for engaging in a self-reflexive process of critiquing one's own consciousness within the context of larger social systems. It effectively confronts the hegemonic notion of the "global village" that benefits all participants and "raises all boats" through capitalist "development."

The critical pedagogy of sustainability also recognizes domination and oppression in many forms—racism, sexism, and domination of nature—as contributing factors to enforced dependency and as oppressive social forces in their own right. Combating these forms of oppression is an important aspect of (re)inhabitation. The revolutionary multicultural pedagogy of McLaren and Farahmandpur calls for counterhegemonic alliances that honor the full complexity and diversity of culture and identity:

> A revolutionary multicultural pedagogy links the social identities of marginalized and oppressed groups—particularly the working class, indigenous groups, and marginalized populations—with their reproduction within capitalist relations of production. It also examines how the reproduction of social, ethnic, racial, and sexual identities, as particular social and cultural constructs as well as shared histories of struggle, are linked with the reproduction of the social division of labor. It therefore moves beyond the often fragmented and atomized entrapments of identity politics, which frequently polarizes differences instead of uniting them around the common economic and political interests of marginalized social groups. (2005, p. 153)

The revolutionary multicultural pedagogy of McLaren and Farahmandpur (2005) recognizes the full social complexity of hegemony within late capitalist globalization, and it articulates with place-centered, counterhegemonic approaches to combating enforced dependency. It does so by recognizing the diversity among the oppressed with regard to particular histories and identities as these have been shaped by class, race, ethnicity, gender, and place and by recognizing that diverse experiences of oppression within the globalized world are systemically linked (McLaren, 2005, p. 87). The pedagogy of McLaren and Farahmandpur (2005) offers an important critical lens for conjoined counterhegemony and (re)inhabitatioin. Their focus on systemic oppression creates a narrative thread of class struggle (2005, p. 173). This narrative calls for unity in difference in counterhegemony and, when viewed through the critical lens of enforced dependency, for (re)inhabitation. Unity in difference (see McLaren & Farahmandpur, 2005, p. 175) forms an essential foundation for the counterhegemonic praxis of the critical pedagogy of sustainability.

McLaren's (2007) articulation of the roles of critical theory and pedagogy today reflects an important emerging vision for a place-centered, counterhegemonic pedagogy of sustainability:

In our pursuit of locally rooted, self-reliant economies; in our struggles designed to defend the world from being forced to serve as a market for corporate globalists; in our attempts at decolonizing our cultural and political spaces and places of livelihood; in our fight for antitrust legislation for the media; in our challenges to replace indirect social labor (labor mediated by capital) with direct social labor; in our quest to live in balance with nature; and in our efforts to replace our dominant culture of materialism with values integrated in a life economy, we need to develop a new vision of the future, but one that does not stray into abstract utopian hinterlands too far removed from our analysis of the present barbarism wrought by capital. (p. 307)

McLaren addresses important areas for sustainability praxis. His emphasis on decolonization of "cultural and political spaces and places of livelihood," on replacing "indirect social labor…with direct social labor," on living "in balance with nature," and on replacing materialism with "values integrated in a life economy" all point toward the need for place-centered critical pedagogy and praxis for sustainability. McLaren's (2007) analysis represents movement toward a synthesis between critical pedagogy and place-based education as envisioned by Gruenwald (2003). He says, "While critical pedagogy offers an agenda of cultural decolonization, place-based education leads the way toward ecological 'reinhabitation'" (McLaren, 2007, p. 4).

Placing (re)inhabitation at the center of counterhegemonic sustainability education and praxis highlights tensions between indigenous scholars and educators and critical theorists rooted in the Western intellectual tradition. According to Grande,

The Western foundation of critical pedagogy…presents significant tensions for indigenous pedagogy and praxis. The radical constructs of democratization, subjectivity, and citizenship all remain defined through Western epistemological frames. As such, they carry certain assumptions about human beings and their relationship to the natural world, the view of progress, and the primacy of the rational process. The implications of such tensions are myriad and significant, giving rise to competing notions of governance, economy, and identity. (2007, p. 320)

According to Grande (2007) the central difference between indigenous and Western critical approaches has to do with the primacy of class struggle within critical theory and the primacy of colonization within indigenous critique. Grande argues that, though indigenous critique and critical theory exist in tension with one another, they are also complementary in important ways:

While critical indigenous scholars do not equivocate the ravages of capitalism, a Red critique of critical pedagogy decenters capitalism as the main struggle concept and replaces it with colonization. Comparatively, the colonist project is understood as profoundly multidimensional and intersectional; underwritten by Christian fundamentalism, defined by White supremacy, and fueled by global capitalism. The fundamental difference shifts the pedagogic goal from the "transformation of exist-ing social and economic relations" through the critique and transformation of capi-talist social relations of production (i.e. democratization) to the transformation of existing colonialist relations through critique and transformation of the exploitive relations of imperialism, (i.e. sovereignty). This in not to say that the politi-cal/pedagogical projects of democratization and sovereignty are mutually exclusive; on the contrary, in this new era of empire, it may be that sovereignty extends de-mocracy its only lifeline. (p. 320)

Given the imperative to (re)establish sustainable human/nature relationships in place in order to ameliorate the effects of rapidly converging socio-ecological crises and move toward sustainable living, Grande's assertion of the primacy of sovereignty to the survival and perpetuation of democratic political processes is insightful indeed.

The critical pedagogy of sustainability recognizes that indigenous cri-tique and critiques based in critical social theory derive from distinct cultural and historical traditions and represent distinct projects in social transfor-mation. This pedagogy proposes, however, that enforced dependency, even in its diverse incarnations and its complex interrelationships with other forms of social oppression, is a shared experience of the vast majority of people worldwide. The critical pedagogy of sustainability, like the pedagogy of McLaren (2007) and McLaren and Farahmandpur (2005), does not seek to subsume indigenous critique, but instead to forge strategic alliances between Western critical educators and indigenous critics and pedagogues. I propose that the theory of enforced dependency represents a vehicle for doing so.

As discussed in chapter five, indigenous culture and language traditions that embody the material, ethical, and spiritual knowledge of how to live in particular places offer important platforms for sustainable (re)inhabitation of place (Armstrong, 1995). In many places where colonization of place and the lifeworld of people has been extensive, few indigenous cultural traditions and little indigenous knowledge remain, and the process of inhabitation must begin anew for all inhabitants. Place-based adaptation, informed when and where possible and appropriate by sustainable indigenous and localized traditions, is central to the critical pedagogy of sustainability. This praxis works to decrease enforced dependency at the same time that it increases diversity and resilience within communities.

Placing decolonization and (re)inhabitation at the center of sustainability education facilitates sustainable human/nature relationships. As I have argued throughout this book, communities rich in social *potentia* that produce for themselves the necessities of life contribute to the dismantling of enforced dependency. They also create balance between human needs and the health of the natural world. Sustainable communities require reciprocating relationships between people and nature in place, relationships that foster the health and integrity of both individuals and the entire human/nature complex that is place.

The theory of enforced dependency highlights the need for conjoined critique and (re)inhabitory praxis. Diverse forms of place-centered living are also required for a sustainable future in which the mobility of people and the long-distance transportation of goods will soon be limited. (Re)localization and concomitant counterhegemonic deglobalization are central foci of the critical pedagogy of sustainability (McLaren & Houston, 2005, p. 182).

Through its place-centered and (re)inhabitory foci, the critical pedagogy of sustainability builds upon the foundations of critical pedagogy and integrates indigenous critical perspectives in ways appropriate to confronting the sustainability crisis. Critical theory typically has not focused on place as a nexus for counterhegemonic, anti-capitalist, sustainability praxis. Instead, critical theorists have focused on class-based analysis and struggle. As Grande (2007) emphasizes, choosing the focus for critique need not be an either/or proposition because class relationships both produce and are produced through enforced dependency and other forms of oppression and exploitation. McLaren and Kumar (2009) note that "colonization, and economic exploitation linked to capitalism, are demonstrated to be co-constitutive of plundering the oppressed...."

Because the conquest of place (which includes the subjugation of land-based cultures and people) is antithetical to the (re)establishment of diverse, resilient, sustainable communities, decolonization of places, peoples, and the lifeworlds of individuals must be a central focus for sustainability-oriented praxis. According to Gruenwald, decolonization and (re)inhabitation form two sides of the same coin in the critical praxis of sustainability (2003, pp. 9–10). Place-centered critical pedagogy focuses on both counterhegemony and (re)inhabitation: the dismantling of one system through the simultaneous building of something new.

The Contradictory Notion of Sustainability as Education

The critical pedagogy of sustainability works through educational
processes that are themselves embodiments of sustainability.
It is, therefore, sustainability as education.

Sustainability educator Rick Medrick has outlined principal values and practices of education as sustainability:

> Education as sustainability explores the theories, processes, and conditions through which individuals, groups, and organizations learn and transform in ways that support a sustainable future…. It is essentially transformative, constructivist, and participatory. It is also integral…in that it seeks to incorporate as many insights and perspectives from as many disciplines as possible to understand events, experiences, and establish contexts…. It is also essential to incorporate education for sustainability into this investigation, exploring human impact on the natural environment as well as the influence nature has on humans. (2005, p. 1)

Medrick's conception of education as "essentially transformative, constructivist, and participatory" (2005, p. 1) correlates with the pedagogy of other well-known critical educators who recognize the value of generating inquiry and class content from among the students themselves, thereby subverting through educational praxis the typical authority structure of the classroom (Brookfield, 1987; Freire, 1970/2000; Mezirow, 2000; Mezirow & Associates, 2000). According to Medrick, education as sustainability strives to integrate disciplinary knowledge and practice, to understand the inextricably intertwined relationships between humans and nature, and to engage in respectful dialogue with diverse people with the understanding that no one person or group has all of the answers to the sustainability crisis. Sustainability as education also embodies a relatively egalitarian conception of leadership, an openness to diverse knowledge and ways of knowing, a prioritization of values in which economic well-being is only one among multiple important values, and a conception of humans as part of the natural world (Medrick, 2005).

Medrick's notion of sustainability as education highlights that outcomes related to *what* we teach derive heavily from *the way* we teach. Students will not learn sustainability praxis through oppressive educational processes. As argued in chapter five, sustainability praxis must promote forms of social organization characterized by authentic, grassroots servant leadership.[1] As a microcosm of society, the educational process must foster such leadership

---

[1] My theories on sustainable leadership are articulated in my article on this topic published in the *Journal of Sustainability Education* (Evans, 2011).

among students. Critical teaching calls upon students to name the world (Freire, 1970/2000, p. 88) and not for the teacher to name it for them (a process that would reinforce hegemonic patterns of domination and oppression). But critical sustainability educators are caught in an untenable bind as they seek to work *both* counterhegemonically *and* democratically with students whose worldviews have been colonized such that their common sense opinions and ideas reflect the status quo (see Brookfield, 2005, pp. 205–209, and Marcuse, 1964). The grading system and other aspects of teacher authority built into higher education only exacerbate this quandary.

The quandary of the sustainability educator is illuminated in Habermas' (1984) theory of ideal speech that is embedded in his theory of critical praxis for democracy (see Brookfield, 2005, chap. 9). At the center of Habermas' (1984) theory is the presumed possibility for people to engage in authentic, ideal speech acts in which they negotiate truth through discourse. Such authentic communication must pass four validity tests: that the claim is true, comprehensible, and sincere, and that it is right for the speaker to make the claim (Carr & Kemmis, 1986, p. 141). The notion of authentic discourse as a foundation for democracy implies that people can forego the use of speech as an instrument of social power, in order to communicate rationally and inclusively in search of truth (Carr & Kemmis, 1986, pp. 140–144).

Using speech as an instrument of domination would be antithetical to educational praxis aimed at eliminating domination and oppression, and authentic educational praxis would presumably involve students in every phase of praxis within a course, including planning curricula. But, in a thoroughly colonized culture, such a process might never become counterhegemonic. Freire states the problem thusly:

> The central problem is this: How can the oppressed, as divided, unauthentic beings, participate in developing the pedagogy of their liberation? Only as they discover themselves to be "hosts" of the oppressor can they contribute to the midwifery of their liberating pedagogy. (1970/2000, p. 48)

The question of how to engage students in authentic counterhegemony in sustainability education remains unresolved, and sustainability educators must negotiate this question as best they can in their day-to-day interactions with students. Below, I will discuss how I have dealt with this question in my teaching.

Sustainability Education in the Classroom

Though teacher authority is virtually unavoidable in higher education, sustainability educators can engage in authentic servant leadership. In doing so, they must foster student agency and avoid, to the extent possible, reinforcing blind compliance with authority.

Because students bring the contradictions of the world with them into the classroom, I believe they would be unlikely to take a strongly counterhegemonic approach if they were to lead in developing the processes of inquiry and the thematic content of my courses. I believe many, if not most, of my students are predisposed to act unconsciously to preserve the status quo with regard to social power and social systems. Unless they are presented with cogently developed counterhegemonic arguments that elucidate unsustainable and oppressive social contradictions, I believe my students are unlikely to critically examine the unquestioned "truths" that support the hegemonic order.[2]

In my experience a handful of students at most in a class of 20 to 35 are hungry from day one to engage in counterhegemonic discourse. Most would rather avoid examining deeply held assumptions, some are frightened and/or deeply disturbed by what they learn through the process of doing so, and a few are outright hostile to critique that calls their views into question. I believe that course readings chosen by the professor, instructor-facilitated class discussions, and structured analytical reflections on course content provide for most students a necessary precursor to critical praxis. As students are presented with alternate ways of viewing what has been taken for granted, they can begin the process of re-visioning and re-creating their worldviews. I find that many of my students (especially those who are freshmen of traditional college age) seem to have been very sheltered, at least regarding how the capitalist world-system works. They typically have not formulated a deeply critical self and social consciousness, a form of personal development that is not generally valued in capitalist society. Brookfield (1987) offers an explanation for why the process of beginning to think critically is both threatening and empowering for such students:

> Asking critical questions about our previously accepted values, ideas, and behaviors is anxiety-producing. We may well feel fearful of the consequences that might arise from contemplating alternatives to our current ways of thinking and living; resistance, resentment, and confusion are evident at various stages in the critical thinking process. But we also feel joy, release, relief, and exhilaration as we break through to new ways of looking at our personal, work, and political worlds. As we

---

[2] See Mezirow (2000) on the concept of disorienting dilemmas.

abandon assumptions that had been inhibiting our development, we experience a sense of liberation. As we realize that we have the power to change aspects of our lives, we are charged with excitement. As we realize these changes, we feel a pleasing sense of self-confidence. (p. 7)

Though I see teacher-led engagement with counterhegemonic thinking as a necessary precursor for student engagement in critical praxis, the process is not democratic. When I engage students in this way, I do my best to help them use written and verbal analysis and reflection to think independently in order to reinforce as little as possible culturally ingrained blind obedience to authority. Following is a description of practices I have employed in both upper- and lower-division courses to help students think both critically and independently.

During the first week or two of class, I flat out tell students that I am not neutral and explain the impossibility of my taking a neutral stance in the classroom. I tell them that the course as a whole presents an argument that I have developed over time through my extensive and conscientious study of pressing socio-ecological issues, and I ask them to seriously consider the arguments presented in the class. I also tell them that I realize I do not have a monopoly on the truth and that I do not expect them to uncritically accept whatever I say. I do my best to model this openness in class discussions and in constructing and grading written assignments.

In my upper-division End of Oil class, for example, I invite students to analyze and question course texts and to listen to diverse views. I point out some of the central claims being made in course texts and ask students to be cognizant of the arguments developed to support these claims. I ask them to consider to what extent the author is/is not successful in developing his/her supporting arguments and to consider how the given claims and arguments apply to real-life situations. I believe this highlighting of claims and supporting arguments helps students understand that truth is contingent and that, based on the soundness of a given discussion and on its applicability to the world as they know it, they may accept all, part, or none of a given argument. Meanwhile, I do my best to expand their worldviews by engaging them through films, reading, and discussion in a counterhegemonic exploration of the legacies of colonialism and the post-war world-system. This exploration addresses the Bretton Woods systems and institutions, neoliberal economics, oil dependency, and fossil-fuel-dependent development patterns, including suburbanization and the Green Revolution.

Throughout this process, I assume little to no counterhegemonic knowledge of history or the social and economic injustices of the capitalist world-system. I hold students responsible for analyzing material only within the framework of ideas developed in the class to date. If a student raises a question or makes a point that will be addressed in a future class reading, I reference that work and say that author X will address that very idea by either supporting or critiquing the idea/question offered by the student. I invite the student to listen to that author's points later in the term and consider them within the framework of her/his evolving knowledge base. This process helps the student see that s/he is capable of asking insightful questions and that s/he is also capable of evaluating responses to those questions.

In drafting essay questions, I am careful not to assume that students agree with particular authors, with particular class content, or with me. I draft these questions in ways that leave openings for students to offer views counter to those presented in class. I remind students, though, that when they present arguments, their work must be supported by more than sheer opinion. Students must present appropriate evidence and develop sound arguments that either concur with or contradict material presented in class.

Working in an atmosphere where student work must be graded creates additional contradictions for critical educators because the practice of assigning grades is highly authoritarian. Furthermore, deciding the relative value of each student's work is a highly questionable process that typically only compounds the cumulative effects of past support/approval or rejection/neglect of students as learners and people. I tell students that I am looking for depth of thought in their essays, but how can I truly determine when a student has made a significant move toward depth of thought as compared to her/his past thought and expression? While it is true that, later in the term, I have a chance to compare each student's written work to her/his previously submitted essays, one term offers me only a small window into the life of a student and her/his personal and intellectual growth. If I give everyone high grades, then students who are motivated by competition are likely to stop applying themselves fully. Therefore, I must do the best that I can in assigning grades through a process of comparing each student's work to his/her past work and to the work of other students in the class.

In order to encourage students' agency in the face of the judgments I am required to make on their written work, I treat the act of grading as, at least in some respects, a conversation. I respond to their work with questions and with comments that question or build upon the claims they make. I do my best to offer new sources for inquiry and to propose new lenses for analysis. This process by no means resolves the contradiction represented by the grading process. It does, however, offer conversation in addition to offering judgment.

Perhaps the most important feature of my critical teaching is that I do not seek power *over* my students. I hope to empower them. My critical teaching is, therefore, akin to the concept of servant leadership (Greenleaf, 1970/1991). Although my pedagogical practices do not allow me to transcend the contradictions inherent in sustainability education in a colonized world, I believe my work calls upon students to engage in important preparatory work toward authentic praxis that they may undertake as part of my course and/or at other points in their lives.

I believe teacher authority can be used toward sustainable ends and be anchored in authenticity rooted in a counterhegemonic reading of the world. Learners can be encouraged to voice their own readings of the world in various contexts such as small-group discussions, full-class discussions, and analytical/reflective writing. Brookfield notes that, given their life experiences and acculturation to capitalist values, students may interpret activities such as class discussions to be a competition rather than an opportunity for authentic communication (2005, chap. 5). Metacommunication about the purposes of course activities may alleviate the drive to compete for some students—or for others, it may simply frame the perceived competition differently. There is only so much the professor can do to lay the groundwork for students to engage authentically in naming the world. The professor can model servant leadership and encourage students to learn to lead.

The organic crisis of capitalism and the sustainability crisis appear to be upon us. These developments represent a historic opportunity—as well as the potential for widespread socio-ecological destruction. As the crises deepen, the critical pedagogy of sustainability can prepare students to engage in much-needed counterhegemonic sustainability praxis. I will continue to engage in the critical pedagogy of sustainability while, at the same time, realizing that I cannot entirely resolve the authority contradictions inherent in its process.

The Importance of Service Learning

Because it can help students make leaps from comprehension to praxis,
service learning is an important aspect of the
critical pedagogy of sustainability.

I have my doubts that a strictly idealist critical pedagogy of sustainability, even one that works powerfully with powerful ideas, can fundamentally alter the worldviews of students. After all, the classroom is only a small part of life in a world where hegemony infuses all aspects of the lifeworld. A course or two may not be enough to provoke radical departures from the way one has interacted with family and friends or to cause a fundamental reorientation of one's values, life goals, career orientation, or lifestyle. In cases where counterhegemonic learning creates dissonance with a student's identity and day-to-day experiences, the student may find it challenging to integrate what s/he is learning in the classroom with life as s/he knows it. When the class ends, budding counterhegemonic aspects of the student's worldview may simply die on the vine.[3]

People also need to be involved with long-term sustainability projects in order to change ingrained cultural patterns, economies, and philosophies. Higher education typically does not offer a venue for such projects. Critical educators in colleges and universities, therefore, face significant challenges in working toward social change that extends beyond the classroom in both space and time.

But, if stimulating sustainability-oriented praxis is the central goal of the critical pedagogy of sustainability, its process must require students to do more than think about agency. For many students who have grown up in suburbia not even knowing their neighbors, involvement in community work of any kind may be unfamiliar and somewhat personally challenging (see Loeb, 1999).[4] Service learning, a pedagogical method that includes conceptual classroom learning, action in the community that is related to conceptual learning, and structured reflection on the relationships between classroom and experiential learning, appears to be effective in helping students overcome this hurdle. Students who experience at least some community engagement as part of a course are also more likely to continue to engage in community work both in- and outside the academy (Astin, Vogelgesan, Ikeda, & Yee, 2001), and these experiences also help students learn to take on more effective leadership roles and to see themselves as more able leaders

---

[3] I believe that dissonance between lived experience and counterhegemonic classroom content occurs most frequently in relatively wealthy societies where repressive desublimation creates widespread support for the status quo (Marcuse, 1964). In such an atmosphere, going along with everyone else makes life seem much easier, even though one's freedom, including freedom of thought, may be progressively circumscribed as a result.

[4] Loeb's (1999) *Soul of a Citizen: Living with Conviction in a Cynical Time* is an important text for helping students move from study to action and for helping them to consider action itself as a continued learning process. Loeb's critique of the "perfect standard" (the idea that activists should be perfectly consistent and correct in their convictions and actions) is a pervasive, action-inhibiting judgment. It is important to discuss this standard with students who engage in social action. Loeb also argues that people must learn activism and leadership step-by-step and with the support of others, another important point to share with students doing service learning work.

(Newman, Bruyere, & Beh, 2007; Astin, et al., 2001).[5] Based on these observed outcomes (Astin, et al., 2001; Newman, Bruyere, & Beh, 2007) and on my experience as a sustainability educator, I suggest that, when students have an opportunity to build personal relationships within a community of practice and to apply their counterhegemonic learning through engagement in community projects and political struggles, they may be more likely to engage in lifelong praxis.

Such was the case with some Environmental Studies students in my Community Internship class. After a great deal of time spent working on an issue or project and a considerable amount of reflection on how that work related to their personal commitments and goals, many students became emotionally and intellectually attached to their work. Even beyond the internship course, these students continued to work with organizations or issues related to their internships, and they maintained relationships with others who were similarly involved.

This depth of attachment can be difficult to foster in courses such as my End of Oil course that have large numbers of students (as many as 35) and a great deal of critical conceptual material to cover. Still, even a brief experience with service learning can benefit these students by providing them with an example of how that work can be rewarding and enjoyable and how it can relate to learning. In chapter nine, we will explore student views on service learning and other aspects of their experience in The End of Oil course.

Local food projects can be excellent vehicles for incorporating service learning in sustainability-oriented courses. These projects

- embody multiple aspects of sustainability-oriented praxis,
- can easily be linked with counterhegemony through class discussion and reflection,
- help generate emotional connection to cyclical and long-term processes of nature by tapping into natural cycles of growth, death, and decay,
- emphasize the importance of (re)inhabitation of place in both its social and environmental aspects,
- can occupy large numbers of people at work at one time, and
- are generally at least somewhat fun and can take place in beautiful places or places that are in the process being made beautiful.

---

[5] Alumni who engaged in sustainability-oriented service learning at Allegheny College also reported enhanced cognitive development and improved communication skills as a result of these experiences (Keen & Baldwin, 2004).

Local food projects are also particularly important to addressing the emerging peak oil and gas crisis which threatens to destabilize industrial food systems, and these projects embody progress toward (re)inhabitation.

My advocacy of service learning is common among sustainability educators (Keen & Baldwin, 2004; Ward, 2006). Still, as Bawden (2004) argues in "Sustainability as Emergence: The Need for Engaged Discourse," institutions of higher learning are typically reluctant to engage with civil society in addressing the challenges of sustainability—challenges that may be the subject of research at these same institutions. He notes that higher education has become abstracted from the communities and societies it serves and that knowledge generated by the academy is often disconnected from values. In response to these failings, Bawden (2004) advocates scholarship that engages directly in sustainability-oriented social change. He calls for a revised and revitalized role for colleges and universities as leaders and participants in communities of practice. M'Gonigle and Starke (2006) advance similar arguments in *Planet U: Sustaining the World, Reinventing the University*. These authors argue that institutions of higher education should serve socially revolutionary roles in service to sustainability.

At its best, sustainability education involves critique as well as praxis within a community setting. As I argue in chapter five, as both context and participant in this transformational process, place provides a container within which systems of reciprocity can be observed, established, and lived and where responsibilities to other people are concrete and immediate rather than abstract (see Armstrong, 1995; Kemmis, 1990; Martinez, 1997; Shuman, 2000; Summers & Markusen, 1992/2003). Place-centered service learning is therefore highly appropriate to the critical pedagogy of sustainability.

Transdisciplinarity

The critical pedagogy of sustainability is inherently transdisciplinary.

After studying a number of sources on transdisciplinarity (Dölling & Hark, 2000; Klein, 2001; Lenhard, Lücking, & Schwechheimer, 2006; Meyer, 2007; Nicolescu, 2002; Schroll & Stærdahl, 2001), I developed the following definition for a faculty committee on interdisciplinary programs at my institution:

> Transdisciplinary studies, research, and action focus attention on thematic threads that inform complex, real-world issues and challenges such as globalization, climate change, and sustainability. Transdisciplinary scholars and practitioners engage with these issues and challenges using integrative approaches to knowledge-making with the aim of transforming the subject(s) of study by informing purposeful human ac-

tivity. Trandisciplinary research draws upon disciplinary methods of knowledge-making as means to generate and synthesize new knowledge, but transcends the disciplines in its drive to approximate the complex reality of its subjects of study. Transdisciplinary work is integrative, socially relevant, and oriented toward problem solving. Therefore, transdisciplinary work engages with human values in producing knowledge and identifying avenues for action.

Thematic threads that draw the attention of transdisciplinary scholars and practitioners run across diverse sectors of society and differing loci of knowledge creation and use. Transdisciplinarians seek to identify, integrate, and act upon points of relationship among centers of knowledge-making as these relate to ideas and phenomena that manifest in complex ways across diverse sectors of society. Transdisciplinary work is relevant to and contextualized within the full complexity of the real world. This work entails an ontological perspective of the world as integrated, complex, and whole.

Trandisciplinarians therefore seek to integrate perspectives and knowledge originating both inside and outside academe and to deal with epistemological questions of the validity of knowledge created in various contexts. Transdisciplinarity therefore implies a critique of the idea that valid knowledge is created solely within disciplinary boundaries and within academe. Since boundaries within knowledge making are both questioned and crossed by transdisciplinarians, trandisciplinary work also implies a critique of the "ivory tower" conception of academic work as a "pure" form of knowledge creation rightly detached from messy real world contexts. Transdisciplinary work, by contrast, seeks explicitly to engage with the real world and derives its character and relevance from this engagement. Transdisciplinary work is distinguished from interdisciplinary work by its engagement with human values within problem solving contexts.

The critical pedagogy of sustainability is solidly transdisciplinary. Because teaching and research that are divorced from their social contexts can easily be directed toward hegemonic ends, the critical pedagogy of sustainability confronts forms of educational specialization that decontextualize knowledge creation and technological innovation.

An aspect of transdisciplinarity that articulates well with counter-hegemony is its concern with epistemology.[6] Questioning epistemologies is central to both transdisciplinary work and counterhegemonic praxis. Critical examination of epistemological frameworks is also essential to sustainability because our very notions of ideas and practices as true and valid may be hegemonically informed. Sustainability educator John Huckle (2004) asserts the importance of what can be described as a transdisciplinary approach to sustainability education:

---

[6] Epistemology is the philosophy of knowledge making. It explores processes of knowledge generation, of determining the relative validity of ways of knowing, and of assessing the validity of articulated knowledge claims.

The key requirement of institutions and courses that seek to educate for sustainability is a philosophy of knowledge that integrates the natural and social sciences and
the humanities, accommodates local knowledge, supports critical pedagogy, and
continues to regard education as a form of enlightenment linked to a vision of a
more sustainable future. (p. 34)

Huckle (2004) explicitly applies critical pedagogy to the context of sustainability education and argues for combining analyses of power, politics,
and governance with ecologically-oriented concepts of sustainability and
environmental health in order to develop a broadly and deeply contextualized
understanding of societies and environments as inextricably interrelated and
interdependent. Precisely because unsustainability is fueled in part by the
compartmentalization of knowledge, by spatial distancing between
knowledge creators and the applications of their knowledge (a key condition
encouraging collective violence) (Summers & Markusen, 1992/2003), by
people's dislocation from and dis-integration with place, by hegemonically
informed worldviews, and by "rationalization" of the means of production,
critical transdisciplinarity like that advocated by Huckle (2004) is an important aspect of the critical pedagogy of sustainability.

The sustainability crisis is the most pressing challenge of our day. It en-
compasses interrelated social contradictions in the realms of governance and
leadership, socio-ecological health and resilience, social justice, and political
economy. It requires the dislocation of the global elite from positions of
concentrated wealth, power, and control and a reorienting of production as a
means to generate use value rather than surplus value. It requires the end of
exponential economic growth and elimination of the socially and environmentally depleting systems of enforced dependency that characterize
neoliberal capitalist globalization. Sustainability requires diverse and locally
adapted (re)inhabitation of place through developing resilient, socio-
ecologically integrated lifeways. If ever there were a time to transcend the
proliferating specialization of learning and knowledge making (Nicolescu,
2002), that time is now. Only through a critical, transdisciplinary effort to
comprehend and act to avert the sustainability crisis can we hope to create
truly sustainable socio-ecological systems and communities.

Nicolescu (2002) builds his definition of transdisciplinarity on insights
from quantum physics that highlight the nonseparable nature of reality as
characterized by both localized and nonlocalized connections among people,
objects, nature, and phenomena. Ours is an intimate world and universe,
whether we recognize it or not. In Nicolescu's view, the insights of quantum
physics point to "one fundamental characteristic of the transdisciplinary

evolution of education: to recognize oneself in the face of the Other" (2002, p. 135).

The critical pedagogy of sustainability—based as it is in a transdisciplinary striving to reverse the damage of capitalism and other forms of exploitation that conceptualize people as objects rather than subjects—is about love. Sustainability calls for us to (re)create intimate connections among people and between people and nature that are relationships of love. If and when we (re)develop lifeways that recognize destruction of the Other, be it destruction of other people or nature, as destruction of ourselves, we will at last be able to authentically love ourselves, each other, and nature.

The critical pedagogy of sustainability, as a transdisciplinary endeavor, may increasingly serve to integrate and (re)contextualize learning in higher education. It is my hope, and the focus of my pedagogical efforts, that higher education does not fail to address the sustainability crisis. If it does, this failure will be embedded in a broader socio-ecological failure of human societies, the dire consequences of which are beyond comprehension.

Conclusions

In this chapter, I have proposed a praxis-oriented critical pedagogy of sustainability, and I have advocated for its application in higher education contexts. I propose that combining the critical pedagogy of sustainability with service learning represents an important avenue for generating agency among students. Engaging students in an in-depth analysis of the contradictions of the late capitalist world-system while calling upon them to engage in action projects that begin to create alternatives to enforced dependency offers opportunities for generating sustainability-oriented social change.

The purposes and goals of this transformative educational praxis depart from those dominant in higher education today. This praxis involves students in the process of naming the world and defining desired action. It seeks to (re)integrate our fractured identities and worldviews. It is counterhegemonic in orientation so that it directly confronts the political economy of late capitalism and its means of production as primary drivers in the sustainability crisis. It seeks to decolonize both our places and our very persons. It takes a transdisciplinary approach to integrating the academic disciplines and seeks to heal dichotomous and destructive fractures within the modern worldview such as the separation of humans from nature. It seeks to authentically reconnect people with each other and with the land. It embodies both sustainable forms of leadership and sustainability *as* education.

In short, actualizing the critical pedagogy of sustainability would mean revolutionizing higher education (see O'Sullivan, 2004, p. 165). The need for this revolution is urgent. If it is diffused or delayed until the dire consequences of socio-ecological and/or economic collapse are upon us, opportunities for higher education to engage in a sustainability-oriented remaking of the world will have vastly diminished.

Chapter 9:

Pedagogy and Praxis in The End of Oil Course

Another world is not only possible. She is on her way. On a quiet day, I can hear her breathing.

—Arundhati Roy

In this book, I have developed the conceptual framework for the critical pedagogy of sustainability. This pedagogy aims to move individuals, communities, and global society toward sustainability as defined in chapter one. The theory of enforced dependency elaborated in chapters three and four is a focal point of this pedagogy because it serves as the touchstone for counterhegemonic critique, on the one hand, and for the praxis of (re)inhabitation, on the other. In the critical pedagogy of sustainability, social critique and action based on that critique comprise a unified educational praxis. This pedagogy emphasizes strategies of (re)inhabitation discussed in chapters five and seven. It advocates building community resiliency through (re)localizing the provision of basic needs while also engaging people in reciprocally nurturing relationships with nature. It also teaches and engages in authentic leadership. In the critical pedagogy of sustainability, these ideas and emphases form the basis for student engagement in service learning experiences.

In this chapter, I will explore the promise and the challenges of implementing this pedagogy in my End of Oil course. Drawing upon final reflective essays written by students in two sections of this course, I will discuss what may and may not be possible in a course that embodies the critical pedagogy of sustainability. I believe offering my reflections upon my own and my students' experiences may prove useful to other educators because my example can help them envision possibilities and constraints of implementing this pedagogy in their own contexts.

I begin this chapter with a brief discussion of the content and pedagogy of The End of Oil course. I then discuss the methods I used to analyze final reflective essays written by students in two sections of the course. These essays serve as a vehicle for generating insights about the potential and challenges of implementing the critical pedagogy of sustainability. Most of this chapter consists of analysis and discussion of the content of these essays. I also offer insights about the implications of my findings for my own pedagogy and for the work of other educators in this chapter and chapter ten.

The End of Oil Course:
An Example of the Critical Pedagogy of Sustainability

I teach The End of Oil at Fort Lewis College, a residential, public, liberal arts college in Durango, Colorado, with an enrollment of approximately 3,700 students. I co-developed this course with my sociologist colleague Janine Fitzgerald in the summer of 2004, and I have taught it 16 times since then. The End of Oil is a four-credit, upper-division, transdisciplinary course in the College's Education for Global Citizenship (EGC) curriculum. Every student, regardless of major, is required complete two EGC courses. These courses call upon students to:

- demonstrate an awareness of the global dimensions of social, ecological, political, economic, or cultural systems.
- critically analyze the global phenomena, problems, issues, or topics that are the specific focus of the course using diverse cultural perspectives and multiple disciplinary frameworks.
- identify possible responses to the global phenomena, problems, issues or topics that are the specific focus of the course. These responses may be enacted by individuals, social networks, movements, organizations, governments or other entities (Fort Lewis College, n.d.).

The typical class size for The End of Oil is 30–35 students. There is little to no budgetary support for field trips or other course expenses, and there is no course fee that would provide funding for materials and/or travel. I have, however, been able to draw upon small grant funds awarded to my Food for Thought program[1] in order to provide material support for student participation in local food projects.

---

[1] Food for Thought is an education and community action project that assists the students and faculty of Fort Lewis College and the residents of the Durango/La Plata County region in meeting the serious challenges of climate change and global peak oil production through creating a stronger and more sustainable local economy, with particular focus on sustainable local food systems. The project teaches students from various classes focused on sustainability how to plant, nurture, and harvest fruit from fruit trees. Food for Thought has been awarded several grants that have enabled the planting and fencing of a 100-fruit-tree campus orchard

Course Content

The theories of sustainability rooted in counterhegemonic social critique and praxis developed in this book serve as the conceptual foundations for The End of Oil course. With regard to energy issues in particular and the sustainability crisis more generally, the course explores the following claims:

- Oil depletion is a geological fact that cannot be remedied through technological means alone.
- Natural limits of physical systems cannot always be surpassed through application of human ingenuity and technology.
- Economic theories and the price of oil have no impact on the thermodynamic laws of physics.
- Most political and economic leaders focus their attention on maintaining and increasing their positions of advantage and not on increasing the security, well-being, or autonomy of the middle classes and the poor.
- Globalized political economy is a system of enforced dependency that systematically concentrates wealth and power in the hands of global hegemons.
- Corporate power and corporate personhood are hegemonic.
- The growth economy and its debt-based monetary system are unsustainable within the system of earth's natural limits.
- The primacy of the bottom line in business thinking is dangerous to public and environmental health.
- Individuals routinely participate in behaviors and processes that are collectively violent.
- The concentration of wealth and power nationally and globally is antidemocratic and both politically and economically unstable.
- Globalized political economy is vulnerable to collapse.
- Neoliberal globalization serves large scale business interests, often at the expense of small businesses, communities, and the environment, and even at the expense of national interests.
- The recent and continuing wars in Afghanistan and Iraq represent, in large measure, geopolitical struggles for U.S. control over energy resources and energy resource transport routes.
- The invasion of Iraq by U.S. forces may have been prompted, in part, by the desire of the U.S. to maintain the petrodollar system.

---

and the purchase of an apple press. Annual fruit tree sales provide a minimal level of support for ongoing activities and the purchase of supplies. The Food for Thought program demonstrates that material support necessary for ongoing student participation in local food projects can be affordable (Evans, 2010).

- Individuals and communities must look out for their own interests in times of energy and economic crisis because leaders are unlikely to look out for us.
- Self-sufficiency and self-reliance (at individual, family, and community levels) is a resiliency strategy in times of instability in the globalized economy.
- (Re)localization, especially with regard to basic necessities, can be a useful long-term sustainability strategy for communities.
- Individual and collective agency can make a difference in the well-being of people and communities as we move deeper into global energy and economic crises.

These claims were developed in some depth in preceding chapters. They are counterhegemonic because they call into question the purported benevolence of globalized political economy and because they stimulate critical thinking.

Pedagogy

I have used a number of texts and films for this course. In addition to texts, I have found that films can carry an emotional immediacy important to student engagement with disorienting dilemmas (see Mezirow, 2000) that challenge their internalized assumptions and worldviews.[2] I have chosen texts that support critical thinking with regard to understanding the causes and consequences of the sustainability crisis. I have also chosen materials that encourage agency aimed at mitigating the crisis and laying the foundations for a more sustainable society. I have also used as a text a table of post-war events and political, economic, agricultural, and international relations developments. I developed this table as a tool for helping students understand globalized political economy as an integrated world-system (see Evans, 2012, January, for a link to this table). This table highlights the development of enforced dependency within the world-system in a way that emphasizes relationships among seemingly disparate phenomena.

In an effort to create a safe space for students to discuss controversial and complex material, on the first day of class, I conduct a ground rules setting activity in which the entire class participates. The main thrust of this exercise is to create a classroom environment where students can delve into discussion of serious, complex, and often disturbing ideas without feeling

---

[2] For a list of texts and films used in this course, see Evans, 2012, February 3.

personally attacked. I believe this ground rules activity sets a tone of individual and collective responsibility and respect.

Because I wish to avoid banking as a learning strategy and to engage students in complex forms of critical thinking, I do not give tests. All assignments involve essay writing, public presentation, and/or taking action based on course content. Essays call for students to respond directly, and in some detail, to course readings and other content, making it difficult for them to do well without attending class and completing readings.

Classes consist of some lecture, small group discussions, full class discussions, showings of films, brief student presentations, and presentations of individual or group action projects.[3] Rather than lecturing, I typically start with discussion questions related to a reading or film in order to encourage students to highlight important ideas on their own. As the class discussion progresses, I respond to questions by providing additional comments that expand upon what the students have brought forward. I also discuss any important points that I think students have missed with regard to discussing a reading or a film and provide additional examples related to material discussed. The active participation of students throughout these discussions encourages their engagement with the material. In these discussions, I treat students respectfully and help them to improve and deepen their understanding of course material, and I avoid making any student look foolish in front of the class. When a student makes a statement that seems flawed, I try to build upon what they have offered or gently contradict it by citing evidence to the contrary.

I require an action project in every End of Oil class.[4] I believe doing so is incredibly important pedagogically because the project calls upon students to engage in action that they see as an outgrowth of the critical framing of the course. Students are also asked to reflect upon the action project in relation to course content, thereby completing a cycle of praxis. The End of Oil

---

[3] Examples of daily class plans can be accessed via my End of Oil course website linked from my website (Evans, 2012, June 28).

[4] Each of the two sections of The End of Oil from which I examined student reflective essays included an action project. These assignments comprised a considerable portion of the graded work for the course (20% in winter 2010 and 15% in fall 2010). In the winter section, students undertook projects of their own choosing, and in the fall, students picked and pressed apples and otherwise assisted with the running of the Durango Apple Days Festival. Assignment sheets for these two projects are available online. See Evans, 2010, February 19, and Evans 2010, fall.

would not embody the critical pedagogy of sustainability without the praxis embodied in the action project.[5]

In winter sections of the course, I typically call upon students to organize and carry out individual or group projects of their choice. Their projects must address one or more of the problems addressed in the course. In the fall, I have called upon students to engage in picking apples, pressing apples into juice, or doing other work in support of a community Apple Days Festival. Students have picked and pressed over the course of one weekend as much as 6,000 pounds of apples donated by local community members.

Teaching The End of Oil is a complex undertaking, in part, because I must regularly revise the materials I use in order to keep up with rapid changes in our world. This brief overview provides a sense of what I teach and how. My most current materials and assignments are also available from my website (Evans, 2012, June 28). The previous chapters of this book also provide insight into my pedagogy.

In order to gain insights on the learning experience from the students' perspectives, we will now examine student reflections on The End of Oil course. I will discuss the methods used to explore the thematic content of student essays, then analyze and interpret the themes that emerge from this exploration. Finally, I will analyze my findings as they relate to the critical pedagogy of sustainability in this course and more generally.

---

[5] Critical educators should harbor no illusions about the ease or rewards of including such projects in their courses. Action projects can be somewhat chaotic and unpredictable, especially when students choose their own projects or when the professor is organizing a great number of students to participate in a group activity. Furthermore, with regard to the faculty member's evaluation as a professor, engaging students in service learning is often considered less valuable than focusing on traditional scholarly research. One must typically be willing to forgo formal recognition for this work, and in some cases, one must be willing to sacrifice professional rewards that might accompany more extensive involvement in traditional scholarship. These can be difficult tradeoffs for faculty members, especially those who are working toward tenure. I am committed to these projects, but I also recognize why others might see them as unworkable in a context where their time and effort are constrained by other personal and professional commitments. Furthermore, in my own case, I must work with a severely constrained budget, and if I want to engage students in a project that requires special equipment, I will likely have to write a successful grant to obtain that equipment. Many faculty members do not face such heavy constraints, but in a climate of economic downturn and shrinking budgets for higher education, I believe more and more faculty members will face such challenges.

Methods of Analysis

I examined the final reflective essays of the 62 students[6] enrolled in the winter and fall 2010 sections of The End of Oil.[7] When I read the essays for the first time, I marked in the margins thematic labels for the content. So that I might consider particular comments for quotation, I also highlighted what I saw as important or representative statements made by students. From this first reading, I created a handwritten table listing the themes that I had identified. I then read the essays again, standardizing the thematic labels and marking with hash marks in a table each time a given theme was referenced.[8] I also made notes in the thematically organized columns about the specific content of student comments so that I could get a feel for some of the subtlety of what students were expressing. I also made notes about important and distinct ideas that were expressed only once or twice so that I could further track the range of ideas expressed by students. In drawing out the thematic content of the essays, I did my best not to read into student comments by drawing connections or making inferences related to course material or by drawing conclusions that were not explicitly stated.

I then transferred the content of the handwritten tables (both the numeric data and the notes in each thematic column) into an Excel spreadsheet for further analysis. I considered those themes that had been mentioned only once or twice about which I had made separate notes, and I was able to fold most of these into the existing themes by making additional content notes. I also combined a handful of themes for which there was a great deal of overlap. I then sorted the themes into larger groupings of related ideas and ended up with 111 themes grouped into 12 thematically linked groupings.

Next, I looked for the strongest themes that emerged from student essays in terms of number of times mentioned and began to organize my discussion of the content of the essays around these thematic anchors. I then reviewed

---

[6] Previously, I had received approval of my research methods from the Institutional Review Board at Fort Lewis College. I received approval from nearly all of my students to quote anonymously from their essays. Note, also, that gendered references to particular students (i.e., "he" and "she") do not necessarily indicate the sex of the student. This ambiguity further ensures student anonymity.

[7] The assignment sheet for these essays is available online. See Evans, 2010, August 20.

[8] It is important to note that I counted each separate time a theme was mentioned, not how many students referenced each theme. If a theme was mentioned in the opening section of an essay, for example, and then again at the end, both mentionings were counted in the table. I chose this approach because I see repetition as emphasis, and I wanted to reveal the relative strength of each theme within the student essays.

the potential quotations that I had marked in student essays and selected many to include in this chapter. From the material I had organized, I then developed my summary and analysis comments that comprise the following sections of this chapter. In an effort to bring student voices to the forefront of my discussion, I integrated many quotations from students. Lastly, I developed my concluding insights derived from this exploration as they relate to my pedagogy in The End of Oil and the critical pedagogy of sustainability more generally.

When reviewing my findings, a few points should be kept in mind. First, The End of Oil is a course I have taught many times and refined over an eight-year period, and I believe the depth of thought evident in my students' final reflections has, by and large, increased along with my level of experience. Therefore, responses by students who took the course recently, cannot be considered as fully representative of student responses across all the sections.

Secondly, I must acknowledge certain challenges to interpreting student essays:

- The content of these essays is likely partly influenced by a desire to please me as the professor, especially because the essays are not anonymous and are graded.
- What I ask for in the essay influences what students write.
- The essays were written immediately upon students' completion of what was likely an intense course experience for many. It is impossible to know how long the effects of taking the course persist in students' thoughts and actions.
- Themes deriving from the most recently covered course material are likely to be emphasized.
- Because students typically want to avoid submitting papers that are too similar, themes covered in other papers are likely to be excluded or downplayed, even if the material was perceived as important to the student.
- Culminating thoughts of students also do not accurately reflect daily challenges of the course with regard to teaching and learning, nor are they likely to emphasize the typical stumbling blocks of students encounter as many of these have been resolved by the end of the term.
- Ideas expressed by students in their final essays may have been learned in ways unrelated to their course experience.

- Themes addressed by students often overlap or are presented in highly integrated ways, making the thematic categorization of my analysis somewhat artificial.

I have done my best to examine what students have said without reading the above limitations into my interpretations, but I believe it is important for my readers to recognize the context within which student essays were generated.

Exploring the Thematic Content of Student Reflective Essays from the End Of Oil:

What Can We Learn about the Critical Pedagogy of Sustainability as Applied in this Course?

Following is my analysis of the thematic content of student reflective essays from my End of Oil course. My discussion is organized according to the themes that emerged from student essays, and my discussion of some prominent themes is also organized according to relevant subthemes. Quotations drawn directly from student papers appear throughout in block format. Paragraph breaks within quoted sections denote different writers.

Counterhegemony

This section addresses counterhegemonic *critique* articulated by students. In later sections, we will explore *actions and strategies* discussed by students, many of which are also counterhegemonic in orientation. Given the course content and its process of engaging students in a Freirean (1970/2000) naming of the world, it is not surprising that a good number of student essays emphasize counterhegemonic thinking. When discussing counterhegemony in student essays, I refer to critical thinking that challenges the prevailing power structure. The central defining practice of critical thinking is questioning received messages and knowledge. Messages critiqued can include social myths conveyed by powerful interests in an effort to mold the behavior, beliefs, and expectations of others into patterns that support hegemonic powers.

Counterhegemony is evident in thinking that, if it became widespread and if people acted upon it, could reduce corporate profits and/or the influence of powerful entities and individuals. For example, beliefs and priorities that consciously challenge the culture of consumerism are counterhegemonic, as is thinking that challenges technological triumphalism (the notion that progress through technological development can address any and all societal and environmental problems). Thinking that questions the efficacy, wisdom, and sustainability of neoliberal globalization, economic growth, or hierarchical leadership is also counterhegemonic, as is thinking that questions the ability of globalization and "development" to remedy concentration of wealth and power, exploitation of people and planet by powerful interests, and other forms of social and environmental injustice.

Student essays evidenced a good deal of counterhegemonic thinking that included well-developed ideas questioning the status quo. It is my hope that the counterhegemony students expressed will serve as a framework for further critique and agency. I examine below the range and depth of counterhegemonic thinking evident in student essays.

Resource depletion. As would be expected with regard to this course, the concept of resource depletion and its impacts was a strong theme across student essays, and most students recognized that the impacts of oil depletion would become obvious during their lifetimes. Acknowledgment that oil depletion is a reality is counterhegemonic in that it calls into question the stability of the current world order. Only one student expressed disbelief in oil depletion. One student stated:

> I would like people to consider the possibility that our natural resources might not be around forever, even during our lifetime.

No technological fix for oil depletion. The idea that technological development alone is not the answer to oil depletion is counterhegemonic. Although a good number of students made direct references to the need for alternative and renewable energy research, development, and implementation (and one student argued that increased development and use of renewable energy technologies would lead to their increased efficiency), no students stated that advances in technology would allow industrial society to continue along its current trajectory. A few students made direct statements noting that oil depletion and/or other large socio-ecological problems cannot be addressed solely through technological development. A few noted that technology would play less of a role in addressing peak oil than they had thought prior to taking the course. At least one student emphasized the importance of examining technological options carefully rather than having faith in their ability to save people from the effects of oil depletion. A number of students also mentioned transportation specifically as a societal and technological challenge, and at least one of these students noted that modern transportation will be especially difficult to maintain. One student discussed further developing rail transportation because of its efficiency.

Student reflections regarding the limited role of technology in addressing oil depletion included the following:

> At the beginning of the course the class covered the topics of global warming and alternative energy sources, but being a scientist I was very obsessed with the technology and development aspect of the issues rather than the implications of the issues themselves. Before this class, I was an advocate of PV cells and I even used to believe that hydrogen might have been the key to losing our dependence on oil. But I had never really put all the pieces together before this class. Although I know chemistry very well, I had not considered what might happen if hydrogen cars existed and if those cars happened to crash into one another. I also did not realize to the extent society has consumed the resources, and I say this in regards to mining. There is a long, energy consuming process associated with the building and developing of the technologies that I had never really taken the time to consider before this class.

> I knew we were close to peak oil production but I believed (like a majority of the population) that alternative energy resources would rescue us from the oil crisis. We then would be able to live our extravagant lives like nothing happened. Alternative energy resources will be able to help us to a certain extent. They will be able to run simple electrical appliances like the light bulb. Oil is really in some ways the perfect energy source since it is energy dense, can be used as a fuel, lubricant, and is used in a variety of other ways. Not a single alternative energy source has quite as many uses as oil.

> One of the biggest things that I learned from this class is that green energy is important, but at the same time the things that we have now, such as solar, wind and other cannot be used as band-aids for something that we use too much of. Nothing is going to yield even close to the amount of energy that oil has for us....We need to stress the importance of finding a way to decrease the amount of energy that we use rather than just finding something that will do the same things that we have now....

> *The Party's Over* by Heinberg gave insight into the technology of energy. In short, it discussed how inefficient the majority of other energy sources are, when compared to oil. This makes replacing oil with another fuel source very difficult.... Being aware of the constraints of current technology is significant since so many people claim that when it comes to the end of oil technology will save us.

These reflections point also to the need for social change in addressing the challenges of fossil fuel depletion.

Dependency. Some students made direct statements about dependency as a concept and construct that can create vulnerability. These statements go beyond general statements of concern about resource depletion and often relate to geopolitical struggle over energy supplies and to issues of security such as national security and food security. Some students expressed a belief that oil companies were involved in keeping the public dependent on their product, and one student noted that the diffuse patterns of settlement that characterize suburbia keep people dependent on driving. Recognition of dependency as a source of personal and societal vulnerability embodies counterhegemonic critical thinking related to oil as a resource as well as to globalization itself. One student noted:

> We have become a society that is unreasonably dependent on others to provide us with even our most basic needs. Over the course of our class it has really become a genuine concern of mine how detrimental this could be to both myself and the world I live in. It scares me to think that if the supermarkets in our town disappeared that it would actually become really difficult for almost everyone in Durango to even stay alive…. This brings me to the changes I want to make in my life. I would like to be less dependent on the system.

The desire to be less dependent on the system, if acted upon over the long term, can manifest as multifaceted and deep counterhegemony.

Sustainability. A good number of students made direct references to sustainability as a broad, integrative, holistic concept. Because it confronts relationships and processes that support powerful and exploitive political and economic interests, sustainability is an inherently counterhegemonic concept and practice. Students' discussion of sustainability went beyond references to various aspects of sustainability discussed in isolation. One student noted:

> I think I have come to the realization that as we are today, humans are not a sustainable species…. Every other species on this planet has found their niche. They have found their balance and are all kept in line by nature itself. If a certain species eats too much of its food source, there will be a big decline in that species population giving the ecosystem a chance to balance back out. It seems to me that we have tried our hardest to find a way around this phenomenon. We are constantly trying to find new ways to provide large populations with processed foods and transport water and resources to areas where human life would not normally be possible.

This comment reflects a deeply critical questioning of the sustainability of the current social paradigm, a form of questioning that calls into question the validity of that paradigm.

Systemic analysis and views. In their essays, many students engaged in systemic analyses that could inform counterhegemonic thinking. These students clearly recognized that our actions take place within a vast web of relationships that span the globe and that our localized actions have far-reaching effects. This recognition can promote counterhegemonic thinking because it encourages one to see that one's actions and the actions of corporations and other entities have very real impacts on the lives of others and on the health of the biosphere. When these relationships remain mystified, it is much easier for individuals and groups to engage in actions that have widespread, damaging effects.

A number of students explicitly recognized that oil affects global systems in myriad important ways and that it makes industrialism possible. Other students spoke of globalization as a complex, integrated system, and one student emphasized that the U.S. plays a large role in what happens globally. Others noted that (re)localization of community economic and cultural life would create systemic ripple effects.[9] Still others noted that, because societies function as integrated systems, we will need to pursue holistic rather than piecemeal approaches to social change. As noted above, a number of students also discussed sustainability as a systemic phenomenon.

A number of students commented specifically on the distancing between cause and effect present in globalized political economy. These students recognized that we make purchases and engage in behaviors that have distant and sometimes far-reaching impacts on people and environments. Some students discussed that these impacts are often obscured through distancing in space and time so that, for instance, our consumption of products appears to us to be an abstract activity, completely divorced from the very specific social and environmental relationships that make that consumption possible. The student quoted here discussed the importance of recognizing complex, systemic relationships with regard to consumption:

> Everything from the clothes that we wear, the oil we consume, the food we eat and the sugar in our coffee, to the resources we use within our houses and the water that comes in the tap and down the sewer, it all has dramatic and systemic effects on others and the environment.

Some students discussed the long supply chains of the global food system as an important form of abstraction while others observed that our use of energy is likely to be higher when the effects of its production are out of sight. One student noted that the distant Middle East wars in Iraq and Afghanistan seem to be little understood in the U.S. Another student observed that people tend to care more for a place that they know rather than the abstract, globalized world. The overall thrust of these comments appears

---

[9] The theme of (re)localization will be discussed in some detail below.

to advocate reducing abstract relationships of production and consumption in order to move toward more intimate and responsible relationships with others and the environment.

Initial recognition of systemic relationships in global political economy can manifest as a shocking, disempowering experience, creating a sense that one is an insignificant cog in a vast machine running out of control. A good number of students commented on the sheer magnitude of the sustainability challenges we face, some noting that the scale and scope of needed changes are daunting, if not overwhelming. On the other hand, quite a few students proposed that small steps taken by many at local levels can have cumulative systemic effects. When paired with the high emphasis on agency in student essays, these comments lead me to believe that most students were not unduly overwhelmed with feelings of hopelessness or helplessness upon the completion of the course. This interpretation is further supported by the expressions of clear-eyed hope in student essays to be discussed below.

Collective violence. Some students specifically directed their systemic analyses at the issue of collective violence by discussing how individuals contribute to collective and large-scale damage to environments and/or injury to other people (see Summers and Markusen, 1992/2003). Concern with issues of collective violence is counterhegemonic in that it represents a questioning of the exploitive practices that reinforce hegemonic power. Some students saw fossil fuel use and, more generally, consumption as collectively violent activities. A number of students expressed specific concern for social justice, and a number noted that they had increased their level of empathy for oppressed people as a result of taking the course.

The following quotations from student essays relate directly to collective violence:

> Ever since I moved to Durango I have begun moving in more of a sustainable fash-
> ion, but I am ready to take it a step further, especially considering all the news, war,
> and politics that we read about and how we, individually, have an impact on the
> harm of the rest of the world. One quote that sticks out to me from this course, al-
> though I don't remember exactly what it said, was something like "while few are
> guilty, all are responsible."[10] While reading about the oil and gas industry and its
> role in our country within politics and the economy I am ready to detach myself
> from that lifestyle because of the harmful effect that my tank of gas has on others
> and the environment.

---

[10] I believe this student is referring to a statement by Rabbi Abraham Heschel that I often use in class when discussing collective violence: "In regard to cruelties committed in the name of a free society, some are guilty while all are responsible" (see Loeb, 1999, p. 11).

The experience that spoke to me the most was viewing *Black Wave*. While I watched this documentary about the Exxon Valdez, I was struck with the realization that the Exxon Valdez was not some sinister entity out in the world doing bad things to the environment, rather as long as I choose to live a life supported by and entrenched in oil, I AM the Exxon Valdez. I can choose to be otherwise and make a difference for myself, my family, community, and the world.

These reflections represent counterhegemonic self-reflexivity.

Political economy. Critical themes related to global political economy were fairly strong in student essays, though not nearly as strong as themes to be discussed below related to agency and to (re)localization as a resiliency- and community-building strategy. Themes I have labeled as related to political economy include concerns students expressed about the profit motive, corporate power and personhood, intellectual property rights, global concentration of wealth and power, various instabilities in the world-system, and geopolitics.

A good number of students commented on the profit motive and the growth economy as distinct social problems. One student noted that a narrow focus on profits leaves other important concerns unaddressed. Others stated that the profit motive can encourage short-term thinking and disregard for energy efficiency and conservation. One student cited war profiteering in the destruction and reconstruction of Iraq and Afghanistan as an example of the amorality of the profit motive. Other students characterized the profit making strategies of the electric utility companies as underhanded, in particular revenue decoupling.[11] Another student noted that competition for profits promotes public deception. The profit motive was also cited as an impediment to sustainability-oriented change.

A good number of students assumed counterhegemonic stances with regard to corporate power and corporate personhood, expressing that they saw these developments as damaging to society:

A corporation can't have the rights of a person, because besides on a financial aspect, it cannot be physically held responsible. Board members that support the demise or compromise the safety of communities (socially or environmentally) don't lose more than their stock value, where if it was as actual individual committing

---

[11] Revenue decoupling is a process based on legislation that "decouples" a utility's profits from its volume of sales. It is a level of profit guaranteed to electric utilities. According to energy policy expert Mark Sardella (personal conversations and presentation in my End of Oil class), such legislation has been passed in many states in the United States, meaning that, when customers succeed through efficiency measures in reducing their electricity consumption, the entire customer base must pay more to compensate the electric utility for lost profits.

these crimes, such leniency would probably not be shown. I am not saying that globalization is entirely horrible, if the goal was to work together for success I think it has the potential to benefit a lot of people, but the intentions are usually profit-based, and not for the common good. Power needs to be divided instead of monopolized, the only way to get the power back to the people is through decentralization.

Corporations are given personhood. As a business student, you would think I would have some knowledge of this, but I had never heard about it until this class. Corporations are given person-like traits and are treated as people, and this is just unethical. I do not think that many people are aware of this, and that is even scarier because as human beings we have a right to know what else the government is calling a "human," and what rights these entities have.

I think that giving a corporation the same rights as an individual human was perhaps one of the biggest mistakes this country has ever made. Certainly when it comes to any law that has been passed, this has to be the worst decision I have ever heard of. Not only does it give hugely unfair amounts of power to corporations, but it also points out how flawed our government has become. It has been some time since America could still be considered a democratic society. I think it is safe to say that we have become perhaps the most well developed plutocracy that has ever been seen.

More than one student saw extensive intellectual property rights granted to large corporations as morally repugnant and emblematic of a dangerous concentration of power in the hands of transnationals. One student noted that intellectual property can interfere with sustainability in the business realm, but that it promotes profits:

One quote that stands out is, "Why exchange each other's cookies when we can exchange recipes." This quote really made me think about how absurd it is to transport things over and over again in order to make a profit when we could just as well explain how to make it to somebody, in turn, effectively removing the shipping process almost entirely. The main issue with this, though is the fact that it is not good business. You make more money by holding on to your business or trade secrets. This creates a dilemma, especially for me considering I am a business man.

This student had been confronted with a disorienting dilemma with regard to his personal plans and practices, an encounter that could possibly lead to further counterhegemonic thinking and/or action.

A number of students also noted that large corporations represent dangerous concentrations of wealth and power and that limited liability contributes to the power and irresponsibility of corporate entities. A few students observed that current political economy can promote monopolies and enforce people's dependency on these monopolies.

A good number of students voiced concerns about instabilities in global political economy and/or about the ultimate unsustainability of globalization. One student expressed a number of concerns in this area and related these concerns to dependency:

> I was enlightened and also frightened to learn how dependent all countries are on each other. We as a nation are slowly foundering due to the bursting credit bubble, and dependency on oil that we cannot support ourselves. We depend so much on other countries such as Japan, China, and the Middle East to either fuel our oil dependency or support our constantly increasing debt issues. These major points that we discussed in class are so important because the future of our planet and country are balancing on a crumbling foundation. We need to be aware of what is going on and we need to be cautious in our next steps in change.

Concerns expressed by students that related to instability in global political economy also included:

- recognition that the continual concentration of wealth and power globally is a destabilizing phenomenon;
- recognition of the potential for a collapse of the U.S. dollar;
- worries about destabilizing levels of national and international debt and about how long other countries would continue to purchase U.S. debt;
- concerns about debt as an unstable foundation for money creation;
- concerns about the potential collapse of the petrodollar system and, more generally, about currency wars; and
- worries about the potential for a near-term great depression, or worse.

The following quotations taken from student papers express such concerns:

> I now hold less faith than before that our world will make a transition into the coming "alternative energy age" that will be smooth and free of suffering. Sure, some countries are phasing in alternative energy so that their transition out of oil dependence will be smooth, but super-power countries like ours and China are not doing enough of this. Considering this and the fact that our economies drastically affect everyone else's, one option that I see on the table is a global depression, the likes of which we have never seen before. This is, however, just a theory, and no one really knows what is going to happen in the coming years, but I see a lot of hardship and struggle for the majority of the world before I see unicorns and rainbows.

> The only reason the economy has been able to produce and grow at the rate it has is because of cheap oil and resources. But when it becomes difficult to drill for oil that is down to a thin layer, that is the end of cheap oil. There will be no more growth and the economy will continue dropping into a depression. This depression is going

to be far greater than anything this country has ever experienced and we are reaching a point where there is no turning back. Each of us needs to be aware that we are about to see a lot of changes in the next few years and know that it's not going to get better on the political and economic level. It is going to get better in our neighborhoods, our homes, and smaller communities. We need to begin a more dependent lifestyle on those that we live near. We need to begin creating a sense of community, working on a barter/trade system, reskilling individuals, and becoming more connected with natural cycles so that when the times get rough because of the political failure, we have what is close to rely on.

This last quotation also emphasizes (re)localization, a counterhegemonic theme that will be discussed in some detail below.

One student voiced specific concerns over the stability of the debt-based monetary system:

Something else has brought us to the position we are in today, that many have an idea about but do not truly understand; the economy. I believe that one of the first things that are most important to understanding our current economic crisis is the fact that I recently became fully aware of with the help of this class. That is the creation of money through debt. If people can understand that the creation of money through debt leads to an unstable economy, I assume that people will be less inclined to generate their own debt, and in the end resolve such an unstable economic problem. However, few are willing to educate others in the matter of economy in this manner. Doing so will cut many business' profits, mainly banks.... In order to change the unstable economy, we need to understand just what makes our economy unstable. If we understand that creating debt leads to an ongoing cycle of debt we can begin to change our ways back to a method of payment that does not rely on debt to work.

This quotation emphasizes several ideas that are counterhegemonic: that the monetary system is unstable, that the educational system tends to support hegemonic interests, and that ordinary people can play a role in changing large-scale systems.

A number of students voiced concerns about a potential political and/or economic collapse, with at least one noting that such a crisis could serve as a catalyst for welcome changes. Others noted that, if we wait too long to address the instabilities within global political economy, we will miss a historic opportunity and ensure a chaotic world-system collapse. One student discussed what she saw as an increasing potential for a political uprising among the people if global political economy continues to slide toward collapse. One student commented on the potential for collapse in a way that integrated many themes into a systemic view of global political economy:

This class has effectively taught me that it is a combination of matters that will bring the drastic change.... The change I keep referring to is, in my opinion, that our modern industrial society will soon end and a more domestic, hopefully sustainable society will prevail. I believe it is due to a combination of peak oil, increasing corporate power, global warming, and poor economic decisions, which are all conveniently related. In short, our extreme consumption of oil had led to both peak oil and global warming; peak oil and corporate power have led to poor international economic practices, while corporate power alone has influenced national economic practices (bank bailouts, the housing bubble, etc.). Corporations now have the ability to blatantly influence politicians, as well, which has and will greatly influence further economic and political agendas. Our society is slowly spiraling out of control but our government and the media combined do an excellent job of keeping this hidden from most of the public.

Geopolitics is an important concept with which to grapple in the process of developing an understanding of global political economy. Quite a number of students discussed geopolitical influences on current national policy, particularly with regard to the wars in the Middle East, and suggested that resource wars would continue and even worsen in the future. One student remarked on the potential collapse of the U.S. empire. Some students assumed critically and historically informed stances with regard to the geopolitics of oil and U.S./Middle East relations:

I was basically unaware of exactly how involved our country and many others are in the oil industry, how important that industry is to the world's economy, and just how far countries (especially ours) are willing to go to keep their power positions in that industry to maintain their status on a global scale.

The emphasis that this course has placed on the interest of the US government in the Middle East as potentially economically driven, has given me a more accurate perception of world events. It is a pretty hefty concept to try and determine whether or not the "War on Terror" is just a way for the American fossil fuel industry and their affiliates to use taxpayer money to secure their financial interests. But, given the track record of big business as far as past violations of human rights it wouldn't be an unreasonable conclusion. After all, with the removal of Saddam Hussein from power comes the need for a completely revamped political power structure to run the country. Clearly the US is not going to help establish a new government that is too disagreeable with US policy. Now the question is how much influence does the fossil fuel industry have in the creation of the new government, and what kind of accommodations will be made in regards to the corporate interest in the resources of the area.

These statements are deeply counterhegemonic in their implication that powerful individuals and entities influence public policy, often behind the scenes, in order to protect their own interests.

A number of students explicitly addressed the petrodollar system and its hegemonic effects in global political economy. These students were able to connect the dots between that system and other destructive policies and behaviors:

> The U.S. has been such a powerful country based on the fact that oil is sold in U.S. dollars. Therefore it promotes consumption within the country in that other countries need to sell their products to us in order to get oil, and we need to keep consuming in order to continue the cycle. And as consumption increases the need for oil increases, which is getting harder to meet as time goes on.

> Because of the way business works, I highly doubt that this rapid consumption of oil will stop. Especially considering that our whole economy is based off of the fact that our currency is backed by oil and oil is traded worldwide in the US dollar. Because of this, it is necessary for us to use oil in order to keep our currency afloat and guarantee our comfortable quality of life.

A good number of students saw a need for political-economy-related changes in the public policy arena. Changes advocated by students included:

- prioritizing cuts to greenhouse emissions;
- prioritizing environmental concerns in decision making;
- encouraging transition away from fossil fuels,[12] in part through creating tax and incentive systems that favor the development of renewables;
- increasing the public accountability of decision makers;
- holding natural gas extraction companies accountable to the public for damages to the environment and to public health; and
- returning power to the people, in part, through devolving political control to the local level.

In sum, a good number of students addressed political economy in their essays, and most of their discussion was counterhegemonic in orientation. Their discussions focused on many important aspects of global political economy, but there was little focus on enforced dependency as a centrally important phenomenon. As we will see below, however, many students did make connections between the critically informed content of the course and (re)localization as an important vehicle for addressing local and global problems.

---

[12] For at least one person, this transition included a focus on nuclear energy. In class, we had discussed the dangers of nuclear energy a number of times, and we had also discussed the nuclear industry as a highly concentrated industry in terms of wealth and power. Still, a number of students supported forms of energy transition that emphasized nuclear options.

Deception, corruption, and manipulation. Recognition that powerful interests use deceptive and corrupt means to maintain their positions of advantage is an important foundation for counterhegemonic thinking and agency. In their essays, many students recognized deception and manipulation in business, media, and government. With regard to deception, some students' comments were quite specific, but a good number of comments also treated the media and government as generally deceptive.

With regard to business practices, students specifically referred to green washing and, more generally, to advertising as deceptive:

> I think another important piece of information was that although corporations make it seem like they are doing good things for the world, they are in fact not. Corporations have been working very hard through marketing and advertising to make themselves seem like heroes. They portray themselves as eco-friendly and most of the time this is just a lie.

Systemic corruption was discussed by a number of students, particularly the corruption of politics by moneyed interests. One student speculated that corporate power had influenced national policy in the recent bank bailouts. Another student asserted that oil companies have been using their political pull to keep people dependent on their product. Several students stated that a great deal of systemic corruption is tied to the fossil fuel industry, directly or indirectly.

Deception of the public by government and the media with regard to the Iraq war was also mentioned a number of times. On a related note, some students observed that the overall geopolitical struggle to control oil and oil transport routes is highly deceptive and corrupt. One student observed:

> An important piece of knowledge that people need to know about is that not everything that our government and media tells us about events is always the entire truth. Most people understand this already. However, they do not fully understand this. For instance, most do not realize that the rise of the Taliban, Osama Bin Laden, and Saddam Hussein to power was through influence of our own government. These facts coming to light for most will help them realize that our only reason for being in the current war is so that we can secure energy reserves.

Some students also noted that the media willingly participate in geopolitical corruption by shaping public opinion to support government agendas that, in turn, support the profits of large corporations. These students pointed to government catering to the fossil fuel industry and catering to corporations that stand to profit from military actions, from the rebuilding of Iraq and Afghanistan, and from the extraction and sale of oil in these locations.

Many students recognized deception, corruption, and manipulation by powerful interests at work within the world-system. This recognition may serve as a platform for further counterhegemonic thinking and action.

Cultural hegemony. As discussed above in relation to several themes, many students demonstrated an awareness of aspects of hegemony at work in the world-system. For the purposes of classifying a student comment as a reference to cultural hegemony, the comment had to explicitly acknowledge that dominant interests manipulate organizations and/or the public, not only for their own gain, but also *in order to make them willing participants in their own oppression.*

Several students noted that access to material comforts and conveniences can effectively quell people's desires to rock the boat, rendering them compliant and passive in the face of power—Marcuse's (1964) notion of repressive desublimation. On a related note, one student stated that people typically think that everything will be alright and that this assumption, though not necessarily correct, is ingrained in people:

> Before I attended this class I had never heard about peak oil and the problems with our world because the common way of thinking is that everything will always be alright. It has been ingrained in my head.

The assumption that everything will turn out alright in the end certainly encourages passivity and discourages agency.

Some students referred to the media as complicit in cultural hegemony:

> I believe that out of any sections of this class to distribute to the masses would be the truth about our media resources and corporations influence within our very own government. I bring these two topics into focus, because of the direct effect it has on every US citizen's beliefs and morals. Throughout classes the unraveling of half truths and misleading facts pertaining to our media was sickening. People should know the truth, whether the news be positive or negative. The artificial spinning of vital information only cages our curiosity and self decisions. Media helps guide US citizens to be herded into majority beliefs.

This statement implies that media deception and manipulation contribute to lack of public awareness and action on important issues.

A student quoted above described the U.S. as a plutocracy while another asserted that most people in the U.S. believe in the American dream because they are prompted by government and media to do so, while in actuality few Americans get to live that dream. One student cited U.S. government fear mongering as a hegemonic tool of social control.

Gramsci argues that an instinctive feeling of being set apart from mainstream society represents a recognition of cultural hegemony that is an early developmental stage of counterhegemony:

> The unity of theory and practice is not just a matter of mechanical fact, but a part of the historical process, whose elementary and primitive phase is to be found in the sense of being "different" and "apart," in an instinctive feeling of independence, and which progresses to the level of real possession of a single and coherent conception of the world. (Gramsci, 1971/1999, p. 333)

One student's discussion of feeling somewhat schizophrenic with regard to the energy crisis resembles this Gramscian consciousness:

> It is really frustrating sort of being stuck in two different mind sets: carrying out regular life in the industrial society, and being aware of the energy crisis and trying to limit consumption within the industrial society. I feel like a lot of people are sort of straddling the fence in this scenario, and I am just as guilty.

As noted above, in an effort to avoid reading meaning into student essays, I set the bar high regarding the relevance of student comments to the theme of cultural hegemony, a factor that may have caused me to underestimate its strength. Students clearly expressed many counterhegemonically-oriented ideas, however, many of which were related to (re)localization and agency (to be discussed below). While I would like to have seen more students voice a clear understanding of the concept of cultural hegemony itself, further emphasizing it as a distinct concept would require me to turn attention away from other course content that I also consider important.

Hegemonic leadership. A few students directly discussed hegemonic leadership as a central challenge to social change. Most of these students stated that they saw leaders and hierarchies as integral factors and agents in generating the current sustainability and economic crises:

> We need to think creatively, and not wait for some cure-all brought to us by the same people who created the problem.

Some of these students suggested that leadership and governance should be focused at the community level. Others emphasized the hegemonic functions of leadership in global industrial society and suggested that individuals and communities cannot wait for current leaders to solve their problems for them but must assume leadership themselves and act on their own behalf:

> I would like others to wake up and realize what's going on in the world, and start to understand how things operate. So many people today have an attitude where they think "nothing is going on" and "everything is OK." We are so conditioned by the news and elsewhere that the people in charge are looking out for our best interests and will make sure we're ok, however that's not the case.

We now turn our attention from exploring counterhegemonically-oriented *critique* evidenced in student essays to examining themes associated with counterhegemonically informed *agency*. The organization of this chapter should not, however, be taken as a suggestion that we should draw a sharp distinction between critique and agency. As argued throughout this book, the two form a unity in praxis.

(Re)localization as a Resiliency Strategy

I argued in chapter five that place-centered sustainability praxis can (re)create lifeways that are resilient and self-sustaining. In student essays, place-centered living as a community building and resiliency strategy was among the strongest themes. I believe this idea represents, at least in part, student responses to 1) a fairly in-depth exploration of modern industrial society as a deeply hegemonic construction and 2) exposure to course material and engagement in action projects aligned with the theme of (re)localization. Many students discussed community resiliency aspects of (re)localization as an embodiment of creative action that grows out of critique and, at the same time, moves beyond it:

> There are so many benefits to thinking local, such as providing jobs and not having to transport everything into town which would cut carbon emissions. Also, with the depletion of oil, I think a stronger sense of community will naturally emerge. Starting to localize now will make it easier in the future when we don't have a choice. I think if we start caring about what community members are producing, it will bring meaningful relationships and a simpler way of life.

In chapter eight, I argued that place-centered service learning experiences can engage students in powerful forms of (re)localization praxis. In their action projects, many The End of Oil students engaged in action projects oriented toward local, sustainable food production and consumption, and all student projects had at least some local community building focus. Several course readings, especially Sharon Astyk's *Depletion and Abundance* (2008) and Richard Douthwaite's "Why Localisation is Essential for Sustainability" (2004) also focused on (re)localization as an individual, family, and community resiliency strategy.

Under the umbrella of (re)localization, students discussed a wide range of topics including the importance of local food, food security, self-sufficiency, self-reliance, and the correlation between a place-centered economy and sustainability. They also noted the value of family and community relationships of interdependence and mutuality to successful living in tough times. The following quotation addresses these themes:

> On an individual level households should grow their own food, medicine and maybe raise some chickens, goats and even bees. Collecting rainwater at a house too is a large contribution.... The community level people should start community gardens and trade systems within their community. Trade goods and services amongst one another instead of depending on outside sources that take money out of the community and send it to larger corporations.

The quotation below treats similar themes and offers an example of conceptual integration of the Apple Days project with the overarching course theme of (re)localization:

> [Apple Days] is providing a community with local resources before there is no other option. These are the changes that I think we most need to see if we as a society are going to be able to make the transition smoothly. I think perhaps the biggest point that needs to be stressed by anyone that supports the idea of local sustainability, is how we cannot wait for things to fall apart. By that point it will be much more difficult to just try and start. Also, making little steps in that direction over the period of a few generations would allow for people to get it right before it becomes essential for their survival.

A number of students specifically recognized local abundance and community assets as foundations for local economies. The following quotation offers an example with regard to Apple Days:

> It took a little while to settle in, but it was amazing to see so much come from nearby areas [for Apple Days]. Most things I look at have stamps saying it was made in another country, and I have very little idea of all the steps it takes to get to me, but in seeing a gigantic pile of apples which had been gathered the day before brought to Durango and quickly changed into juice and other things I was stunned.

In addition to the broad theme of (re)localization present in student essays, a number of distinct subthemes also emerged. We will now explore these subthemes.

Localized, diversified energy systems. With regard to (re)localization as a resiliency strategy, some students specifically emphasized the importance of energy system diversification and localization:

Learning about the various forms of clean and renewable sources of energy we have I think we can implement them in areas that they can prosper, so instead of looking at the renewable sources on a national level, we need to focus more on where the different types will be more effective in local regions. For example we can use solar power in high sun areas such as Arizona, Texas, and southern California because for most of the year they have almost unlimited sunshine. In other areas that are very windy we can use wind turbines to generate electricity.

Place-centered leadership. One student offered reflections on place-centered leadership that derived directly from his experience working at Apple Days. These reflections demonstrate recognition of important differences between grassroots servant leadership for sustainability and leadership as commonly practiced within the context of the national economy:

I saw lots of similarities and differences in terms of how the US economy runs during the Apple Days Festival. Everyone who contributed to the festival had a specific duty for a few hours and in the process contributed to something larger than themselves. The US economy runs on a similar outline. There was however a large difference between the Apple Days Festival and the US economy. There was no hierarchy in terms of most important person or job during the festival. Everyone who did contribute had equally important jobs that helped make the festival a success. I liked that system because it made everyone feel appreciated and proud that they contributed to the activity. This way of thinking is not how corporate America works however. Executives think that they're the most important members of the company when the workers are the ones doing the hard work to make the corporation a success. Corporations also aren't local enterprises since a large percentage of their revenues go back to corporate headquarters for profit. When they take money back to corporate headquarters, money is lost from the local economy.

The value and power of community. (Re)localization as a sustainability-oriented strategy was emphasized almost equally in the two sections of The End of Oil discussed here, even though the Apple Days project involved much more community interaction than did the individualized/small group projects. Students who participated in Apple Days, however, emphasized the value of community relationships in the context of depletion more heavily than did students in the winter section of the course. The following quotation emphasizes how one student viewed the community-building aspects of the Apple Days project and how he related these aspects to the larger conceptual framework of the course:

The sense of community is great when participating in the [Apple Days] festival, although most of the time unspoken, there is a feeling in the air that we know we're doing something good and useful. This might even be an intrinsic, biological induced feeling—one that we get from growing food, practicing husbandry, or farming or hunting—practices that are inherently wholesome and good. It's also important because we are taking advantage of our local food supply. As has been mentioned in the class, not only is shipping food around the world extremely wasteful, it's not as healthy as eating locally. And when the time comes, local food might be ALL we have to rely on and the sooner we get started organizing and harvesting, the better off our community will be. In the future, all of these aspects will need to be relied on—community, local food and resources, and know-how.

Many other students commented directly on the value of community building and community relationships to sustainability and resiliency:

A major change in my worldview involves the value of individualism. Brought up in a society that values individualism, I have learned to rely on and help myself to get where I want to go in life. Now however, I realize that in order to accomplish my goals (especially with the coming crisis), I need to learn how to rely on and help others as well. Society as a whole will have to become more interdependent as well.

This is one of the most important things that I would like others to know, that local community is extremely important, and the American culture needs to move back towards a society of neighbors rather than people who live next to each other.

In response to the question on what kinds of actions people should take, I say reach out. Find community, seek educators that can teach how to farm, hunt, understand local plants, preserve and can food, have potlucks, begin practicing skills that can be traded for other goods, start playing music, begin producing instead of consuming, start growing your own food (AS MUCH AS POSSIBLE), start learning how to build with recycled and natural resources, and most importantly start slowing down, bring more love into your life, and begin to create a lifestyle that is not so reliant on unsustainable energies.

The power of community to come together to accomplish large tasks was mentioned many, many times by students. This theme was strongest in the section of the course that participated in Apple Days. The student who wrote the following captures this theme quite nicely:

I think what I found most rewarding out of the entire project, was to see how few people it actually took to produce such a huge amount of apple picking. Sure the trees had already been planted and we did have the assistance of a dump truck, but when it came down to it the majority of the work was done by good old fashioned manual labor. It was kind of crazy to think that with only the manpower from two classes we were able to pick a couple thousand pounds of apples in a single day..... This got me thinking about how little it might actually take to grow your own food especially if every time it was ready to be harvested, a couple dozen people went out and helped you pick or dig up your food.

Themes related to teamwork and cooperation more generally were also slightly more emphasized among the students who participated in Apple

Days while the value of interpersonal relationships was emphasized almost equally in each section. The potential of a crisis to either bring people together or incite competition was also mentioned.

Human relationship with the environment. In chapter five, I argued for a holistic ontology of place. A number of students specifically addressed human relationship with the environment in their essays. In the following quotation, the student connects the idea of working in harmony with the environment to a practice of healthy living characterized by loving relationships:

> We need to start working in harmony with the environment instead of against it. However, as I write these thoughts and ideas, I am very aware that I can preach to no one and that I can only educate through my actions. I can start doing these things individually and hope that those practices spread. I can create a life around me that is beautiful, thriving, and sustainable with lots and lots of love, so that whether or not we go through this crisis and changing times I will be living a lifestyle that is healthy and ready for anything.

Generalized concern over environmental damage caused by people was also a fairly strong theme. A small number of students made direct references to modern industrial society behaving as if it is disconnected or perceiving itself as disconnected from nature, and a small number of students directly expressed a desire for personal and/or societal reconnection with nature. A handful of students discussed humans and nature as inextricably interconnected—as a human/nature complex. One student noted, "What we do to the environment, we in return are only doing to ourselves." Reciprocity was also mentioned a few times as an important aspect of the human/nature relationship.

Human health. Students also expressed concerns about potential negative impacts to human health and about potential changes to healthcare precipitated by the energy crisis. One student noted that modern medicine is extremely reliant on fossil fuels and that this reliance would translate into big healthcare challenges for the future in an energy-constrained world. Some students cited possible improvements to human health in a low-energy world where people would have to be more active and would have to produce and eat more local food. The human health theme was moderately strong in student essays, though it was not discussed in detail in class. A chapter of Sharon Astyk's *Depletion and Abundance* (2008), a required text for both sections of the course, provided the main treatment of health topics. Though this theme was not exclusively addressed by students within the context of (re)localization, often it was, and Astyk most certainly drew this connection.

Agency

Agency emerged as among the strongest themes in student essays. Their reflections ranged from direct discussion of the importance of taking action to discussion of specific actions planned or already undertaken in response to taking The End of Oil. In many cases, the theme of (re)localization was interwoven with themes of individual and collective agency. Representative general statements on the value of agency included the following:

> I believe that the most important message that was covered in this class was that it is possible to make a difference. Although this class did not cover happy topics it was presented as a problem that we can work to fix, not a hopeless situation that allows us to give up personal responsibility.

> Maybe I'm being naïve but I still feel like we have a chance to make a difference if we stop being selfish and take the steps to do so. A lot of people say that when you're in college that you feel like you can change the world and that that feeling fades, as you get older, because then you see how the world really works. Well maybe what needs to change is how this world works.

> I do agree that it is truly remarkable to live in this time of great advances in knowledge and technology, but I have come to the conclusion that there is also quite a heavy responsibility involved in living during this time. I believe that we must all recognize the problems of the end of oil and accept the challenge to become committed to doing our part to effect positive change. Our survival, and the survival of our children, and their children, depends on our actions now!

Not all statements about the efficacy of agency were positive, however. At least one student discussed social change as difficult with regard to giving up certain luxuries:

> It will be a hard switch because there are some things that I really like doing that use fossil fuels. Flying is one of those things, it is a passion of mine and it will be hard to give it up.... I want to be able to travel the world and see its beauty, but if we continue with our habits there will not be much to see.

A few respondents voiced an unwillingness to take responsibility for needed changes, and a couple of people noted that they were unwilling to make sacrifices in their chosen lifestyles in a society filled with others who were busy consuming and otherwise enjoying the temporary benefits of living unsustainably. A few students seemed a bit conflicted about recognizing the need for sustainability-oriented social change while remaining unwilling to make changes in their lives. This quotation from a student essay provides an example:

> I know that there are problems with the world today, and, as I said, I feel that most could be solved with a simple reduction of the human population. Alas, I am merely one of the masses, and I am no martyr. I certainly will not be the one giving up any luxuries for the good of the planet. That is not to say that I don't turn the light off when I leave the room, or park my car just for the sake of walking. It's just that I drive my car almost daily, and I deem it basically a necessity to do so in order to live the way I want to live. Until regular people like me stop consuming, I have a distinct feeling that things will get much worse before they get better.

Under the umbrella of agency as a broad theme, a number of distinct subthemes emerged that we will now explore.

Relevance/practicality of the course. A good number of students commented on the relevance or practicality of the course with regard to their education and their lives, some of them stating that everyone should learn the material covered in the class. One student noted that what she learned could be applied to many aspects of life other than her career:

> I believe that these classes are necessary in our education, as it strengthens our knowledge of the world, not just that of our career choice.

Urgency with regard to social change. Reflections on agency also included a focus on the urgency of needed social changes. This urgency comprised a very strong subtheme running through student essays. Reflections on this theme typically focused on the narrow window of time available to address socio-ecological problems together with the severity and/or scope of the problems we face. Some students noted that long-term intergenerational effects of the sustainability crisis will stem from our actions, or lack thereof, in the present. The following quotations offer examples of student reflections on urgently needed social change:

> We have to come together as a worldwide community, and treat this situation like what it is, a disaster. We should treat this situation like a meteor hitting the earth, because it's going to have just as much of an impact.

> The longer we wait to make the drastic changes we need to on a global scale to stop the quick depletion of the planet's finite resources, the more abrupt and shocking the change will be.

Awareness of socio-ecological problems as a precondition for agency.
Expressed need for this awareness was a moderately strong theme that
emerged from student essays. It is evident in the following statements:

> Overall this class project and the class in general, has been very beneficial to my
> education and to my life. I believe that it is a class that everyone should take, in
> schools all over the country. I feel that the most important thing that people can do
> is get educated on these topics. The first step in making a difference is knowing
> what the problem is and deciding to make changes in our lives that will contribute to
> a solution.

> Knowing the true extent of corruption, or at least the true extent of the potential for
> corruption that is prevalent in today's social systems, is important in order to com-
> bat said corruption.

With regard to energy specifically, several students mentioned having a
newly acquired energy consciousness that had prompted them to take at least
small actions to conserve energy.

Reflections on specific actions planned or taken. Discussion of specific
actions students were already undertaking or planning to undertake in their
lives also figured prominently in their essays, as the following examples
demonstrate:

> This course has given me the firm nudge I needed to stop being a passive environ-
> mentalist and become an active environmental citizen.... I can no longer buy any-
> thing without serious consideration as to whether it's necessary and where the thing
> I'm purchasing came from as well as the real cost to our planet. I made the decision
> to use that criteria for my purchasing decisions back in October and a strange thing
> happened. Since the time I chose to look at purchase through that lens, nothing has
> met the criteria and aside from food, I have purchased nothing. At the risk of sound-
> ing dramatic, everything I see appears to be literally covered in oil and I cannot bear
> the idea of my purchase being responsible for the manufacture of one more unneces-
> sary plastic object. Oddly enough, I've also found that the quality of my life has not
> suffered; on the contrary it actually seems better. By choosing to open my eyes and
> be aware of the impact of my choices it's as if my mind blooms ideas about ways to
> use and reuse what I have as well as creative ideas for procuring what I don't have
> from other sources that don't require buying newly manufactured goods.... This is
> what much of Sharon Astyk's book was about; the ways that average people can
> make a difference by how they choose to live their day to day lives.

> The goal of my future in terms of family is to prepare my children to become self-
> reliant and to make choices that will increase their success in the face of a changing
> world. Many of these preparations still need to be researched in order to educate my

children, but End of Oil class has adequately inferred the importance of these skills
for the future.

A good number of specific actions planned or undertaken by students in
response to taking the course focused on conservation of energy and other
resources and on reducing waste. In fact, such discussion comprised one of
the strongest themes emerging from student essays. The following quotation
serves as an example:

> Conserving has been a good policy I have been adopting and trying to wean into a
> way of life without some of the daily processes that I have been so accustomed to.
> Conservation can only go so far because we still are going to run out of resources in
> the future. That's why I have been trying to not give myself an option to cheat.
> Walking up to campus instead of driving has been a hard task I've been working on
> but it's always easier to make an excuse to drive. So I made it a goal to walk to
> campus next semester at least 3–5 days out of the week.

With regard to conservation of energy and other resources, students often
mentioned the value of individual change as an initial step in social change.
They also often pointed to the value of cumulative lifestyle changes in
realizing widespread social change. Student comments in these areas ranged
from general statements about heightened awareness of personal energy use
or about the need to conserve or simplify lifestyles to very specific ideas
about how to conserve. In these discussions, students emphasized recycling
and reuse of various products and/or packaging. One student who expressed
feeling overwhelmed and somewhat hopeless with regard to the scale and
scope of sustainability challenges saw value in taking small steps to reduce
her carbon footprint:

> I have always kind of felt helpless when it comes to making a difference. Through
> the course I have had discouraging thoughts about the future and what lies ahead, it
> seems that the problems that have come forth, such as pollution, natural resource
> depletion, overpopulation and over consumption, are too developed and progressed
> to be reversed. I have however, been inspired to continue to make little changes in
> my life to reduce my carbon footprint.

The theme of recycling was weaker among student essays from the fall
2010 section of the class, but very strong in essays from the winter 2010
section. This difference may derive from the difference in the action project.
The individualized and small group projects completed by students in the
winter 2010 class often focused on various forms of recycling and reuse of
materials and products. One student commented on recycling as a theme in
student projects by saying that "recycling was not really mentioned through-

out the class but it was a huge part of our projects." This theme might also have been stronger in student essays from the winter section because students presented their projects to the full class at the end of the term, shortly before student reflective essays were due.

With regard to specific actions taken, some students discussed that they were producing and/or consuming more local food, or that they had plans to do so. The following quotation presents an example:

> Another change that I found weird about myself for thinking, was actually wanting to have my own garden. I say weird because never before in my life have I considered this something that I would desire. Nonetheless I have decided that I think it would be important for me to someday grow at least a portion of my own food. Now I realize that this is a little unreasonable at my current point in time but someday when I am shopping for my very own piece of property, this will definitely be something that I will be keeping in mind.

More than one student related her/his local food agency directly to the Apple Days experience:

> Personal experiences are always more empowering to me than is a book or teaching on a subject. I find that opportunities like [Apple Days] are very empowering, especially if you are aspiring to become a local food producer. Seeing the community support at the Apple Days Festival was great, it was like a community within a community. I will possibly reach out to those involved in the festival to collaborate, get ideas, or just simply support each others' endeavors. I am sure many of the faces at the Apple Days festival are also those at the farmers market, green business roundtables, and community gardens.

> I will use this experience [Apple Days] in the future and even continue participation. I would like to volunteer next year. It is inspiring to see people really coming together and prove that we can really live locally! It also made me want to learn more about bees and beekeeping, wild and edible plants, and even write songs to share my beliefs and knowledge of issues today and that we can be strong communities and make positive change.

Some students discussed how the course had influenced their purchasing decisions. The following quotations offer examples:

> When we discussed in class the idea of purchasing new vehicles, I had the idea that one day I would be wealthy, own a nice, expensive muscle type car. Now, I have the vision of buying a car that will get me from A to B with minimal cost, last for more than a couple years, and is gas mileage effective. I have used the same car for the last 5 years, and wish to continue that trend. Not only is it better, as purchasing a "better," and more gas efficient car sometimes actually is not worth it as in order to

build such a car it still requires a manufacturing company to use oil for shipping and creating the product.

It may sound silly, but I quit using straws because they are a complete waste of plastic. They may keep my teeth from getting a little chilly but they are used only once before being dumped in a landfill for millions of years. I can name hundreds of other useless plastics we consume daily that are simply, a luxury. I also quit buying new clothes, 80 percent of what I wear is second hand. Not only am I helping the environment by reusing but it makes my style unique and cheap.

I look at many products both large and small scale, meaning I consider the impact of the creation of the product, as well as things like how long it will last me and what I'll have to do with it when it has fulfilled its purpose.

Related to the theme of taking action, a few students also discussed the value of and/or challenging nature of manual labor in an energy-constrained world.

Career influences. A small number of students noted that the course had in some way influenced or altered their career plans by focusing their attention not only on making money, but on how to earn a living sustainably. One student stated:

As for my future plans, I am an entrepreneur and really plan on starting a company at some point after graduation. These company ideas have been based on ideas of globalization and even moving to Europe and starting a skate shop. With rapid depletion of resources, mostly oil, these dreams of easy transportation and traveling have come to a speed bump. They have not stopped me but I am shifting my business ideas towards things that are local and can help the community without exploiting resources and people.

With regard to career decisions, the following quotation from a student paper demonstrates a clear recognition of the contradiction between a capitalist economy and a sustainable one and how that conflict translates into issues of personal integrity in business practice:

It is extremely hard to come up with a business plan that is profitable AND energy efficient. I think one of the hardest things I will face in the future, as a business man, will be to meet the need of my customers without sacrificing my integrity.

It remains to be seen whether and how this student might resolve this conflict.

Collective agency. As discussed above with regard to (re)localization, the theme of collective agency in community is very strong in student essays. Students also addressed forms of collective agency that were not specifically related to localized communities, and this theme is a moderately strong one emerging from student reflections. Following are examples of student writing on this theme:

> This is something I think that everyone has to understand, the power is in the people, rather than the government, and if we, the citizens decide that we want something changed, then we can organize to have something changed and the government will have to listen.

> We have the ability right now to shape our future. We can either continue on the course we are on now and deplete our limited resources as rapidly as possible until we are faced with a likely deadly and abrupt change or we can try to make the transition easier and less drastic. Out future is completely dependent on society's collective effort.

One person referred to prominent social movements in the history of the U.S. to make the point that collective agency is very challenging but a necessary and meaningful form of resistance to domination:

> The eight hour work day, suffrage, civil rights, etc. had to be fought for, tooth and nail, to make progress. As long as we sit idly by and hope someone else takes care of it, our rights WILL be taken away. Our rights to clean air, water, food, and even expression are being eroded and it seems that nobody cares.

Educating others. Educating others with regard to peak oil and sustainability was a very strong theme in student essays. The strength of this theme may derive in part from the essay prompt which asked students to discuss important ideas from the course that they would like others to understand.

Many noted that they had discussed course content with others and had even taken action in collaboration with roommates or family members in response to what they had learned in the course. Some commented that it is difficult to educate others who tend not to believe in the urgent need to address pressing socio-ecological problems. Several students commented on the value of teaching by example, for instance, through engaging in community gardening. Some mentioned recommending the course to others or that it was recommended to them. Others discussed feeling more informed themselves as a result of taking the course and, therefore, noted that they felt better prepared to educate others. Apple Days was mentioned as a form of community education, and one student also emphasized that participating in the event made her feel more optimistic about the possibility of educating others. A couple of students who developed and participated in educational public performances for their small group action project expressed a desire to continue to educate others through artistic expression. Several students commented specifically on the need to have an open minded approach to educating others and/or that they had appreciated this kind of approach in the classroom. The following quotation from a student paper emphasizes some of these themes:

> I feel that the most important information that can be distributed from End of Oil is the simple fact that hydrocarbon energy sources are finite and as such, they will be used up. I have discussed this fact with several of my neighbors and it seems like the idea was new to them. I don't feel that it is necessary to preach or warn of the impending difficulties, only to educate and let people make the preparations that they feel are the most important. I hope I can relay this information in a manner that is similar to Tina's teaching style, inform but don't infer.

Praxis. A small number of students articulated a clear vision of agency as praxis, though they did not use the term praxis in their reflections. The following quotations offer examples:

> The single most important recurring thread to me was that the point of this course is that we all must not only recognize the problem, but we all must also take an active role in constructing some positive interventions. This is a personal duty to ourselves, to our community, and to the world—today, tomorrow, and in the future.

> Though educational techniques have been, and will continue to be questioned by scholars, many agree that experiential activities aid in the effectiveness of a lesson. These activities allow people to engage in an experience, reflect on it, understand its meaning, and apply it to their own life.

A good number of students were, at the time they wrote their reflections, already taking action based on what they learned in the course, and some had plans to do so. Many students also recognized the value of critically informed action to mitigating the sustainability crisis. It is my hope that many students who take the class will continue to act on what they have learned.

The Learning Experience

We will now explore themes drawn from student essays that relate to their experience as learners in The End of Oil. This section is divided into several subsections representing subthemes.

Integration of the action project with the conceptual content of the course.
Although I have often wondered how well students integrate their action

project experiences with their conceptual learning in the course, a good number of students demonstrated fairly strong integration in their reflections. Examples of how students saw the Apple Days project as integrated with course content include the following:

> Conservation was stressed throughout our course as well as in class readings. During Apple Days we learned how not to waste any resource, without us those apples would probably be sitting on the ground rotted. Even the pulp from the pressed apples, which may seem like trash to most people, was dried and used as sugar. Since it was right in town, we were all able to walk/bike there saving fuel. We also used hand presses for the apples instead of electrical, making it even more beneficial to conservation.

> It is really nice when you can discuss issues like local food, sustainability, renewable resources, and come together as a class and community and see how we can tackle some of these issues. I feel like learning in the classroom and then putting it all together and being part of a community event like Apple Days is an awesome way to incorporate class material....

Two students who completed vermiculture projects in winter 2010 offer examples of integration of individualized or small group projects with course content:

> By using this type of composted materials [worm compost] to add to a garden, then there would be less of a need for the companies that sell us food, both produce and animal products, which would in turn lead to a decrease in power for oil companies, which would in turn lead to a decrease in the amount of control that oil, coal and natural gas have in our politics both nationally and globally. This would be a healthy alternative for the world.

> I loved the project that I did for this class. Doing a worm bin was one of the best parts of my semester. It is easily relatable to this class because soon everybody is going to be required to use all of what they have. This is an easy thing to do with food scraps that will help you grow more food in the future. In the future I think that we will not be throwing anything away. Resources will be so scarce that we will be able to find a practical purpose for all of these things that we just throw away right now. People in the future will look back at this time in history and just wonder at the excess that we must have been living in to be able to throw so many useful things away. I will definitely use this knowledge that I have gained from this project in the future. I am sure that I will have a worm composter going for the rest of my life. Like I said in my presentation it is kind of like having a bunch of little pets. You start to really look after them and get to know what kinds of food they like and things like that.

One student reflected upon the course theme of reciprocity in human/nature relationships as this theme relates to vermiculture and gardening.

This student did not complete a vermiculture project herself. Her reflections were derived from listening to the presentation given by the student quoted above and from completing a container garden project herself.

> One [student] stated, "The worms turned into little pets and I grew attached to them." It is pretty interesting how a person does become attached to something such as a worm or a plant. I actually became attached to my plants. I think we become attached to what we are taking care of because not only does the plant or worm rely on us to take care of it but we rely on it to take care of us as well.

One student who completed a small group project noted that not all aspects of the project were sustainable, but acknowledged that this work represented steps in the right direction:

> Areas of our [multi-person bike building] project that are not and were not sustainable did and do exist.... If we had more time, it would have been possible to do the grinding and cutting using energy-free tools like saws and files, but welding is an integral part of the process and, to my knowledge, is impossible without electricity and expensive gases. Also, as with all bikes, ours is subject to wear, tear, and degradation, mostly in the wheels. Tubes and tires (both petroleum products) will need to be replaced over time. Of course, for the amount of electricity that went into the production of the bike, it seems to be far worth the resource. Our bike, besides the replacement of tires and tubes, should be able to survive for our entire lives as it is very simple and structurally sound.... This project is tied in wonderfully with the course as it was a hands-on experience that made us think about things like sustainability and how we could use our trade skills to influence and participate in our community.... We already have had several people around town express interest in getting similar bikes made for them and are ready and willing to pay/trade us for our services.

The above reflection integrates nicely with the overarching course theme of sustainability. It also points to this student's abilities to think critically and consider ways to improve upon personal and collective action, a process of reflection that is a core component of praxis.

Many students were able to integrate their project experiences with central course themes and other conceptual content.

Agency in action project choice. An aspect of the projects that I have found challenging has been balancing student agency in the choice of their projects with time efficiency for me as the professor. Students who completed individualized or small group projects seemed, by and large, to be very satisfied with what they had done. The atmosphere in class during the presentations was energizing, and many students expressed verbally and in their reflective essays a good deal of satisfaction with their chosen work. One student commented explicitly on the value of personal choice with regard to the project:

> This project greatly enhanced my learning in this course because having this opportunity to work with the bees has fulfilled a desire that I have had for years. I have learned so much about bees, beekeeping, and the importance of them recently and that experience alone has made me incredibly grateful for this course. I really appreciated that we had the choice to work on the project that we wanted to because I was much more engaged in this project probably more so than I would have been any of the other ones.

Organizing student participation in Apple Days each fall takes less class time than does organizing and presenting individualized or small group projects. Though students participating in Apple Days generally do not seem to feel as personally empowered at the conclusion of the project work as do those who undertake individualized or small group projects, Apple Days participants often refer to the experience as exhilarating and empowering in terms of engaging the power of community.

Critical thinking. Critical thinking is a foundational activity for teaching and learning that is transformational (Brookfield, 1987, 2000, 2005; Freire, 1970/2000; Mezirow, 2000). Some students discussed directly how the course had called upon them to engage as critical thinkers while also commenting on the value of critical thinking as a practice in daily life. The process of questioning authority, powerful entities, news and other information, and one's personal assumptions was the central activity mentioned in students' discussions of critical thinking. One student noted that deception by powerful interests reduces critical thinking among the public. One student discussed the power and importance of critical thinking to social change:

> On an individual scale, the strongest actions I could suggest at this point would be to learn open-minded thinking, and to learn as much as possible about the truth of the world. The news spits things at us, and a lot of people just take it for granted that it's the truth. Individuals need to take responsibility for the things they know because otherwise things will remain the same. As for larger entities, they need to be made to serve the people again instead of using them.

Worldview change. Change in worldview was a fairly strong theme in student essays. The strength of this theme is likely due, in part, to my specifically asking students to discuss whether, why, and how their worldview night have changed as a result of taking the course.

A good number of students noted that their worldviews had not really changed, but rather, had deepened or been reinforced through the course:

> This class did not do a whole lot to change my world view, I'll admit, what it did do though, was to inform it. I now feel like I have a good understanding about what is going on in my own world, and I feel like the feelings that I have about oil, conservation, the importance of community and family, along with a drive to preserve the environment as we know it, are all now valid and legitimate.

Worldview changes described by students as distinct departures from previous ways of seeing the world ranged from those that affected a fairly narrow area of belief or perception to changes that were deeply transformative. The following quotation describes a worldview change within a fairly narrow but important area:

> My view of the world has also changed throughout this class, mostly because I am a business major. In all of my business classes we were taught how to make the most profit and be the most efficient. Although I was aware of the evils of corporations, I did not have much knowledge on those evils until this class. Therefore my view of the world has changed because the corporations in America are not looking quite as glorious as they once looked.

With regard to transformative changes in one's worldview, I refer to transformation as defined by Mezirow (2000). According to Mezirow, transformation occurs as a result of a student's encounter with a deeply disorienting dilemma that calls foundations of his/her belief system into question. The encounter is, not only intellectual, but emotional, and the student can experience one or more negative emotions as a result, including feelings of betrayal, guilt, shame, anger, or disappointment. Transformation occurs when the student significantly reshapes her/his system of beliefs to incorporate the new information that had comprised the disorienting dilemma.

A handful of students in each section discussed worldview changes on this order. Some of these students referred to the class as an "eye opener." Many students reported feeling strong emotions in connection with taking the course, both positive and uplifting and negative/depressive, though negative emotions were mentioned much more often than positive ones. I believe these negative emotions (and sometimes positive ones as well) may result from students encountering disorienting dilemmas, but I cannot assume that all of these encounters resulted in transformation. The following quotations indicate transformative potential with regard to worldview:

> I left class after watching some of the films we viewed very depressed, but in a way I think it was a good thing.

After this semester, I feel I have gone from having a very naïve view on things to feeling incredibly disappointed at the way this world is run. Disappointment is almost an understatement as there has definitely been a shock from what I've learned. The world truly operates on a cash flow and bottom line, and individuals do not matter except when they further growth. I'm concerned that I'm a cog in this giant money press and I'm none the wiser. It is a very heavy thought and will take time to fully comprehend. That said, the rose-shaded glasses are definitely fractured.

Sure, a lot of what we cover in class seems to be somewhat depressing as it divulges an unclear forecast for our future as individuals, communities, and nations, but at the same time, knowing these things and preparing for the changes to come is the best way to live great lives in the future.

First off I will say I was really excited for this class going into it. Now I feel excitement to be motivated to make changes to my life and help change community life to fit a sustainable practice. I feel this motivation comes from a very strong disheartening feel I got from a lot of material presented in the course. Disheartened due to the severity and reality of the issues presented in the class.

We are at the forefront of a major, drastic, life changing occurrence and everybody has different views on how to deal with it and what the cause is. Anxiety, excitement, curiosity, pressure, confusion, determination; these are just a few of the thoughts and emotions that accompany the change mentioned above. This class has been very enlightening and helpful in finding my own view. It is still in pieces but beginning to fall into place and education is the only thing to assist in this process.

Over the course of this semester I have been tested, quizzed, been angry, sad, happy, disappointed, depressed, and complacent, but of all the courses I have taken this class came to mind most often at home or when spending time with friends and family. There are times when I disagreed with the course teachings, wondering if they were truth, or whether just as much of a fabrication as the other side of the argument, however, in doing that, I was able to learn more about what personal opinions I have in relation to the themes of the course, and why I have those opinions. This course opened my eyes further to the dire need for a change in our world. That without change we, as a human race, are heading towards a future that cannot be sustained, and that we may not be able to survive through.

All in all, I think this class has really challenged my view on how I as an individual fit in this world. What my role should be as a person as well as what everyone else's role should be in relation to each other. We need to start relying on each other more on a smaller scale. Right now our consumer-driven society is causing us to try and push further and further away from each other and it all has to do with not having to rely on your neighbors for anything. Why would I ever want to rely on my neighbor for anything when it is so much easier to go to the store and buy something I need for myself? We are headed in a scary direction and I only hope that we can get shaken from our current path before we run out of time.

One student linked personal transformation with the course's emphasis on action:

> Of all the EGC courses that I have taken here at Fort Lewis, I think it's safe to say that this is the only one that has actually made me reconsider the way I want to live my life. We covered a lot of material over the course of the semester and it really opened my eyes to some things going on around me that never even get discussed with any real thought. If that were all however, this course would not be different from any of the others that I have taken. I think what sets this class apart is because instead of just learning about how terrible things are in the world and how everyone is contributing to global problems, this class actually took it a step further and addressed potential solutions.

One student discussed fairly extensively what seemed to be personal transformation resulting from taking The End of Oil, but he contradicted these statements later in his essay, leaving the impression that he remained conflicted with regard to the course experience. The student offered these thoughts at the beginning of his essay:

> Basically our way of living has made the future of our world a question mark. This has been the theme of the course End of Oil from day one. It has been a scary and interesting ride with everybody pointing fingers at every issue. I have learned more than I ever dreamed about learning throughout this semester about the world we live in and our oil dilemma. This knowledge has been bitter sweet to swallow due to the frightening facts. This class has made me think on my own and question everything from people's opinions to the facts themselves. There is no doubt in my mind that this class has changed me as a person and has made me think of ways I can be a better human on earth.

This same person stated the following near the end of the essay:

> What do I do with these facts, and how do I use them in the future? It is scary to think that the world as we know it could end tomorrow. I am not going to change the way I am living my life now to perhaps prolong this inevitable event. I guess if everybody is contributing to the consumption I will too because I am not going to inconvenience my lifestyle by riding a bike instead of driving a car or spend my hard earned money on expensive technologies that would be sustainable when I could use those funds to go on vacation or make a large purchase. Sure people could call me dumb for doing what the corporation wants me to do, and not caring about our children's earth. I am not going to live in constant worry about what can happen when no one else will stop consuming. No one else is putting in any effort to stop the end of oil. This sounds horrible for me to say, but this class has been great for me even though it has not changed the way I live. It has however affected my thinking. It has showed me the bubble that we are living in and that all bubbles pop at some point. A great example is the housing bubble and crash we witnessed where a few, educated people knew it was going to happen. With the End of Oil course I

now know that the depletion of this resource is imminent. I know the facts. When the time comes I will be stocked up and ready.

These last quotations remind me that choosing, as a critical pedagogue, to participate in rocking people's worldviews is a heavy responsibility with an unpredictable outcome, and that the full outcome for individual students may take time to emerge. But, as an educator who has very good reasons to believe that the way people are living constitutes personal, societal, and ecological suicide, it would also be a heavy responsibility *not* to rock people's worlds. Still, as critical pedagogues, we must exercise care in working with students involved in learning deeply disturbing material while also realizing that knowing how much criticality is too much for any one student at any particular time is not entirely possible.

We now turn from exploring the learning experiences of students in The End of Oil to a discussion of how the course's content and learning processes influence students' ideas about hope.

Clear-Eyed Hope

None of the students in the two sections discussed here expressed feelings of total hopelessness or helplessness in their final essays though, as noted above, a small number of students expressed a reluctance or unwillingness to take responsibility for social problems of sustainability and resource depletion. Many students addressed the theme of hope indirectly in their essays. By including discussion of actions they were taking or that they planned to take, they implied that these actions were worthwhile and that there must be some hope for themselves, their communities, and the world. A good number of those who discussed hope directly expressed what I call "clear-eyed hope" —a notion that there is value in doing the right thing, even in the context of knowing the daunting problems we face and recognizing that the outcome of our struggles for a better world is uncertain.[13] This notion of hope is dis-

---

[13] This form of hope is eloquently articulated by Václav Havel (1990, p. 180). He states:

> Hope…is not the same as joy that things are going well, or willingness to invest in enterprises that are obviously headed for early success, but, rather, an ability to work for something because it is good, not just because it stands a chance to succeed. The more unpropitious the situation in which we demonstrate hope, the deeper that hope is. Hope is definitely not the same thing as optimism. It is not the conviction that something will turn out well, but the certainty that something makes sense, regardless of how it turns out.

cussed in The End of Oil,[14] but students are not required to address it in their final essays.

Some students expressed clear-eyed hope quite eloquently in their own terms, and it is my hope that this idea will stick with them in the long term as the sustainability crisis deepens. One student discussed how guest speaker Riki Ott[15] had led her toward a sense of clear-eyed hope when she was feeling powerless after learning about corporate power and influence. Her reflection shows that exposure to the idea of agency and to examples of agency in others can contribute to students developing a sense of clear-eyed hope:

> Unfortunately my worldviews have changed for the worse. I feel like I lost a bit of faith in humanity because of the way governments are run and how the power of corporations has such an impact on life. After indulging in the reading for this class, it was slightly frightening to find out the motives behind these major corporations that people have no control of modifying. For instance, major oil companies invest in government interests, which many people do not even know about. Although I became aware of the lack of control citizens have, the talk with Riki Ott helped open my eyes to the amount of people who really do want a change in the way our society is functioning.

At least one student directly expressed a sense of clear-eyed hope as an outcome of his agency in the course action project:

> [Apple Days] was the contrast to all the negative things that were brought to light. Given the material, if all that had been presented were the bad, in my current place of understanding I would have simply shut down or even dropped out from it being too intense to process. Essentially, Apple Days provided me with the positive ideals I needed to absorb the information.

The following quotations from student essays express a sense of clear-eyed hope as an outcome of the full course experience:

---

[14] I reference Havel's notion of hope in class discussion. In fall 2010, Orr's *Down to the Wire* (2009) was also used as a course text, and Orr repeatedly discusses a notion of hope akin to that of Havel.

[15] Riki Ott is a grassroots environmental and political activist who works on issues of oil pollution and its impacts on human and environmental health. She is also participating in an effort to rescind corporate personhood in the U.S. and on other related issues. Her work figures prominently in the film *Black Wave* (Cornellier & Carvalho, 2008) about the impacts of the Exxon Valdez oil spill on both people and the environment. I showed this film in both the winter and fall 2010 sections of the class.

I like the way that [Heinberg, author of *The Party's Over*] approached the subject [of peak oil], because it wasn't about how humanity is in huge trouble and there is nothing we can do about it, rather it was about how we need to accept the position that we are in and figure out a way to make things better for ourselves now and in the future. This is also one of the core messages in my favorite book from the course, that being *Depletion and Abundance*. Sharon Astyk is one of the single best writers that I have encountered that can convince me that there is hope in this world. She gives her readers hope rather than leaving them feeling helpless. I can't tell you how valuable I felt that book was, because she is living a real world example of how to combat peak oil. It's from her book that I have taken examples of how to change my day to day life.

I thought by the end of this class I would feel depressed, and of course I came out of class some days ready to curl up into a ball and cry. But now that the class has ended, I am left with more inspiration and hope than anything else. I see the energy crisis as a challenge, and I know it will not be easy. Now that I have more knowledge on the energy crisis, thanks to this class, I feel like the transition to a new lifestyle will be easier, and I am willing to embrace this new lifestyle when the time is right.

If there is one thing I most cherish in taking away from this course, it is the hope that sustainability on a personal level is plausible, and that the chances of finding a community/rural neighborhood that feels the same is as well. As Orr mentions [in *Down to the Wire*], optimism is thinking positively while hope is reached after critically examining information. This class has given me hope on this scope.

We have now concluded our exploration of student reflections on aspects of their learning experience in the course. We will now consider the themes related to the course as a form of praxis.

The End of Oil Course as Community Praxis

Although the concept of praxis itself as developed in this book was not a strong theme in student essays, some specific reflections related to praxis were quite powerful, and they demonstrated that the course is having some impact beyond the classroom.

One student discussed how he had been very impressed with the actions of his friends who had taken the course in a previous term. He stated that, when he returned to Durango in the fall, he saw that these friends had been very busy over the summer gardening and growing their own food. He was impressed that, when they were cooking, instead of going to the supermarket, they went into their garden to get some of the food they had grown.

Another student discussed the impact of the course on her roommate who had taken it in a previous term. She also discussed how the course was continuing to influence both her and her roommate in their thinking and actions:

> I remember my roommate taking this course last year. She used to come home and tell me how crazy it is going to be when we run out of oil. That was the semester I learned plastic bags were actually made out of oil, and that those bags were polluting our oceans. Since I love the ocean and all it is, we switched to re-usable bags permanently. This class had a significant impact on my roommate, and because it seemed so radical I decided to check it out.... Over this semester, I have concluded that this course is radical indeed. Through the lens of peak oil, the way I perceive my entire life has changed. I have been forced to ask myself, "What does happen when oil runs out?" Life as we have come to know it in America will change forever. Our cars won't run, there will be no plastic, ultimately we won't be able to operate at the intense capacity that we do now. Capitalism will change, and our communities will have to look to themselves in order to survive. The entire American ideal of super-independence will be ultimately destroyed. As a people we will have to learn how to communicate with each other and care for our environment.... In a very interesting way, I find the problem of peak oil a good thing. Personally, it has caused me to step back and ask myself, "What is truly important in life?" After some serious consideration, and the help of Sharon Astyk [author of *Depletion and Abundance*], I feel it is people. Life is all about the people around you and the relationships you build. We only have a limited amount of time on this earth, but in America we miss it because we get so lost in money and power.

This student and her friend have persisted in acting on what they learned in the course beyond the duration of the course itself. This student is also voicing a fairly deep and comprehensive analysis of the global implications of oil depletion and considering priorities for individual and collective action. Although she cannot say exactly where her insights may lead her, she is beginning to engage in praxis.

This same student commented on the impact Apple Days had had on her thinking prior to taking the course:

> I remember the first year we had the event [Apple Days] in Buckley Park. A friend and I happened to stumble upon the huge pile of apples that people were collecting. We actually had no idea what was going on, so we grabbed an apple and started to play hacky-sac with it. Pretty soon we saw people pedaling in towing large amounts of apples behind them. It is a scene I will never forget. Never before had I seen so many people come together to celebrate something as simple as apples. At that moment, I fell in love with the idea of simplicity.

I hope that Apple Days has also helped community members to recognize the value of local food, to see the joy that celebrating it can bring to a community, and to realize the possibility for community building that engagement with local food represents.[16]

As discussed above with regard to agency and community building, a good number of students recognized the value of the action projects, not only to themselves, but to the wider community. This recognition points to the potential of the course as a vehicle for community praxis.

We have now concluded our exploration of the themes that emerged from student reflections on The End of Oil course. In the following chapter, I will offer my conclusions about how this exploration has informed my pedagogy and about how my findings, interpretations, and conclusions might contribute to the critical pedagogy of sustainability as practiced by others in other contexts.

---

[16] I do know that the festival has been well attended and that many community members have expressed a pleased bewilderment when they learn that the festival is not a business venture, but an event emerging from the collective generosity and will of participants themselves.

Chapter 10:

Conclusions on Implementing the Critical Pedagogy of Sustainability in the End of Oil Course

I want to be with you.
You cant.
Please.
You cant. You have to carry the fire.
I don't know how to.
Yes, you do.
Is it real? The fire?
Yes it is.
Where is it? I don't know where it is.
Yes you do. It's inside you. It was always there. I can see it.

—Cormac McCarthy

After carefully examining students' final reflective essays from my End of Oil course, I conclude that much of what I have proposed throughout this book, with regard to both the conceptual framework and the processes of the critical pedagogy of sustainability, is to some extent possible to achieve in a single college course. At the same time, this pedagogical praxis is challenging for the professor in terms of her/his time commitment. Practitioners are also likely to find that institutional recognition/support and budgetary support for this pedagogy is less than forthcoming.

Outcomes for students are also incomplete and tenuous. Reflecting on my own learning process over the years, the incompleteness of student outcomes is not surprising. I consider my own learning, even as it relates to issues covered in some depth in this text, to be incomplete. I do believe I offer my students a significant foothold in developing counterhegemonically-oriented thinking and praxis and that some students will continue to develop their worldviews and their actions in this direction.

Analyzing students' reflective essays has also led me to conclusions about specific aspects of The End of Oil course. I argued in chapter eight that critical pedagogues must be concerned, not only with what we teach, but with how we teach. I argued that courses must call upon students to name the world and that professors must leave room for students to engage in independent meaning making in response to encounters with disorienting dilemmas. Striking difficult balances between more comprehensive coverage of complex issues and problems, on the one hand, and taking the time necessary for students to voice their perspectives and engage in projects of their own design, on the other, is challenging. This balance is vitally important, however, because sustainability-oriented educators must work against reinforcing patterns of passivity in the face of daunting socio-ecological problems. The importance many students ascribed to action projects with regard to their learning and agency inspire me to continue requiring service learning experiences in The End of Oil. I believe, as some students confirmed in their essays, that these experiences also serve as vehicles for moving toward clear-eyed hope in the face of the critical content of The End of Oil that is quite heavy and dark.

My analysis of student essays also calmed an ongoing concern I have had about the conceptual integration of the action project with overarching course content. I have 30–35 students in each class, I cover a great deal of material (three to four books plus additional articles, book chapters, films, and websites), and I require students to write approximately six papers per term. Therefore, the action project can receive only so much of my and my students' attention. Although integration between the project and the conceptual framework for the course was described as difficult in a couple of instances by students in both sections of the course, a good number of students were able to articulate in some detail and with some enthusiasm how the project(s) related to the course. Although balancing comprehensiveness of coverage and agency-oriented projects will remain a challenge, the ability to integrate action projects into a larger conceptual framework demonstrated by many students encourages me to continue requiring service learning in my sustainability-oriented classes.

With regard to large-group, community-oriented projects such as Apple Days versus individualized or small group projects, I don't know that one or the other is best; they are simply different in their outcomes with regard to the student learning experience. Both sorts of projects offer ample opportunities for students to integrate their actions with the conceptual content of the course, and I will likely continue to require either one or the other kind of project depending on my time constraints and community needs.

In chapter eight, I advocated systemically-oriented critique of global political economy as a key aspect of the critical pedagogy of sustainability. Given the amount of class time devoted to illuminating the inner workings of neoliberal globalization, I had hoped for greater emphasis in student reflections on specific aspects of globalized political economy. Given the complexity of the issues covered (and the fact that many of these issues were covered early in the term), however, my findings are not entirely surprising. Judging from the content of student reflections, it seems likely that, even though a great deal of detailed material on globalized political economy was presented and discussed, most students drew very general conclusions about the enforced dependency and corruption that permeate the system. It seems students may have learned just enough about globalized political economy to see the system as a generally bad and risky deal. Given the emphasis students placed on agency and community resiliency, generalized conclusions about the problems of the world-system may have prompted them to seek alternatives to globalization.

With regard to enforced dependency specifically, students did not demonstrate an ability to apply the fully developed concept as a basis for their analyses. But many students did demonstrate systemic understanding of social power and of complex interrelationships within the world-system. Some students recognize dependency as a potential source of vulnerability for individuals, communities, and nations. Helping students to also recognize the enforcement aspects of enforced dependency would help them deepen their counterhegemonic thinking with regard to globalized political economy. If I want my future students to more fully conceptualize enforced dependency as an important aspect of globalized political economy, I will need to require readings that focus more explicitly on it as a distinct concept. It is still my goal to help students to unite enforced dependency–based critique and (re)inhabitory praxis in the form of service learning projects.

I may also need to be more explicit about conceptualizing various forms of political economy (using labels such as subsistence economy versus surplus production or pre-capitalist, capitalist, and post capitalist political economy) so that students have an analytical heuristic that can be used to integrate and articulate their understanding of political economy. What I have done is to contrast localized economies with globalization, and it seems that students have been able to use these categories reasonably effectively as analytical lenses on their community and the world. They have also been able to use these conceptualizations as a foundation for planning and, at times, engaging in counterhegemonic action.

The emotional aspects of taking the course have also proven to be an impetus for actions undertaken or planned by some students. A good number of students referred to negative emotions they experienced as a result of taking the course as uncomfortable but important to their learning. They also often linked their discussion of these emotions with discussion of personal actions they were taking or planning to take. Some students expressed a distinct sense of empowerment and/or satisfaction that emerged from their action project, and a few directly mentioned being inspired to engage in further action.

Even though students describe a broad range of emotional responses to The End of Oil, the course remains popular each term it is offered. It usually fills within two days of the start of early registration. Even when a second section is offered, the course still fills very early, despite that many students appear to be generally aware of the course content before enrolling. I conclude that many students are interested in the counterhegemonic content, even if they are unaware at the start of the term that much of this content will be difficult to grapple with, both intellectually and emotionally. I do not plan extensive changes to the course with regard to its intellectual and emotional content.

In an overarching sense, as student reflective essays demonstrate, The End of Oil is a challenging course for students. The course material is challenging for students to integrate. Furthermore, the course's duration does not approach the length of time necessary to developing the in-depth rela-tionships among people and with places that are foundational to sustaining deep processes of sustainability-oriented social change. It is not surprising, therefore, that systemic thinking expressed by students in their reflective essays is not as broad or deep as I would hope—I am also hoping for a lot. Although courses and programs can help students learn to engage in criti-cally informed praxis, the critical pedagogy of sustainability is really a lifelong orientation and process of learning that can begin inside the walls of the academy but, ultimately, must live outside of classrooms and educational institutions in the lives of individuals and communities.

I want to be clear that I hold no illusions about the ability of my course to easily transform large numbers of students into lifelong advocates of sustainability. There are very real limits to one course in teaching systemic critique and action. I also acknowledge that the essays I have examined were written at the culmination of an intense course experience and that, for some students, the course may turn out to be only an isolated episode in their lives, the immediacy of which will fade over time. Still, I believe the course does create an opening for some students to transform their worldviews and their lives in sustainability-oriented ways.

In closing, I would like to focus on clear-eyed hope as an important theme that emerged in students' writing. This hope, I believe, is akin to love. When we love others, we have to be honest about their shortcomings in relation to our own expectations, love them anyway, and create relationships that help us and those we love to grow. The same goes for our intimate relationships with place that should also be clear-eyed and reciprocally nurturing. We should not expect our places to conform to our preconceived visions or to yield material wealth beyond their ability to renew themselves. When we approach the world with clear-eyed hope—with love—we must first do our best to see the world clearly, then seek to integrate ourselves within the world and, at the same time, to change the world in ways that enhance the health and vitality of ourselves, others, and nature. Ultimately, the critical pedagogy of sustainability is about love for others and the world expressed in a deeply open and creative sense.

Epilogue

Reflections on Occupying Education

because we must
because the alternative is unthinkable
because failing to is unlivable
because even the faintest glimmer of light has its own fleeting magic
with which we must have patience
so that we might be touched
so that we might ever-so-briefly
    bask in Nature's luminance.

—Bill Devlin

My critical pedagogy of sustainability is my response to the sustainability crisis. It is my attempt to rise to the challenge of making teaching and learning relevant, responsible, and practical in a time when ecologies and economies are perched on the brink of collapse. It is my call to others to join forces—to occupy education by confronting the forces that reduce it to an instrument of perpetuating the status quo.

My pedagogy hinges upon my critical social theory of sustainability and my theory of enforced dependency. It rejects the unsustainable growth imperative of modern industrial capitalism, and it is counterhegemonic because hegemonic social power drives socio-ecological breakdown. The critical pedagogy of sustainability promotes social justice as integral to sustainability. It is rooted in a holistic ontology of place that seeks to (re)integrate human communities with nature in healthy, reciprocating interrelationship. It is also particularly concerned with developing resilient socio-ecologies by (re)localizing the provision of basic human needs. It is inspired and informed by diverse critical theoretical traditions and by cultural traditions of healthy, place-centered living. The critical pedagogy of sustainability engages with the deep social contradictions of the late-capitalist world with the hope of transforming human-to-human and human-to-nature relationships. It is a deeply challenging praxis of clear-eyed hope that sees sustainability resting on a far-off horizon that we may never reach—but it is moving anyway, with its eye on that horizon.

The critical pedagogy of sustainability is indebted to many who have engaged in a Freirean naming of the world (Freire, 1970/2000, p. 88) with the hope of changing it—and it is my naming of the world. It is my hope that this work will inspire other educators working in formal and informal settings and in communities who will engage in sustainability-oriented praxis that is finely tuned to their own socio-ecological and historical contexts. It is also my hope that their stories of sustainability praxis will be communicated broadly and that these stories will ignite hope and agency in growing numbers of people and communities worldwide. This book is my story of a shared experience of meaning making and action in the world. We need more such stories, rooted in diverse epistemologies and experiences of the world, in order to inspire and motivate sustainability-oriented social change.

Lastly, I acknowledge that we are in desperate times, and it is not at all clear that we will make it to a new horizon of living sustainably. It is, however, never too late to act with integrity and do the best we can. I hope that others can learn something from my example that will be useful in their own educational and community contexts.

Afterword

Richard Kahn
Antioch University, Los Angeles

If the doors of perception were cleansed every thing would appear to man as it is, infinite. For man has closed himself up, till he sees all things thro' narrow chinks of his cavern.

—William Blake (2000, p. 201)

I can choose. I have to choose. I have to make my mind up whom I will take into my arms, to whom I will lose myself, whom I will treat as that vis-a-vis, that face into which I look, which I lovingly touch with my fingering gaze, from whom I accept being who I am as a gift.

—Ivan Illich (1997, p. 22)

If we live truly, we shall see truly.

—Ralph Waldo Emerson (1987, p. 39)

It is an honor to have the opportunity to provide some final remarks on behalf of Tina Lynn Evans's *Occupy Education: Learning and Living Sustainability*. Significantly, this book beneficially lends itself to the further recharacterization and reinhabitation of a sustainability education-oriented theoretical space that has been emerging academically over the last couple decades, one in which critical pedagogues and other antioppression educators have begun to forge transformative solidarity with environmental educators, on the one hand, and indigenous scholars, on the other, in a kind of counter-hegemonic bloc of ideological alliance.[1] Moreover, all these efforts have

---

[1] This important conversation has moved in all directions and has numerous origins. I do not mean to suggest at all that such work is the result of promethean efforts within critical pedagogy solely. While critical pedagogues like Peter McLaren, Antonia Darder, Bill Ayers, Marta Baltodano, Shirley Steinberg, and Curry Malott have openly worked in ways to allow for greater alliance as iterated here, the undertakings of critical environmental educational researchers and ecojustice theorists such as Marcia McKenzie, David Greenwood, Connie Russell, Pauline Sameshima, Chet Bowers, Madhu Prakash, and Rebecca Martusewicz, and of indigenous scholars like Sandy Grande, Eve Tuck, Four Arrows, Peter Cole, and Dolores Calderon, have been no less crucial. Indeed, especially as regards the steps taken by indige-

increasingly taken place in an ongoing dialogue with nonacademic knowledge workers, and with a wide-range of social and environmental justice, as well as sovereignty, activists, many of whom themselves have often been out ahead of the professoriate in affiliating intersectionally by enmeshing themselves in a diverse movement of movements. In my own work, I have named this emergent, evolving, dialectical, and democratic multitude for planet Earth: the ecopedagogy movement. But whatever it is that we are called and however we identify ourselves, certainly myriad capacities for sustainable human and organizational development continue to unfold across society and around the world in much needed, but altogether unguaranteed, ways.

This means that despite the progress of our struggle, it remains limited and prone to contradictions. No beloved community has yet arisen to stop the occult horrors of militarism, industrial capitalism, and racist colonialism (as well as the various conjoining and non-derivative forms of oppression—such as patriarchy, ableism, and speciesism) that clearly constitute through and through the nightmare that presently weighs upon the brains of the living. No vast and inclusive proletarian base has hitherto come to know and trust in itself that it is capable of abolishing the dominator culture of a white suprem- acist affluent class. A dream of a totally liberated and just "planetary com- munity" is a vanguard phrase spoken by only a relative few intellectuals.[2] The educational Left is more collegial than perhaps ever before, but profes- sional altercations over the parceling of academic real estate (Agger, 1990) remain more common than performances of the kind of resilient grassroots service leadership that this book argues is axiomatic to a thriving communi- ty-in-place. When it comes to learning and living sustainability, those of us

---

nous academics to openly trust potential white allies, it could be said that within this solidary network some partnerships thus far undertaken are likely more generous, extraordinary, and worthy of acknowledgment than others. Finally, it bears emphasizing that the names mentioned here are hardly a complete list of the actors contributing to this overall work, to which a great many have increasingly shown noteworthy commitment.

[2] In my opinion this dialectical, moral and postsecular space is a preferable dream to complex secular spaces like the "glocal," "alter-global," or "international," which it shares affinities with all the same. Outside educational research, theorists such as Leonardo Boff, John Brown Childs, Ervin Laszlo and Sean Kelly have made important contributions to the study of planetary community. In education proper, one can look to Edmund O'Sullivan, Moacir Gadotti, and my own formulations of this idea—thus, a general theoretical contribution of ecopedagogy. Within ecopedagogical literature, specifically, it may be that thus far only I have made a particularly overt demand for methodological decolonization though. To this end, I argue that the "planetary" can only be achieved (minimally) through a postcolonial struggle for "planetarity" (see Kahn 2011; 2010). This links to Evans's "re-localization."

in higher education are not yet a true *collegium*. All too often we are simply the estranged labor of colleges, which are themselves competitively operating as biopolitical institutions for a neoliberal academic enterprise that serves at the behest of greater masters still.

In spite of this, the relationships girding ecopedagogy and related radical sustainability movements are currently strong and viable, but they too are also under the continued provocation of various hegemonic forms of cultural enclosure (e.g. cognitive, affective, spiritual, biological). When accomplished, even partly, these enclosures function to dehumanize us as critical scholars, citizens, and friends capable of loving kindness in these movements, and in this way (as this book painstakingly clarifies) re-enforce our dependency upon the irrational and violent commoditization of the global industrial complex. Thus, *Occupy Education* provides a timely reminder that when folk commit to caring for their collective subsistence and aspire to maintain a re-localized economy of gift-giving on behalf of the larger commonwealth, a revitalized ecology of body/mind/spirit is cultivated that represents a qualitative alternative to the status quo as well as a moment of organized resistance to the domination of nature writ large. Further, by modeling both how and why sustainability advocates should deploy a multiperspectival critical social theory of sustainability education, one based solidly in an analysis of global political economy, Evans herself reaches out diplomatically to bring differing audiences together around common themes. In this way, *Occupy Education* provides a crucial transdisciplinary mediation of discursive debates now underway that can (beyond its other contributions):

1. help Marxist and other socialist colleagues to recognize and value, however latent it may be for some, the primary ecological dimension of their work; and
2. demonstrate for environmental educators and researchers the stark necessity of their learning to theorize and teach against the destructive socio-cultural systems of domination, as these ultimately structure the damaged relationships that characterize so much of contemporary global life, even as they attempt to colonize the rest.

I cannot stress enough that *Occupy Education* is a movement-building text. Therefore, the book rightly ends on a note of hope—what it calls "clear-eyed hope"—a demand that we distinguish human from inhuman historical tendencies, and then strive consciously to join with the former against the latter. For sure, this is a quasi-universal method active within a plethora of critical educational and political projects, so much so that it can begin to sound like a thoughtless mantra at times. Yet, if the propagation of hope can sometimes ring false, seem naïve and coldly abstract, by contrast clear-eyed hope attempts to stare concretely into in the face of planetary endgame and act with bold integrity. Whereas Derrick Jensen (2006) has in this same way suggested that present circumstances dictate that hope be jettisoned altogether in order to allow for people to act as directly and fully within the state of emergency as possible, I personally find a pedagogy of clear-eyed hope appropriate and preferable to his call for hopelessness. There is too much humanity left to uncover through the prophetic imagination[3] and in the utopian visions contributed to society by denunciatory/annunciatory regimes of truth to abandon a methodology of hope now…just because things appear so awfully bleak. Likewise, as Weber and Mumford famously suggested, it is the potentially terrible cult of political administrators and scientific technocrats, those that enforce megamachinic laws of social efficiency through their instrumental rationality, who can generate mass anomie by erasing traditional ties to the land and others, and who drive in the shadowy logic of modernity that continuously threatens to end in fascism, which ultimately arrives with the last bugle call of a disenchanted world spirit.

Jensen may be correct that warriors don't hope, they fight. Still, at this time, a hopeless critical pedagogy potentially courts the formation of a nihilist sensibility that would fold the revolution back into the establishment and seemingly play more into the tactical hands of the oppressors than the oppressed. Ironically, then, it is just because there continue to be significant numbers of people for whom hope is not an option that hope lives on. As W. E. B. DuBois came to conclude, it is from "those who suffered most and have the least to lose that we should look to for our steadfast, dependable, and uncompromising leadership" (James, 1997, p. 28). Their leadership is another name for our hope.

If, therefore, I were to raise a question about clear-eyed hope, it would not be about the nature of its existence but rather about the clarity of its eyes. For me, hope—in an age of planetary ecocrisis, for an existence increasingly characterized by geographies of genocide, ecocide, and zoöcide—emerges from an uncompromising sadness akin to the prophet Jeremiah. In this way, hope can flow like the streams of tears that often flood uncontrollably from the red, swollen eyes of those deep in mourning. This hope feeds back into our critical theory and educational activity, nourishing and redoubling its

---

[3] In our book, *Education Out of Bounds* (2010), Tyson Lewis and I define, theorize, and exemplify this imaginative force at length as a revolutionary zoömorphic love that invitationally prepares for the Other in the form of a critical and creative posthumanism.

negativity. So, as the mainstream routinely intimates a desire for a pedagogy that can help contribute to students developing clear critical thinking, a radical hope for sustainability counsels the need for disciples of the wild jeremiad instead. It upholds people's right to practice the dignified non-erasure of their suffering. It calls forth their sustained wails, and unending invectives against the powers-that-be, as well as for them to learn self-respect through their prolonged cries for a continuance of a just war that seeks to defend the peaceable, the good, and the beautiful from wrongful violation. This is the hope of the Great Refusal.

As with Evans, while I can imagine "sustainability" *beyond* this type of refusal, I cannot do so *without* it. Additional reasons, more than those that have already been provided up unto this point, hardly need be given. Suffice it to say that those interested in any form of sustainability education in the twenty-first century will need to dwell seriously on the ways in which it has become an integrated part of the dominant ideology. From this, critical research, theory, and pedagogical campaigns must more broadly take up the rigorous investigation of how sustainability education has, via an institution-al inversion, come to represent a treacherous counterrevolutionary movement to those who stand on the side of bio- and cultural diversity, who value participation in the commons, and whose processual homecoming amounts to the further humanization of the world.[4]

How many times have we heard the question as educators and activists, "What does sustainability mean to you?" Almost invariably, we hear a general type of answer emerge in response—a positive vision, one almost pastoral in its romance of the bountiful kingdom of heaven on earth that may be right now available to us (as if)...and this from the less reactionary quarters! From another more malevolent set of industry spokespersons, sustainability equates only with the affirmation of disaster capitalism. For them, it is an opportunity for training large segments of the population to believe not only in the relevance of ecocrisis, but in the corporate state's ability to properly manage it, and then to awe people through its self-professed ability to turn the lead of social problems into the gold of improved standards of consumer living. The Big Lie of such public relations alchemy, as we know, works in part because of the creation of previously unimag-inable technologies along with the spectacular cultural turns based upon upholding them as magical, compelling, and highly relevant to people's

---

[4] This work is underway in various parts of the world and foundational research by Bob Jickling, Edgar González-Gaudiano, and Arjen Wals (to name but a few notable figures as a kind of primer) has been published on the topic.

lives. But spellbinding the populace to expend their labor power on behalf of running another lap on the global capitalist treadmill of production is really just a snake rapidly consuming itself from the tail on up.[5]

By turn, it ought to become a ubiquitous part of our parlance that the etymological foundation of the concept of sustainability is in fact connected to uttering statements of opposition within the legal tradition. Hailing from the judicial arguments made during the beginning of the last century, sustainability was therein invoked for the first time when a member of counsel disagreed for the record with court procedure or opinion. "I object," an attorney rose and said, and when said objection was found to have merit, a judge responsibly affirmed, "Objection *sustained*." At a minimum, then, sustainability activists and educators need to recover this original sense of legitimate protestation alive in the idea of sustainability. They must more actively vocalize their reasonable rebuttals of social protocols that they find to be clearly illegal. They should stage interventions into and disrupt tactical attempts to utilize ambiguous forms of precedent toward the re-entrenchment and instantiation of hegemonic positions for the modernization and programmatic development of vernacular life spaces.

It follows, then, that those of us interested in learning and living sustainability should be unequivocal in the recognition that there is a vast amount to disagree with and reject in how educational circles are advancing sustainability right now. For instance, take the recent White House Summit on Environmental Education held on April 16, 2012, championed as the first of its kind by presiding administrator of the Environmental Protection Agency (EPA), Hon. Lisa P. Jackson. According to the event's website[6]:

> The White House Summit on Environmental Education convened a diverse group of stakeholders to discuss the importance of environmental education and the core concepts and principles that contribute the most to environmental literacy. The Summit featured two panels of environmental education leaders, remarks from several Administration officials and a panel on the Federal government's on-going commitment to the field of environmental education.

A diverse group of committed stakeholders indeed! Offering a prevailing message that government agencies need to come together to maximize

---

[5] My apologies to snakekind all—who deserve far, far better than this comparison. My point in deploying it is that if perceived from an alternative standpoint, the same sets of behaviors it images may not represent an absolute death-drive but rather the symbolic renewal and promise of an ouroboros.

[6] See http://www.epa.gov/education/eesummit.html.

resource opportunities—thereby essentially creating a Department of Homeland Sustainability—so that the state can best assist corporations to work productively with schools and nonprofits in "collaborative partnerships" for environmental education, authorities testifying at the Summit included Secretary of Education, Arne Duncan, as well as top representatives of multinationals like NASCAR, Disney, Toyota, Samsung, and the Girl Scouts of the USA.

Also present was a chief columnist for *Time* magazine (an organ of Time Warner Inc.), whose wise counsel was that environmental education's twenty-first century challenge was decidedly not to work against accountability and privatization schemes, but rather to facilitate the accountable integration of public and private interests for the promotion of "sustainability" ventures that could generate much needed jobs. Seconding this were the authors of *The Failure of Environmental Education (And How We Can Fix It)*, a popular 2011 book that propagandizes for the creative destruction of the educational system toward a business-friendly takeover of the curriculum. According to these authors, in their expert opinion the future of the American people (and hence the world) desperately requires a form of environmental literacy that can help foster entrepreneurial desires to capitalize on green economic trends. Of course, this conclusion met with rapturous applause in the Summit and hearty endorsements by governmental appointees and their *Forbes* 500 counterparts.

Lastly, and perhaps most shamefully, the President of the National Environmental Education Foundation (NEEF), and the Executive Director of the North American Association for Environmental Education (NAAEE), additionally offered their own commentaries. Much to the delight of Secretary Duncan and the others, NEEF declared it was using the Summit as an opportunity to coordinate a National Environmental Education Week dedicated to the greening of STEM (Science, Technology, Engineering, and Mathematics) programs—a preeminent fascination of neoliberal, human capital ideologues. NAAEE, for its part, openly gushed about the organization's enthusiasm to be a part of what it saw as an exciting collaborative and philanthropic enterprise. Meanwhile, an article on the Summit in *Education Week* exposed a great deal of the event's utter hypocrisy. As reported, it turns out that even as the attendees mused on the necessity of their coming together as never before to support environmental literacy, plans had actually been quietly afoot within the Obama administration for some time to gut and slice in half the EPA's paltry Office of Environmental Education (the initial thought was to kill it outright), as well as to cancel the place-based, experiential learning initiatives of the Bay Watershed Education and Training program, along with the availability of $8 million of environmental literacy grants that had previously been offered annually by the National Oceanic and Atmospheric Administration (Robelen, 2012).

My hope is that readers of *Occupy Education* will understand that a chronicle as I have just outlined equates with sustainability being altogether stood upon its head. For me, this is undeniable and definitive evidence that both major political parties in the United States have now put ecoliteracy (conceived as a democratic mandate) under direct siege by a market fundamentalism that has proved brazenly hostile to democracy throughout the past. In consort together, Democrats and Republicans have unhappily launched the public good of environmental education for sustainability far over the event horizon of fiscally conservative rhetoric and into the authoritarian black hole of Orwellian truthspeak.

This type of sustainability education they imagine—one that makes environmental education into an unfailing servant for the economic machinations of corrupt government, various transnational corporations, and a power lunch club of large-scale Washington Consensus nonprofits—requires our disgust and rage. Moreover, if accountability and standards are indeed to remain the norms of our everyday lives under such technocapitalism, then we must make sure that the institutional leaders who work gladly to neutralize more liberatory sustainability standards are themselves held accountable by as much indignant criticism as possible. As Howard Zinn told Henry Giroux before he died, "we are in a situation where mild rebuke, even critiques we consider 'radical' are not sufficient" (Giroux, 2010). Hence, if the guillotine is no longer in style, then I say let reputations fall and be forever tarnished such that stoning would be a welcome reprieve for those found questionable. For far too long pronouncements like those that typified the White House Summit on Environmental Education have been allowed to occupy our wider educational policy-making. What has resulted is the Anthropocene. Decolonizing the same at least means receiving formal apologies from those that are institutionally responsible, while securing their outright removal from office whenever and wherever it can be meaningfully achieved.[7]

In saying this, it bears emphasizing that engaging in a thoroughgoing Gramscian "war of position" over sustainability, in which dominant ideolog-

---

[7] To his credit, Alan Reid (Editor of *Environmental Education Research*) issued a quick call for comment on the White House Summit and attempted to organize a symposium of minds to evaluate and engage it as a site of possible intervention. For my part, I replied by asking that the minutes of such discussion record that I greeted the occasion of the Summit's success with a loud Rabelaisian fart.

ical interests are attacked across civil society, does not preclude auxiliary preparation for a concurrent "war of maneuver" and potential confrontation with the armed troops or police when the time shall prove necessary. It is these guardians of property, after all, who effectively put the brute force behind the larger plan of enforced dependency that is one of *Occupy Education*'s key theoretical objects. Though certainly beyond the purview of this book and of only indirect consequence to its discussion, sustainability educators must try to find ways to push themselves to grapple with the strategic role that is played by the security apparatus in maintaining the unsustainable norms of the hegemony. For even to put forth the sorts of radical educational ideas and vehement conclusions about the nature of power in society that Tina Lynn Evans does generally comes with an intimidating warning of repressive costs for having done so—these can range from being the victim of personal and professional surveillance, to undergoing processes of suspension or firing, and beyond. As we move forward, both activists and scholars alike who support the greater integration of the social and ecological justice movements should thus be considered to be as threatened a species as polar bears. However, rather than allow this conclusion to make us cower in fear, we should try consciously to build esteem as a community around the insight that the transformative counterhegemonic work we undertake (in some cases anyhow) is effective enough to warrant being considered legitimately dangerous. Yet, for that very same reason, those of us doing it must learn to take especially good care of one another, document the rectitude of our critical paranoia, account for the stressful and traumatic conditions in which we try to foment friendly relationships and seek change with our lives, and be especially vigilant as to our conduct and its likely repercussions.

References

Agger, B. (1990). *The decline of discourse*. London, UK: Falmer Press.

Blake, W. (2000). *The selected poems of William Blake*. Hertfordshire, UK: Wordsworth Editions, Ltd.

Emerson, R. W. (1987). *The essays of Ralph Waldo Emerson*. Cambridge, MA: Belknap Press of Harvard University Press.

Giroux, H. A. (2010). Howard Zinn: A public intellectual who mattered. *Truthout* (January 28). Online at: http://archive.truthout.org/howard-zinn-a-public-intellectual-who-mattered56463?utm_source=twitterfeed&utm_medium=twitter.

Illich, I., Brown, J. & Mitcham, C. (1997). Land of found friends: Conversation between Ivan Illich, Jerry Brown, and Carl Mitcham. *Whole Earth Review*, 90 (Summer), 22.

James, J. (1997). *Transcending the talented tenth: Black leaders and American intellectuals*. New York: Routledge.

Jensen, D. (2006). *Endgame, volume 1: The problem of civilization*. New York: Seven Stories Press.

Kahn, R. (2010a). *Critical pedagogy, ecoliteracy, & planetary crisis: The ecopedagogy movement*. New York: Peter Lang.

Kahn, R. (2010b). Love hurts: Ecopedagogy between avatars and elegies. *Teachers Education Quarterly* 37 (4), 55–70.

Lewis, T., & Kahn, R. (2010). *Education out of bounds: Reimagining cultural studies for a posthuman age*. New York: Palgrave Macmillan.

Robelen, E. (2012). Obama sends mixed signals on environmental education. *Education Week* (April 25). Online at: http://blogs.edweek.org/edweek/curriculum/2012/04/obama_sends_mixed_signals_on_s.html.

Bibliography

Achbar, M., & Simpson, B. (Producers), Bakan, J., Crooks, H., & Achbar, M. (Writers), Achbar, M., & Abbott, J. (Directors). (2005). *The corporation* [Motion picture]. United States: Zeitgeist Films.

Allen, T. F. H., Tainter, J. A., & Hoekstra, T. W. (2003). *Supply-side sustainability*. New York: Columbia University Press.

Antonio, R. J. (1981). Immanent critique as the core of critical theory: Its origins and developments in Hegel, Marx and contemporary thought. *British Journal of Sociology, 32* (3), 330–35.

Araghi, F. A. (1995). Global depeasantization. *The Sociological Quarterly, 36*, 337–368.

Armstrong, J. (1995). Keepers of the earth. In Rozak, T., Gomes, M. E., & Kanner, A. D. (Eds.), *Ecopsychology: Restoring the earth, healing the mind* (pp. 316–324). San Francisco, CA: Sierra Club Books.

Astin, A. W., Vogelgesang, L. J., Ikeda, E. K., & Yee, J. A. (2001). How service learning affects students: Executive summary. Los Angeles: Higher Education Research Institute, UCLA.

Astyk, S. (2008). *Depletion and abundance: Life on the new home front*. Gabriola Island, BC, Canada: New Society.

Avramovic, D. (1986). Depression of export commodity prices of developing countries: What can be done? *Third World Quarterly, 8*, 953–977.

Barbour, I. G. (1993). *Ethics in an age of technology: The Gifford lectures, 1989–1991, Vol. 2*. San Francisco: HarperSanFrancisco.

Barlow, M., & Clarke, T. (2002). *Blue gold: The fight to stop the corporate theft of the world's water*. New York: New Press.

Bawden, R. (2004). Sustainability as emergence: The need for engaged discourse. In Corcoran, P. B., & Wals, A. E. J. (Eds.), *Higher Education and the Challenge of Sustainability* (pp. 21–32). Dordrecht, The Netherlands: Kluwer Academic Publishers.

Bell, M. M. (2004). *Farming for us all: Practical agriculture & the cultivation of sustainability*. University Park: The Pennsylvania State University Press.

Bennet, T. S. (Writer/Director). (2007). *What a way to go: Life at the end of empire* [Motion Picture]. Pittsboro, NC: VisionQuest Pictures.

Bentz, V. M., & Shapiro, J. J. (1998). *Mindful inquiry in social research*. Thousand Oaks: Sage Publications.

Berkes, F. (1999). *Sacred Ecology: Traditional ecological knowledge and resource management*. Philadelphia: Taylor & Francis.

Berman, A. (2010, July 19). Interview with Art Berman, part 1. The Association for the Study of Peak Oil and Gas, USA. Retrieved March 3, 2012, from http://www.aspousa.org/index.php/2010/07/interview-with-art-berman-part-1/

Berry, W. (1987). *Home economics: Fourteen essays*. San Francisco: North Point Press.

Black, S. (Producer/Director). (2001). *Life and debt* [Motion Picture]. New York: New Yorker Films.

Bond, G. D. (2004). *Buddhism at work: Community development, social empowerment and the Sarvodaya movement*. Bloomfield, CT: Kumarian Press.

Booth, D. E. (2002). *Searching for paradise: Economic development and environmental change in the mountain West*. Lanham, MD: Rowman & Littlefield.

Bradshaw, Y. W., & Huang, J. (1991). Intensifying global dependency: Foreign debt, structural adjustment, and third world underdevelopment. *The Sociological Quarterly, 32*, 321–342.

Braverman, H. (1974). *Labor and monopoly capital: The degradation of work in the twentieth century*. New York: Monthly Review Press.

Brookfield, S. D. (1987). *Developing critical thinkers: Challenging adults to explore alternative ways of thinking and acting*. San Francisco: Jossey-Bass Publisers.

———. (2000). Transformative learning as ideology critique. In Mezirow, J., & Associates, *Learning as transformation: Critical perspectives on a theory in progress* (pp. 125–48). San Francisco: Jossey-Bass.

———. (2005). *The power of critical theory: Liberating adult learning and teaching*. San Francisco: Jossey-Bass.

Bullard, R. (1993/2008). Confronting environmental racism. In Merchant, C. (Ed.), *Ecology* (pp. 265–276). Amherst: Humanity Books.

Buttel, F. H. (1980). Agricultural structure and rural ecology: Toward a political economy of rural development. *Sociologia Ruralis, 20* (1/2), 44–62.

Cahill, D. (2009, January). The end of neoliberalism? *Z Magazine*, 35–38.

Cajete, G. (2001). Indigenous education and ecology: Perspectives of an American Indian educator. In Grim, J. A. (Ed.), *Indigenous traditions and ecology: The interbeing of cosmology and community* (pp. 619–638). Cambridge: Harvard University Press.

Calthorpe, P. (1993). *The next American metropolis: Ecology, community, and the American dream*. New York: Princeton Architectural Press.

Campbell, C. J. (2008, October). *Peak oil: A turning point for mankind*. Retrieved June 23, 2010, from
http://www.aspo-spain.org/aspo7/presentations/Campbell-Mankind-ASPO7.pdf

Campbell, C. J., & Strouts, G. (2007). *Living through the energy crisis: Preparing for a low-energy world*. Skibbereen, Cork, Ireland: Eagle Print.

Capra, F. (2002). *The hidden connections: A science for sustainable living*. New York: Anchor Books.

Carlsson, C. (2008). *Nowtopia: How pirate programmers, outlaw bicyclists, and vacant-lot gardeners are inventing the future today!* Oakland: AK Press.

Carolan, M. S. (2006). Do you see what I see? Examining the epistemic barriers to sustainable agriculture. *Rural Sociology, 71*(2), 232–260.

Carr, W., & Kemmis, S. (1986). *Becoming critical: Education, knowledge, and action research*. London, England: Falmer Press.

Chiras, D., & Wann, D. (2003). *Superbia! 31 ways to create sustainable neighborhoods*. Gabriola Island, BC, Canada: New Society.

Clark, W. R. (2005). *Petrodollar warfare: Oil, Iraq and the future of the dollar*. Gabriola Island, BC, Canada: New Society.

Cooke, M. (2005). Avoiding authoritarianism: On the problem of justification in contemporary critical social theory. *International Journal of Philosophical Studies, 13*(3), 379–404.

Cornellier, R., & Carvalho, P. (2008). *Black wave* [Motion Picture]. Montreal, Quebec, Canada: Macumba.

Daly, H. E. (1999). Uneconomic growth in theory and in fact. The first annual Feasta lecture. *FEASTA Review, 1*. Retrieved June 25, 2010, from http://www.feasta.org/documents/feastareview/daly1.pdf

Daly, H. E., & Farley, J. (2004). *Ecological economics: Principles and applications*. Washington, DC: Island Press.

Dant, T. (2003). *Critical social theory: Culture, society and critique*. London: Sage.

Darley, J. (2004). *High noon for natural gas: The new energy crisis*. White River Junction, VT: Chelsea Green.

Dasgupta, P. (2001). *Human well-being and the natural environment*. Oxford: Oxford University Press.

Davis, M. (2006). *Planet of slums*. New York: Verso.

Deffeyes, K. S. (2001). *Hubbert's peak: The impending world oil shortage*. Princeton, NJ: Princeton University Press.

Desmarais, A. A. (2009, May 20). Building a transnational peasant movement. *NACLA*. Retrieved April 2, 2012, from: https://nacla.org/node/5831

Devall, B. (1980/2008). The deep ecology movement. In Merchant, C. (Ed.), *Ecology: Key concepts in critical theory* (pp. 149–163). Amherst: Humanity Books.

Di Chiro, G. (1996). Nature as community: The convergence of environment and social justice. In Cronon, W. (Ed.), *Uncommon ground: Rethinking the human place in nature* (pp. 298–320). New York, NY: W.W. Norton.

Diamond, J. M. (1999). *Guns, germs and steel: The fates of human societies*. New York: W. W. Norton & Co.

Dölling, I., & Hark, S. (2000). She who speaks shadow speaks truth: Transdisciplinarity in women's and gender studies. *Journal of Women in Culture and Society, 25*(4), 1195–1198.

Dormael, A. van (1978). *Bretton Woods: Birth of a monetary system*. New York: Holmes & Meier Publishers.

Douthwaite, R. (1999a). *The ecology of money*. Schumacher Briefing no. 4. Totnes, Devon, England: Green Books.

———. (1999b). *The growth illusion: How economic growth has enriched the few, impoverished the many and endangered the planet*. Gabriola Island, BC, Canada: New Society.

———. (2004). Why localisation is essential for sustainability. In Douthwaite, R., & Jopling, J. (Eds.), *Growth: The Celtic cancer. FEASTA Review*, No. 2, 114–124.

Dussel, E. (2008). *Twenty theses on politics*. Durham, NC: Duke University Press.

Dyer, J. (1998). *Harvest of rage: Why Oklahoma City is only the beginning*. Boulder: Westview Press.

Evans, T. L. (2010). Food for thought local food program. Retrieved January 1, 2011, from http://www.fortlewis.edu/EnvStudies/food_for_thought.asp

———. (2010, February 19). Action project. End of Oil. Retrieved June 30, 2012, from http://tinalynnevans.com/oil/w2010/project.html

———. (2010, August 20). Final reflection paper, the end of oil. Retrieved June 30, 2012, from http://tinalynnevans.com/oil/f2010/reflection.html

———. (2010, fall). Apple Days Festival community based project. Retrieved, June 30, 2012, from http://tinalynnevans.com/oil/f2010/apple-days.html

———. (2011). Leadership without domination? Toward restoring the human and natural world. *Journal of Sustainability Education, 2*. Retrieved March 8, 2012, from: http://www.jsedimensions.org/wordpress/content/leadership-without-domination-toward-restoring-the-human-and-natural-world_2011_03/

————. (2012, January). The power of oil: Relationships among world changing ideas, institutions and technologies. Retrieved June 30, 2012, from
http://tinalynnevans.com/oil-table.pdf
————. (2012, February 3). End of oil course texts, past and present. Retrieved June 30, 2012, from http://tinalynnevans.com/end-of-oil-texts.html
————. (2012, June 28). Tina Evans' web page. Retrieved June 30, 2011, from
http://tinalynnevans.com/
Feasta (The Foundation for the Economics of Sustainability). (2004, June). Curing global crises: Let's treat the disease not the symptoms [Electronic version]. Dublin, Ireland: Feasta. Retrieved January 22, 2009, from
http://www.feasta.org/events/debtconf/sleepwalking.pdf
————. (2008, May). Cap & share: A fair way to cut greenhouse emissions [Electronic version]. Dublin, Ireland: Feasta. Retrieved December 31, 2008, from
http://www.feasta.org/documents/energy/Cap-and-Share-May08.pdf
Feit, H. A. (2001). Hunting, nature, and metaphor: Political and discursive strategies in James Bay Cree resistance and autonomy. In Grim, J. A. (Ed.), *Indigenous traditions and ecology: The interbeing of cosmology and community* (411–452). Cambridge: Harvard University Press.
First National People of Color Summit. (1991/2008). Principles of environmental justice. In Merchant, C. (Ed.), *Ecology* (pp. 265–276). Amherst: Humanity Books.
Folke, C., Hahn, T., Olsson, P., & Norberg, J. (2005). Adaptive governance of social-ecological systems. *Annual Review of Environment and Resources, 30*, 441–473.
Fort Lewis College. (n.d.). *EGC Course Development Guidelines*. Retrieved January 1, 2011, from http://explore.fortlewis.edu/faculty_staff/egcpaperwork.asp
Foster, J. B. (2009). *The ecological revolution: Making peace with the planet*. New York: Monthly Review Press.
Fox, J. E., Gulledge, J., Engelhaupt, E., Burrow, M. E., & McLachlan, J. A. (2007). Pesticides reduce symbiotic efficiency of nitrogen-fixing rhizobia and host plants. *Proceedings of the National Academy of Sciences, 104*(24), 10282–10287.
Freire, P. (1970/2000). *Pedagogy of the oppressed*. (30th Anniversary Ed.). New York: Continuum International Publishing Group.
Garcia, D. K. (Writer/Director/Producer). (2004). *The future of food* [Motion picture]. Lily Films.
Goldschmidt, W. (1947). *As you sow*. Gencoe, Illinois: Free Press.
Gomes, M. E., & Kanner, A. D. (1995). The rape of the well-maidens: Feminist psychology and the environmental crisis. In Rozak, T., Gomes, M. E., & Kanner, A. D. (Eds.), *Ecopsychology: Restoring the earth, healing the mind* (pp. 111–121). San Francisco: Sierra Club Books.
Gonzales, T. A., & Nelson, M. K. (2001). Contemporary Native American responses to environmental threats in Indian Country. In Grim, J. A. (Ed.), *Indigenous traditions and ecology: The interbeing of cosmology and community* (pp. 495–538). Cambridge: Harvard University Press.
Gramsci, A. (1971/1999). *Selections from the prison notebooks*. New York: International Publishers.
————. (1926–1943/1996). *The prison notebooks*. New York: Columbia University Press.
Grande, S. (2007). Red Lake Woebegone: Pedagogy, decolonization, and the critical project. In McLaren, P., & Kincheloe, J. L. (Eds.), *Critical pedagogy: Where are we now?* (chap. 13). New York: Peter Lang.

Greene, G. (Director). (2004). *The end of suburbia: Oil depletion and the collapse of the American dream* [Motion picture]. Toronto: The Electric Wallpaper Co.

Greenleaf, R. K. (1970/1991). *The servant as leader*. Indianapolis: The Robert K. Greenleaf Center.

Greider, W. (1997). *One world, ready or not: The manic logic of global capitalism*. New York: Simon & Schuster.

Grim, J. A. (Ed.). (2001). *Indigenous traditions and ecology: The interbeing of cosmology and community*. Cambridge: Harvard University Press.

Grossman, R. (Producer/Director). (2005). *Homeland: Four portraits of native action* [Motion Picture]. Oley, PA: Bullfrog Films.

Gruenwald, D. A. (2003). The best of both worlds: A critical pedagogy of place. *Educational Researcher, 32* (4), 3–12.

Guggenheim, D. (Director). (2006). *An inconvenient truth* [Motion Picture]. Hollywood, Paramount.

Habermas, J. (1984). *The theory of communicative action*. (T. McCarthy, Trans.). Boston: Beacon Press.

Hall, C. A. S., & Klitgaard, K. A. (Eds.). (2012). *Energy and the wealth of nations: Understanding the biophysical economy*. New York: Springer.

Harvey, D. (1989). *The condition of postmodernity: An enquiry into the origins of cultural change*. Oxford, England: Basil Blackwell.

Harvey, D. (2005). A brief history of neoliberalism. New York: Oxford University Press.

Havel, V. (1990). *Disturbing the peace: A conversation with Karel Hvížďala*. (P. Wilson, Trans.). New York: Knopf.

Heinberg, R. (2005). *The party's over: Oil, war and the fate of industrial societies*. (Rev. ed.). Gabriola Island, BC, Canada: New Society.

Held, D. (1980). *Introduction to critical theory: Horkheimer to Habermas*. Berkeley: University of California.

Henzell, P., & Rhone, T. D. (Producers/Directors). (2002). *The harder they come* [Motion Picture]. Santa Monica, CA: Zenon Pictures.

Hirsch, R. L., Bezdek, R., & Wendling, R. (2005, February). *Peaking of world oil production: Impacts, mitigation, & risk management*. A report sponsored by the U.S. Department of Energy. Retrieved January 14, 2009, from
http://www.netl.doe.gov/publications/others/pdf/Oil_Peaking_NETL.pdf

Homer-Dixon, T. (2006). *The upside of down: Catastrophe, creativity, and the renewal of civilization*. Washington, DC: Island Press.

Hopkins, R. (Ed.). (2005). Kinsale 2021: An energy descent action plan, version.1. Retrieved September 3, 2008, from
http://transitionculture.org/wp-content/uploads/KinsaleEnergyDescentActionPlan.pdf

———. (2008). *The transition handbook: From oil dependency to local resilience*. Fox Hole, England: Green Books.

Horkheimer, M., & Adorno, T. (1944/1972). *Dialectic of enlightenment*. New York: Herder and Herder.

Huber, P. (2002). The energy spiral. *Forbes, 169*(8), 102.

Huckle, J. (2004). Critical realism: A philosophical framework for higher education for sustainability. In Corcoran, P. B., & Wals A. E. J., (Eds.), *Higher education and the challenge of sustainability* (pp. 33–47). Dordrecht, The Netherlands: Kluwer Academic Publishers.

Hudson, M. (2005). *Global fracture: The new international economic order*. London: Pluto.

Intergovernmental Panel on Climate Change [IPCC]. (2007). *Climate change 2007 synthesis report*. Retrieved January 15, 2009, from http://www.ipcc.ch/pdf/assessment-report/ar4/syr/ar4_syr.pdf

International Commission on the Future of Food and Agriculture [ICFFA]. (2008). *Manifesto on climate change and the future of food security*. Retrieved September 10, 2009, from http://www.arsia.toscana.it/petizione/documents/clima/CLIMA_ING.pdf

International Consortium of Investigative Journalists [ICIJ]. (2003). *Water barons: How a few powerful companies are privatizing your water*. Washington, DC: Public Integrity Books.

International Labor Organization [ILO]. (2005, February 14). *New ILO report sees weak global job growth in 2004, says European job growth remains static*. Retrieved October 14, 2007, from http://www.ilo.org/global/About_the_ILO/Media_and_public_information/Press_releases/lang--en/WCMS_005183/index.htm.

International Society for Ecology and Culture [ISEC]. (Producer). (1993). *Ancient futures: Learning from Ladakh*. Bristol: International Society for Ecology and Culture.

Jackson, W. (1996). *Becoming native to this place*. Lexington: University Press of Kentucky.

Jacobs, J. (1969). *The economy of cities*. New York: Random House.

———. (2000). *The nature of economies*. New York: Modern Library.

Jameson, F. (1991). *Postmodernism, or, the cultural logic of late capitalism*. Durham: Duke University Press.

Jarecki, E. (Writer/Director). (2006). *Why we fight* [Motion Picture]. Culver City, CA: Sony Pictures Home Entertainment.

Jensen, D. (2000). *A language older than words*. New York: Context Books.

———. (2004). *The culture of make believe*. White River Junction, VT: Chelsea Green.

Kahn, R. (2010). *Critical pedagogy, ecoliteracy, & planetary crisis: The ecopedagogy movement*. New York: Peter Lang.

Kaplinsky, R. (2005). *Globalization, poverty and inequality: Between a rock and a hard place*. Malden, MA: Polity.

Keen, C., & Baldwin, E. (2004, April). Students promoting economic development and environmental sustainability: An analysis of the impact of involvement in a community-based research and service-learning program. *International Journal of Sustainability in Higher Education, 5*(4), 384–394.

Kemmis, D. (1990). *Community and the politics of place*. Norman: University of Oklahoma Press.

———. (2001). *This sovereign land: A new vision for governing the West*. Washington, DC: Island Press.

Keynes, J. M. (1941/1980). Proposals for an international currency union. In Moggridge, D. (Ed.), *The collected writings of John Maynard Keynes*, vol. XXV (pp. 42–66), London: MacMillan.

Kitchen, R., & Bartley, B. (2007). *Understanding contemporary Ireland*. London: Pluto.

Klein, J. T. (2001). Interdisciplinarity and the prospect of complexity: The tests of theory. *Issues in Integrative Studies*, (19), 43–57.

Kneafsey, M., Cox, R., Holloway, L., Dowler, E., Venn, L., & Tuomainen, H. (2008). *Reconnecting consumers, producers and food: Exploring alternatives*. Oxford, United Kingdom: Berg.

Korten, D. C. (2001). *When corporations rule the world* (2nd ed.). Bloomfield, CT: Kumarian.

Kovel, J. (2002). *The enemy of nature: The end of capitalism or the end of the world?* London: Zed Books.

Kropotkin, P. (1902/1989). *Mutual aid: A factor of evolution.* Montréal: Black Rose Books.

Kunstler, J. H. (2005). *The long emergency: Surviving the converging catastrophes of the twenty-first century.* New York: Atlantic Monthly.

LaDuke, W. (1999). *All our relations: Native struggles for land and life.* Cambridge: South End Press.

Laszlo, E. (2006). *The chaos point: The world at the crossroads.* Charlottesville, VA: Hampton Roads.

Lenhard, J., Lücking, H., & Schwechheimer, H. (2006). Transdisciplinarity: Expert knowledge, mode-2 and scientific disciplines: Two contrasting views. *Science and Public Policy, 33*(5), 341–350.

Leonardo, Z. (2004). Critical social theory and transformative knowledge: The functions of criticism in quality education. *Educational Researcher 33*(6), 11–18.

Li, Minqi (2008). *The rise of China and the demise of the capitalist world economy.* New York: Monthly Review Press.

Loeb, P. R. (1999). *Soul of a citizen: Living with conviction in a cynical time.* New York: St. Martin's Griffin.

Lomborg, B. (2001, August 16). Running on empty? *The Guardian.* Retrieved June 23, 2010, from http://www.guardian.co.uk/education/2001/aug/16/highereducation.climatechange

Ludwig, A., & Blum, F. (Producers), Opitz, F. (Writer/Director). (2006). *The big sellout* [Motion picture]. San Francisco: California Newsreel.

Lynch, M. (2001, September). *Closed coffin: Ending the debate on "the end of cheap oil": A commentary.* Retrieved June 23, 2010, from http://sepwww.stanford.edu/sep/jon/world-oil.dir/lynch2.html

———. (2003). The new pessimism about petroleum resources: Debunking the Hubbert model (and Hubbert modelers). *Minerals & Energy, 18,* 21–32.

Lyson, T. A. (2004). *Civic agriculture: Reconnecting farm, food, and community.* Medford, MA: Tufts University Press.

Mandel, E. (1972/1975). *Late capitalism* (J. De Bres, Trans.). London: NLB.

Manley, M. (1987). *Up the down escalator: Development and the international economy: A Jamaican case study.* Washington, DC: Howard University Press.

Mann, C. C. (2002, March). 1491. *Atlantic Monthly, 289,* 41–53.

Marcos. (2001). *Our word is our weapon: Selected writings.* J. P. de León (Ed.). New York: Seven Stories.

Marcuse, H. (1964). *One-dimensional man.* Boston: Beacon Press.

———. (1972/2008). Ecology and revolution. In Merchant, C. (Ed.), *Ecology: Key concepts in critical theory* (pp. 67–70). Amherst: Humanity Books.

Martinez, D. (1997). American Indian cultural models for sustaining biodiversity. In Proceedings of the sustainable forestry seminar series, Oregon State University, 1995. *Special forest products: Biodiversity meets the marketplace.* Retrieved September 2, 2007, from http://www.fs.fed.us/pnw/pubs/gtr63/gtrwo63g.pdf.

———. (2010, May). The value of indigenous ways of knowing to Western science and environmental sustainability. *Journal of Sustainability Education.* Retrieved July 16, 2010, from http://www.journalofsustainabilityeducation.org/wordpress/content/the-value-of-indigenous-ways-of-knowing-to-western-science-and-environmental-sustainability_2010_05/

Marx, K. (1844/1964). *The economic & philosophic manuscripts of 1844.* New York: International Publishers.

Mathews, F. (1992/1999). Ecofeminism and deep ecology. In Merchant, C. (Ed.), *Ecology: Key concepts in critical theory* (pp. 235–245). Amherst: Humanity Books.

McChesney, R. W. (1999). *Rich media, poor democracy: Communication politics in dubious times.* Urbana: University of Illinois Press.

McLaren, P. (2005). *Capitalists and conquerors: A critical pedagogy against empire.* Lanham, MD: Rowman & Littlefield.

———. (2007). The future of the past: Reflections on the current state of empire and pedagogy. In McLaren, P., & Kincheloe, J. L. (Eds.), *Critical pedagogy: Where are we now?* (pp. 1–5). New York: Peter Lang.

McLaren, P., & Farahmandpur, R. (2005). *Teaching against global capitalism and the new imperialism.* Lanham, MD: Rowman & Littlefield.

McLaren, P., & Houston, D. (2005). Revolutionary ecologies: Ecosocialism and critical pedagogy. In McLaren, P., *Capitalists and conquerors: A critical pedagogy against empire* (pp. 166–85). Lanham, MD: Rowman & Littlefield.

McLaren, P., & Jaramillo, N. (2007). *Pedagogy and praxis in the age of empire: Towards a new humanism.* Rotterdam: Sense Publishers.

McLaren, P., & Kincheloe, J. L. (Eds.). (2007). *Critical pedagogy: Where are we now?* New York: Peter Lang.

McLaren, P., & Kumar, R. (2009 Feb. 19). Being, becoming and breaking-free: Peter McLaren and the pedagogy of liberation. *Radical Notes.* Retrieved May 13, 2009, from: http://radicalnotes.com/content/view/88/39/

McLaughlin, N. (1999). Origin myths in the social sciences: Fromm, the Frankfurt School and the emergence of critical theory. *Canadian Journal of Sociology, 24*(1), 109–139.

Meadows, Donella, Randers, J., & Meadows, Dennis. (2004). *Limits to growth: The 30-year update.* White River Junction, VT: Chelsea Green.

Medrick, R. (2005). Sustainability as education. Unpublished manuscript, Ph.D. Program in Sustainability Education, Prescott College, Prescott, AZ.

Merchant, C. (1996). Reinventing Eden: Western culture as a recovery narrative. In Cronon, W. (Ed.), *Uncommon ground: Rethinking the human place in nature.* New York: W. W. Norton & Co.

———. (Ed.). (1999). *Ecology: Key concepts in critical theory.* Amherst: Humanity Books.

———. (Ed.). (2008). *Ecology: Key concepts in critical theory.* (2nd ed.). Amherst: Humanity Books.

Meyer, M. (2007). Increasing the frame: Interdisciplinarity, transdisciplinarity and representativity. *Interdisciplinary Science Reviews, 32*(3), 203–211.

Mezirow, J. (2000). Learning to think like an adult: Core concepts of transformational theory. In Mezirow, J., & Associates, *Learning as transformation: Critical perspectives on a theory in progress* (pp. 3–33). San Francisco: Jossey-Bass.

Mezirow, J., & Associates. (2000). *Learning as transformation: Critical perspectives on a theory in progress.* San Francisco: Jossey-Bass.

M'Gonigle, M., & Starke, J. (2006). *Planet U: Sustaining the world, reinventing the university.* Gabriola Island, BC, Canada: New Society.

Miller, A. (1999). Economics and the environment. In Merchant, C. (Ed.), *Ecology: Key concepts in critical theory* (pp. 78–87). Amherst: Humanity Books.

Moore, J. L. (1995). Cost-benefit analysis: Issues in its use in regulation [Electronic version]. Congressional Research Service Environment and Natural Resources Policy Division Report for the U.S. Congress (ENR Publication No. 95–760). Retrieved December 30, 2008, from http://www.ncseonline.org/NLE/CRSreports/Risk/rsk4.cfm?&CFID=7004999&CFTOK EN=5009675

Morrow, R. A., & Brown, D. D. (1994). *Critical theory and methodology. Contemporary social theory, vol. 3*. Thousand Oaks, CA: Sage Publications.

Moyers, B. (Host). (2002). *Bill Moyers reports: Trading democracy* [Motion picture]. Princeton, NJ: Films for the Humanities & Sciences.

———. (Writer), & Casciato, T. (Writer). (2006). *Is God green?* [motion picture]. United States: Films for the Humanities & Sciences.

Murphy, C. (1999). *Cultivating Havana: Urban agriculture and food security in the years of crisis*. (Development Report No. 12) Food First, Institute for Food and Development Policy. Retrieved September 10, 2009, from
http://www.foodfirst.org/pubs/devreps/dr12.pdf

Nabhan, G. P. (2009). *Where our food comes from: Retracing Nikolay Vavilov's quest to end famine*. Washington, DC: Island Press.

Næs, A. Deep ecology. (1973/2008). In Merchant, C. (Ed.), *Ecology: Key concepts in critical theory* (pp. 143–147). Amherst: Humanity Books.

Navdanya. (2009). Organization website. Retrieved September 22, 2009, from
http://www.navdanya.org/

Nelson, R. K. (1983). *Make prayers to the raven: A Koyukon view of the northern forest*. Chicago: University of Chicago Press.

Newman, P., Bruyere, B., & Beh, A. (2007, July). Service-Learning and natural resource leadership. *Journal of Experiential Education, 30*(1), 54–69.

Nicolescu, B. (2002). *Manifesto of transdisciplinarity*. Albany: State University of New York Press.

Norberg-Hodge, H. (1991/1992). *Ancient futures: Learning from Ladakh*. San Francisco: Sierra Club.

Norberg-Hodge, H., Merrifield, T., & Gorelick, S. (2002). *Bringing the food economy home: Local alternatives to global agribusiness*. London, England: Zed Books.

O'Connor, J. (1991/2008). Socialism and ecology. In Merchant, C. (Ed.), *Ecology: Key concepts in critical theory.* (pp. 189–197). Amherst: Humanity Books.

Odum, E. P., & Barrett, G. W. (2005). *Fundamentals of ecology* (5th ed.). Belmont, CA: Thomson Brooks/Cole.

Ollman, B. (1971). *Alienation: Marx's conception of man in capitalist society*. New York: Cambridge University Press.

O'Sullivan, E. (2004). Sustainability and transformative educational vision. In Corcoran, P. B., & Wals A. E. J., (Eds.), *Higher education and the challenge of sustainability* (pp.163–80). Dordrecht, The Netherlands: Kluwer Academic Publishers.

Orr, D. (2009). *Down to the wire: Confronting climate collapse*. New York: Oxford University Press.

Pagden, A. (1982/1988). *The fall of natural man: The American Indian and the origins of comparative ethnology*. Cambridge: Cambridge University Press.

Patel, R. (2007, February 2). *Tortilla prices: An analysis*. Retrieved January 14, 2009, from
http://stuffedandstarved.org/drupal/node/96

Perkins, J. (2004). *Confessions of an economic hit man*. San Francisco: Berrett-Koehler.

Persaud, R. B. (2001). *Counter-hegemony and foreign policy: The dialectics of marginalized and global forces in Jamaica*. Albany: State University of New York Press.

Pfeiffer, D. A. (2006). *Eating fossil fuels: Oil, food and the coming crisis in agriculture*. Gabriola Island, BC, Canada: New Society.

Pittman, J. (2007, August). *Whole systems design and living sustainability.* Presentation given at the orientation to the Prescott College Ph.D. Program in Sustainability Education, Prescott, AZ.

Ploeg, J. D. van der. (2008). *The new peasantries: Struggles for autonomy and sustainability in an era of empire and globalization.* London, England: Earthscan.

Polanyi, K. (1944/1957). *The great transformation: The political and economic origins of our time.* Boston: Beacon Press.

The power of community: How Cuba survived peak oil [Motion picture]. (2006). Yellow Springs, OH: Community Service, Inc.

Pretty, J. (2002). *Agri-culture: Reconnecting people, land and nature.* London, England: Earthscan.

———. (2005). *The pesticide detox: Towards a more sustainable agriculture.* London, England: Earthscan.

———. (2007). *The earth only endures: On reconnecting with nature and our place in it.* London, England: Earthscan.

Pro Publica. (2008–2012). Fracking: Gas drilling's environmental threat [Series]. Retrieved March 3, 2012, from http://www.propublica.org/series/fracking

Proudhon, P. J. (1890/1966). *What is property? An enquiry into the principle of right and of government.* New York: Howard Fertig.

Reisner, M. (1993). *Cadillac desert: The American West and its disappearing water* (Rev. ed.). New York: Penguin.

Robbins, R. H. (1999). *Global problems and the culture of capitalism.* Boston: Allyn and Bacon.

Roberts, P. (2004). *The end of oil: On the edge of a perilous new world.* Boston: Houghton Mifflin.

Rowbotham, M. (2000). *Goodbye America! Globalization, debt and the dollar empire.* Charlbury, Osfordshire, England: Jon Carpenter Publishing.

Rozak, T., Gomes, M. E., & Kanner, A. D. (Eds.). (1995). *Ecopsychology: Restoring the earth, healing the mind.* San Francisco: Sierra Club.

Russo, R. (2010). *Jumping from the ivory tower: Weaving environmental leadership and sustainable communities.* Lanham, MD: University Press of America.

Salamini, L. (1974). Gramsci and Marxist sociology of knowledge: An analysis of hegemony-ideology-knowledge. *The Sociological Quarterly, 15,* 359–380.

Salmon, E. (2000). Kincentric ecology: Indigenous perceptions of the human-nature relationship. *Ecological Applications, 10,* 1327–1332.

Scheer, H. (2002). *The solar economy: Renewable energy for a sustainable future.* London, England: Earthscan.

———. (2007). *Energy autonomy: The economic, social and technological case for renewable energy.* London, England: Earthscan.

Schroll, H., & Stærdahl, J. (2001). The concept of the *Journal of Transdisciplinary Environmental Studies.* Retrieved May 12, 2009, from http://www.journal-tes.dk/PDFfiler/concept.pdf.

Shepard, P. (1995). Nature and madness. In Rozak, T., Gomes, M. E., & Kanner, A. D. (Eds.), *Ecopsychology: Restoring the earth, healing the mind* (pp. 21–40). San Francisco: Sierra Club.

Shiva, V. (2000). *Stolen harvest: The hijacking of the global food supply.* Cambridge, MA: South End Press.

————. (2005). *Earth democracy: Justice, sustainability, and peace*. Cambridge: South End Press.

————. (2008). *Soil not oil: Environmental justice in a time of climate crisis*. Cambridge: South End Press.

Shuman, M. H. (2000). *Going local: Creating self-reliant communities in a global age*. New York: Routledge.

Simmons, M. R. (2005). *Twilight in the desert: The coming Saudi oil shock and the world economy*. Hoboken, NJ: John Wiley & Sons.

Spretnak, C. (1997). *The resurgence of the real: Body, nature, and place in a hypermodern world*. Reading, MA: Addison-Wesley.

Stiglitz, J. E. (2002). *Globalization and its discontents*. New York: W. W. Norton.

————. (2009, January). Capitalist fools [Electronic version]. *Vanity Fair*.

Summers, C., & Markusen, E. (1992/2003).The case of collective violence. In Ermann, M. D., & Shauf, M. S. (Eds.), *Computers, ethics, and society* (3rd ed.) (pp. 214–231). New York: Oxford University Press.

Sveiby, K., & Skuthorpe, T. (2006). *Treading lightly: The hidden wisdom of the world's oldest people*. Crow's Nest, New South Wales, Australia: Allen & Unwin.

Tainter, J. A. (1988). *The collapse of complex societies*. New York: Cambridge University Press.

Tar, Z. (1977). *The Frankfurt School: The critical theories of Max Horkheimer & Theodor W. Adorno*. New York: John Wiley & Sons.

Terry, R. W. (1993). *Authentic leadership: Courage in action*. San Francisco: Jossey-Bass.

United States Energy Information Agency. (2008, August 22). How dependent are we on foreign oil? Retrieved January 14, 2009, from
http://tonto.eia.doe.gov/energy_in_brief/foreign_oil_dependence.cfm

United States Treasury Department. (2012, January 31). Monthly statement of the public debt of the United States. Retrieved March 5, 2012, from
http://www.treasurydirect.gov/govt/reports/pd/mspd/2012/opdm012012.pdf

Vegh, I. de. (1943). The international clearing union. *The American Economic Review, 33*, 534–556.

Vidal, J. (2010, March 10). Billionaires and mega-corporations behind immense land grab in Africa. *Alternet*. Retrieved June 25, 2010, from
http://www.alternet.org/story/145970/billionaires_and_mega-corporations_behind_immense_land_grab_in_africa

Wallerstein, I. (1974). *The modern world-system: Capitalist agriculture and the origins of the European world-economy in the sixteenth century*. Studies in Social Discontinuity. New York: Academic, 1974.

————. (1976). A world-system perspective on the social sciences. *British Journal of Sociology, 27*, 343–352.

————. (2003, July–August). U.S. weakness and the struggle of hegemony [Electronic Version]. *Monthly Review*.

————. (2005). After developmentalism and globalization, what? *Social Forces, 83*, 1263–1278.

————. (2006, July–August). The curve of American power. *New Left Review, 40*, 77–94.

————. (2007, spring). Precipitate decline: The advent of multipolarity. *Harvard International Review*, 50–55.

————. (2008a, February 4). 2008: The demise of neoliberal globalization. *YaleGlobal Online*. Retrieved December 23, 2008, from
http://yaleglobal.yale.edu/article.print?id=10299

————. (2008b, August 4). Theory talk #13—Immanual Wallerstein on world-systems, the immanent end of capitalism and unifying social science. *Theory Talks*. Retrieved December 23, 2008, from http://www.theory-talks.org/2008/08/theory-talk-13.html

Walliman, I. (1981). *Estrangement: Marx's conception of human nature and the division of labor*. Westport, CT: Greenwood Press.

Ward, H. (2006). *Acting locally: Concepts and models for service-learning in environmental studies*. Sterling, VA: Stylus.

Windle, P. (1995). The ecology of grief. In Rozak, T., Gomes, M. E., & Kanner, A. D. (Eds.), *Ecopsychology: Restoring the earth, healing the mind* (pp. 136–145). San Francisco, CA: Sierra Club Books.

World Bank (2010). *About us*. Retrieved June 30, 2010, from
http://web.worldbank.org/WBSITE/EXTERNAL/EXTABOUTUS/0,,contentMDK:2065
3660~menuPK:72312~pagePK:51123644~piPK:329829~theSitePK:29708,00.html

Zohar, D. (1990). *The quantum self: Human nature and consciousness defined by the new physics*. New York: Quil/William Morrow.

Index

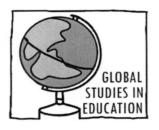

A.C. (Tina) Besley, Michael A. Peters,
Cameron McCarthy, Fazal Rizvi
General Editors

Global Studies in Education is a book series that addresses the implications of the powerful dynamics associated with globalization for re-conceptualizing educational theory, policy and practice. The general orientation of the series is inter-disciplinary. It welcomes conceptual, empirical and critical studies that explore the dynamics of the rapidly changing global processes, connectivities and imagination, and how these are reshaping issues of knowledge creation and management and economic and political institutions, leading to new social identities and cultural formations associated with education.

We are particularly interested in manuscripts that offer: a) new theoretical, and methodological, approaches to the study of globalization and its impact on education; b) ethnographic case studies or textual/discourse based analyses that examine the cultural identity experiences of youth and educators inside and outside of educational institutions; c) studies of education policy processes that address the impact and operation of global agencies and networks; d) analyses of the nature and scope of transnational flows of capital, people and ideas and how these are affecting educational processes; e) studies of shifts in knowledge and media formations, and how these point to new conceptions of educational processes; f) exploration of global economic, social and educational inequalities and social movements promoting ethical renewal.

For additional information about this series or for the submission of manuscripts, please contact one of the series editors:

A.C. (Tina) Besley: tbesley@illinois.edu
Cameron McCarthy: cmccart1@illinois.edu
Michael A. Peters: mpet001@illinois.edu
Fazal Rizvi: frizvi@unimelb.edu.au

Department of Educational Policy Studies
University of Illinois at Urbana-Champaign
1310 South Sixth Street
Champaign, IL 61820   USA

To order other books in this series, please contact our Customer Service Department:

(800) 770-LANG (within the U.S.)
(212) 647-7706 (outside the U.S.)
(212) 647-7707 FAX

Or browse online by series:
www.peterlang.com